Shipping Business and Maritime Economics

An Annotated International Bibliography

James McConville and Glenys Rickaby

MANSELL

First published 1995 by
Mansell Publishing Limited, *A Cassell imprint*
Wellington House, 125 Strand, London WC2R 0BB, England
387 Park Avenue South, New York, New York 10016–8810, USA

British Library Cataloguing-in-Publication Data
Shipping Business and Maritime Economics:
Annotated International Bibliography
I. McConville, J. II. Rickaby, Glenys
016.3875

ISBN 0–7201–2180–9

Library of Congress Cataloging-in-Publication Data
McConville, J.
Shipping business and maritime economics : an annotated international bibliography / James McConville and Glenys Rickaby.
p. cm.
ISBN 0–7201–2180–9 : $140.00
1. Shipping – Bibliography. 2. Merchant marine – Bibliography.
I. Rickaby, Glenys. II. Title.
Z7164.S55M38 1995
[HE735]
016.3875 – dc20 94–23314
CIP

Phototypeset in Great Britain by Intype
Printed and bound in Great Britain by
Biddles Ltd, Guildford and King's Lynn

Contents

Preface xiii
Bibliographical notes and abbreviations xvi

1 **Reference books** 1
Dictionaries; bibliographies; yearbooks; atlases; encyclopedias; handbooks 1
Research 9

2 **General** 10
Textbooks 10
Economics 12

3 **Maritime development (post-1950)** 16
Shipping development 16
Shipowners; entrepreneurs; companies 20

4 **Demand** 23
General 23
General 23
Forecasting 24
Economic theory 25
General 25

5 **Supply** 26
General 26
General 26
Productivity 28
Congestion 28

Safety 28
Routeing 30
Economic theory 33
General; shipping problems 33
Ship price; second-hand 34
Lay-up 35
Ship cost 35
Operational research 37
Speed 37
Elasticity 38
Shipowning; management 38
General 38
Enterprises 41
Ship management 42
Tankers 42
Finance, cost and investment 43
General 43
Ship finance 44
Bunker costs 47
Operating costs 48
Investment 49
Shipbuilding and repair 51
General 51
Government 56
National shipbuilding 57
Conversion; repair; drydocking 61
Safety 62
Loss and scrapping; demolition 62
General 62
Market 63
Scrapping activity 63

6 International seaborne trade 64
General 64
General 64
Freight and transport costs 67
Bulk wet 68
Bulk dry 69
Bulk 'mini' and 'neo' 75
Other 76
Economic theory and modelling 79

General 79
Tariffs 80
Economic nationalism 81
Forecasting 81
Balance of payments 83
General 83

7 **Market structures** 85
General 85
Theory 85
Optimum ships 87
Freight rates; markets 89
Shipping enterprise 92
Intermodal and multimodal 93
Transhipment 95
Dry bulk trades 96
General and theory 96
Freight rates; markets 101
Demand 104
Supply 104
Regulation 106
Wet bulk trades (including combined carriers) 106
General and theory 106
Freight rates; markets 112
Demand 115
Supply 116
Pollution 118
Product carriers 120
Regulations 124
Liner trades 125
General and theory 125
Liner code 137
Liner conference and non-conference 141
Containers 150
Ro-ro 160
Conventional cargo liners 161
Cruise trade 161
Reefers 162
Coastal; short-sea trades 164

General and theory 164
Bulk trades 167
Ferries and ro-ro 168
Hovercraft and other types 171
Rescue systems 171

8 Ports 172
General 172
General 172
United Kingdom 174
Western Europe and Scandinavia 178
Central and Eastern Europe and Middle East 180
North America 181
Japan 183
Socialist Asia 183
Africa 183
Central and South America 186
Asia, Australasia and Oceania 186
Economic theory and price 188
General 188
Optimal cargo handling 197
Terminals and productivity 202
Port purpose and congestion 204
Port and regional development 207
Safety and pollution 213
Finance and investments 214
Finance 214
Subsidies 215
Investments 216
Manpower and management 218
General 218
United Kingdom 221
North America 224
Other countries 227
Management 228

9 Sea canals and waterways 231
Panama 231
Suez 232
Barge systems 234

Sea and inland waterways 234
The Great Lakes 236

10 Human resources in shipping 237
General 237
General; cost and wages; contracts 237
Officers 242
Ratings 242
Education and training 243
On-board systems 245
Trade unions; professional associations; collective bargaining 249
Theory 252
General 252
Communication 255
Job satisfaction 256
Productivity 258
Differential subsidies 258

11 Management 259
General 259
Ship management 261
Education and training 263
Export and freight forwarding 264
Shipbrokers 267

12 Law 268
General 268
General 268
United Kingdom 277
Western Europe and Scandinavia 278
Central and Eastern Europe and Middle East 279
North America 280
Socialist Asia 283
Africa 283
Central and South America 284
Asia, Australasia and Oceania 284
Carriage of goods 285

Contract 285
Sale of goods 298
Charterparties 299
Bills of lading 305
Freight fixtures 310
Laytime 310
Demurrage 312
Agencies 312
Liabilities 312
Maritime law 316
Admiralty law 316
Registration 327
Flags of convenience 336
Pilotage 336
Salvage 337
Limitation of liabilities 342
Maritime liens 350
Collisions 352
General average 356
Ship sale and purchase 358
General 358
Sale of vessel 362
Arbitration 363
Seaworthiness 368
Criminal activity 370
Fraud 370
Piracy 372
Terrorism 375
Criminal law 378

13 Marine insurance 380
General 380
General 380
Premiums 389
Frauds 389
Protection and indemnity (P & I) 389
Brokers' education 391
Hull 391
General 391
War risk 392
Cargo 393

General 393
Containers 394
Freight 394

14 Flag and fleet development 395
General 395
General policy and theory 395
United Kingdom 396
Western Europe and Scandinavia 399
Central and Eastern Europe and Middle East 400
North America 405
Japan 406
Socialist Asia 406
Africa 407
Central and South America 408
Asia, Australasia and Oceania 408
Open Registry 411
General 411
Flagging out 416
Argument for and against 416
Trade union activity 418

15 The State and policy 420
General 420
General policy and theory 420
United Kingdom 427
Western Europe and Scandinavia 430
Central and Eastern Europe and Middle East 431
North America 432
Japan 437
Socialist Asia 437
Africa 438
Central and South America 439
Asia, Australasia and Oceania 439
Protection 441
General and theory 441
Cargo preference; reservation 443
Subsidy 444
Trade embargo 445

Flag preference 445
Deregulation 446
Defence 447
General 447
Military logistics 449

16 International policy and regulation 450
General 450
General 450
International Maritime Organisation 452
Inter-Governmental Maritime Consultative Organisation 454
General Agreement on Tariffs and Trade 454
European Economic Community; European Community; European Union 454
Codes, conventions and reports 461
General 461
United Nations Conference on Trade and Development (not Liner Code) 462
International Labour Organization 465
New Organizations 465

Name Index 466
Subject Index 486

Preface

Accessing relevant information sources and literature is an essential but often time-consuming task in undertaking academic and business research. Furthermore, the search may often produce what is being looked for in a narrow sense rather than stimulating wider and more pertinent approaches.

This bibliography is intended to facilitate literature searches for those engaged in the practice and study of shipping business in general, and maritime economics in particular. There is no detailed bibliography existing for either shipping business or maritime economics; for the first time this volume brings together material both specifically academic and relevant to practitioners of shipping business. It is hoped that this work will provide an opportunity for increased cross-fertilization between academic researchers and 'the real world'.

The definition of shipping business which has been used as a basis for the selection of the works cited is a wide one, but excludes the nautical aspects of ship operations. The business of shipping is concerned with the supply of ships and port services; the behaviour of owners and operators and their interaction with the demanders of such services; the charterer, shipper and passenger. Maritime economics is the study of the economic criteria which govern the activities of those directly involved in, or concerned with, regulating the carriage of goods and passengers by sea or waterway. It is a discipline and a wide area of interest with imprecise boundaries.

It is ingenuous to assume that any bibliography can claim to be a definitive work. It is essential, therefore, to set parameters which make the task of compilation possible without unnecessarily limiting its usefulness. The guiding principle has been to include significant contributions to the subject published in English during the last twenty years, although seminal works published prior to 1970 have also been cited. It is fortunate that the lingua franca of shipping is English since important works in other languages are available in English editions. The bibliography comprises books, journal articles, reports, monographs, conference proceedings and theses. Publications wholly of a statistical type have been excluded as these are covered elsewhere (see p. 3, entry 20). Ephemera and the maritime press have also been excluded.

Gathering the material has illustrated both the extent to which a bibliography must be publication led, and the existence of important lacunae in the published literature. For example, in maritime economics there is a paucity of analysis on elasticity, a concept which is vitally important to business decision-making. It is not entirely facetious to suggest that those planning to undertake relevant shipping business research would do well to leaf through this volume to identify where the most rewarding gaps lie.

Most entries in the bibliography have a 30- to 50-word annotation so that users can get the most efficient access to relevant and apposite sources for their needs. It has not been possible to examine all the material included, and theses are only annotated from the mid-1980s. The contents are divided into 16 categories; these are subdivided into sections and may be regarded as freestanding in terms of identifying areas of interest. For example, containerization is a section in its own right but it will also be included, where relevant, in other categories or sections such as ports and intermodalism.

The compilation of this bibliography has been part of the research programme of the Centre for International Transport Management of London Guildhall University. The Centre was established with funding from the educational charity, The London School of Foreign Trade. This project could not have gone forward without their generous support.

The successful completion of a work of this kind does not merely depend on the authors. It involved the support of our own institution as well as contributions from members of the research team. At the outset there was blithe ignorance of the sustained physical and intellectual outlay that would be required, and the joint efforts of all involved are gratefully acknowledged.

First, our thanks must go to the London Guildhall University and particularly to the computer staff at Moorgate and the University Librarians who gave their unstinting support. We are also indebted to the staff of the libraries at the International Maritime Organisation, the Chartered Institute of Transport and the Institute of Chartered Shipbrokers for the assistance which was so willingly given.

Special mention must be made of the contributions of John Makin of Nottingham and Susan Hawker. Susan's scholarly assistance, particularly in the area of maritime law, was invaluable and we are grateful to the Research Committee of the University's Faculty of Business Studies for their decision to fund her appointment as a research assistant. The final, vital, link in the production of the finished manuscript was Patricia Mullarkey who showed unflagging energy and cheerfulness at all times.

While every effort has been made to ensure accuracy, perfection – always to be aimed at – is, unfortunately, rarely achieved. In the time-honoured manner we take full responsibility for all errors, omissions and shortcomings in this edition. Our hope is that this volume will make a contribution to the further study of, and research into, shipping business and maritime economics.

James McConville
Centre for International Transport Management
London Guildhall University

Glenys Rickaby
Librarian
London Guildhall University

March 1994

Bibliographical Notes and Abbreviations

A number of publications have changed their titles over the years and this has not been specified in every entry. For example the Chamber of Shipping merged with the British Shipping Federation to become the General Council of British Shipping in 1975. In 1991 it reverted to an earlier title, the Chamber of Shipping. On each occasion the publications were continued by the new organization. Other examples are Fearnley and Egers Chartering Company Ltd which has become simply Fearnleys, whilst H. P. Drewry have modified their title simply to Drewry.

AAR	Against all risks
ANCOM	Andean Common Market including Bolivia, Colombia, Ecuador, Peru and Venezuela
ANL	The Australian National Line
ASEAN	Association of South-East Asian Nations
BIFFEX	Baltic International Freight Futures Exchange
BIMCO	Baltic and International Maritime Council
B.V.	Bureau Veritas Classification Society (France)
CBI	Confederation of British Industry
CD	Construction–differential subsidy
C & F	Cost and freight
CENSA	Council of European and Japanese National Shipowners' Associations
CFB	Creative Fares Board
c.i.f.	Cost, insurance and freight
c.i.f. and e.	Cost, insurance, freight and exchange
C.I.I.	Chartered Insurance Institute
CIM	International Maritime Committee
C.I.T.	Chartered Institute of Transport
CLC	Civil Liability Convention

CMEA	Council for Mutual Economic Assistance
CMI	Comité Maritime International (France)
CMR	Convention of Contract for the International Carriage of Goods by Road 1965
CNAA	Council for National Academy Awards
CNCP	Combined Naval Commercial Port
COGSA	Carriage of Goods by Sea Act
D.N.V.	Det Norske Veritas – Classification Society
d.w.t.	Deadweight tonnage
ECMT	European Conference of Ministers of Transport
ECE	Economic Commission for Europe (USA)
EEC	European Economic Community
EEZ	Exclusive Economic Zone
ESCAP	Economic and Social Commission for Asia and the Pacific
ETA	Expected time of arrival
F.A.S.	Free alongside
FMC	Federal Maritime Commission (USA)
f.o.b.	Free on board
FOC	Flag of convenience
FONASBA	Federation of National Associations of Ship Brokers and Agents
FOS	Free on steamer or Floating oil storage
FOSFA	Federation of Oils, Seeds and Fats
FPA	Free of particular average
FPS	Fuzzy probability simulation
FTA	Free trade agreement
G.A.	General average
GAFTA	Grain and Feed Trade Association
GATT	General Agreement on Tariffs and Trade
GCBS	General Council of British Shipping
G.L.	Germanischer Lloyd Classification Society (Germany)
g.r.t.	Gross registered tonnes
HMSO	Her Majesty's Stationery Office

H.R.	Hellenic Register of Shipping Classification Society (Greece)
IAME	International Association of Maritime Economists
I.A.P.H.	International Association of Ports and Harbours
ICC	Institute cargo clauses
I.C.C.	International Chamber of Commerce
ICHCA	International Cargo Handling Co-ordination Association
I.C.S.	Institute of Chartered Shipbrokers
ILA	International Longshoremen's Association
I.L.O.	International Labour Organization
IMB	International Maritime Bureau
I.M.O.	International Maritime Organisation
IMCO	Inter-Governmental Maritime Consultative Organization (now I.M.O.)
INCOTERMS	International commercial terms
INTERTANKO	International Tanker Time Charterparty
IPP	Improving Port Performance
IR	Industrial relations
ITF	International Transport Workers' Federation
KGB	Soviet state security service
LASH	Lighter aboard ship
LDC	Less developed country
LNG	Liquefied natural gas (carriers)
LOF	Lloyd's Open Form (standard form of salvage agreement)
LOT	Load on top
LPG	Liquefied propane gas (carriers)
MARAD	Maritime Administration (USA)
MAR	Form of policy
MARPOL 73/78	The International Convention for the Prevention of Pollution from Ships (1973 modified 1978)
MIT	Massachusetts Institute of Technology
MNE	Merchant Navy Establishment Scheme
MOT	Ministry of Transport
MTO	Multimodal transport operator

NEDO	National Economic Development Office
NFS	Not for sale
NIC	Newly industrialized country
NK	Nippon Kaisi Kyokai Classification Society (Japan)
NRT	Net registered tons
NSFU	National Sailors and Firehands Union
NUMAST	National Union of Marine, Aviation and Shipping Transport Officers
N.U.S.	National Union of Seamen (now R.M.T.)
NVOCC	Non-vessel owning common carrier
N.Y.K.	Nippon Yusen Kaisha (Japan Mail-Steamer Company)
NYPE	New York Produce Exchange
O.B.O.	Oil/bulk/ore carriers
O.B.O.	Ore/bulk/oil carriers
OECD	Organization for Economic Co-operation and Development
O.O.	Oil/ore carriers
Pan.Lib.Hon.	Panama, Liberia, Honduras – Flags of convenience countries
P & I	Protection and Indemnity (Club)
P. & O.	Peninsular and Oriental Line
PR	Polish Register Classification Society
R.I.	Registo Italiano Classification Society
R.M.T.	The National Union of Rail, Maritime and Transport Workers
Ro-ro	Roll-on roll-off type vessel
SDR	Special Drawing Rights
SG form of policy	Basic contract for marine insurance – now little used
SHACDEM	A series of econometric models
SITPRO	Simpler International Trade Procedures Board
SMA	Society of Maritime Arbitrators (New York)
SNAME	Society of Naval Architects and Marine Engineers
SOLAS	Safety of Life at Sea (1974 Convention)
TCM	Combined Transport of Goods

TEU	Twenty-foot equivalent unit (containers)
T & GWU	Transport and General Workers Union
TRC	Tax Reserve Certificates
TRINMAR	Training Development in Maritime Transport
UCC	Uniform Commercial Code
ULCC	Ultra-large crude carrier
UNCITRAL	United Nations Commission on International Trade Law
UNCTAD	United Nations Conference on Trade and Development
UNITAR	United Nations Institute for Training and Research
USMC	United States Maritime Commission
UWIST	University of Wales Institute of Science and Technology
VLCC	Very large crude carrier
WA	With average
WMU	The World Maritime University
W. S. rates	Worldscale

1 Reference Books

Dictionaries; Bibliographies; Yearbooks; Atlases; Encyclopedias; Handbooks

1 **Aldcroft, Derek H.** and **Mort, Derrick.** *Rail transport. Sea transport. (Reviews of United Kingdom statistical sources. Vol. 14).* Oxford and New York: Pergamon Press, 1981. 268 pp.

Reference to sources of statistical material, both official and non official. Operating and financial statistics. Sea transport – demarcates sea transport from non-seagoing transport activities. Differentiates UK activities from other nations.

2 **Association of Singapore Marine Industries.** *Singapore marine industries directory.* Singapore: Association of Singapore Marine Industries.

A comprehensive listing of the services and facilities provided by the marine industry. Write-ups on related government bodies and statutory boards.

3 **Bakr, M.** (ed.) *Elsevier's maritime dictionary.* New York: Elsevier, 1987. 664 pp.

4 **Billy, George J.** *Sources of information in transportation. Part 3. Shipping.* 4th edn. Illinois: Vance Bibliographies, 1990.

Annotated bibliography of general sources; cargo and computers; economics; engineering; laws and regulations; navigation and seamanship; container, cruise, LNG and tanker ships. Also covers statistical sources, directories, periodicals, conferences and dictionaries.

5 **Bolton, Rita V.** and **Bryan, F. J.** *Marine transport: A guide to libraries and sources of information in Great Britain.* London: Witherby & Co., 1983. 76 pp.

A guide which lists British libraries and sources of information in the field of marine transport, including current and historical material.

6 Branch, Alan E. *Dictionary of commercial terms and abbreviations.* London: Witherby & Co., 1984. 458 pp.

A dictionary of commercial terms and abbreviations. Covers entire spectrum of the business sector embracing terms in accountancy, advertising, banking, commerce, computers, economics, insurance, international trade, law, marketing, personnel management, real estate, selling, statistics, tourism and transport.

7 Branch, Alan E. *Dictionary of shipping: International trade terms and abbreviations.* 3rd edn. London: Witherby & Co., 1986. 587 pp.

Correct and effective use of shipping, international trade terms and abbreviations. Third edition covers international marketing; international banking; cargo handling; shipping; ports; documentation; chartering, terms found in Far Eastern markets. International trade and shipping organizations. World currencies, ports.

8 Branch, Alan E. *Multi-lingual dictionary of commercial international trade and shipping terms: English, French, Spanish and German.* London: Witherby & Co., 1990. 253 pp.

Overall some 2,200 terms have been assembled to cover those used in the daily conduct of international trade and commerce.

9 Branch, Alan E. and **Abu Hakmeh, Juma.** *Dictionary of English–Arabic commercial, international trade and shipping terms.* London: Witherby & Co., 1988. 257 pp.

10 Brodie, Peter R. *Dictionary of shipping and shipbroking terms: French–English and English–French.* London: Lloyd's of London Press, 1980. 195 pp.

French–English/English–French translation. Terminology used in chartering and operation of tramp and liner ships. Abbreviations for main shipping phrases and terms.

11 Brodie, Peter R. *Illustrated dictionary of cargo handling.* London: Lloyd's of London Press, 1991. 175 pp.

Focuses on the visual aspects of cargo handling and shipping. Definitions, descriptions and illustrations of the many forms of cargo-handling equipment, vessel types, technical sections of merchant ships, containers, ports and facilities and equipment.

12 Brodie, Peter R. *New foreign language shipping directory.* London: Lloyd's of London Press, 1988.

13 Brodie, Peter R. (ed.) *Dictionary of shipping terms.* London: Lloyd's of London Press, 1985. 155 pp.

Provides a comprehensive coverage of the words, terms, phrases and abbreviations used in the international carriage of goods by sea and shipping activities.

14 Brodie, Peter R. (ed.) *Dictionary of shipping terms: French–English and English–French.* London: Lloyd's of London Press, 1989. 206 pp.

15 Brodie, Peter R. (ed.) *Dictionary of shipping terms: Greek–English and English–Greek.* London: Lloyd's of London Press, 1989. 174 pp.

Greek–English/English–Greek translation for those who use/operate cargo ships both tramp and liner. Areas covered – voyage/time charters; documentation, including bills of lading; ship types and their gear; ports and equipment and facilities; cargo and packing; geographical and weather features.

16 Brodie, Peter R. (ed.) *Dictionary of shipping terms: Italian–English and English–Italian.* London: Lloyd's of London Press, 1988. 182 pp.

17 Brodie, Peter (ed.) *Dictionary of shipping terms: Spanish–English and English–Spanish.* London: Lloyd's of London Press, 1988. 216 pp.

18 Brown, Robert H. *Dictionary of marine insurance terms and clauses.* 5th rev. edn. London: Witherby & Co., 1989. 632 pp.

Incorporates all important terms and abbreviations for the marine insurance industry, including many shipping and commercial terms.

19 Bryon, R. V. and **Bryon, T. N.** *Maritime information: A guide to libraries and sources of information in the U.K.* London: Witherby & Co, 1993.

A directory listing major maritime holdings of 500 libraries, museums, record offices, institutions and organizations.

20 Calvert, J. and **McConville, J.** *The shipping industry statistical sources.* London: City of London Polytechnic, 1983. 337 pp.

Provides access to the wealth of statistical information available on the shipping industry. The major areas of the shipping industry are covered. Subject headings are subdivided into books, and journals and annuals. Each entry gives publication details and list of statistics contained therein.

21 *Containerisation international year book.* London: National Magazine Co.

A guide to port facilities, terminals, and container traffic statistics. The directory contains a register of over 4,800 container-carrying vessels.

22 Couper, Alastair (ed.) *The Times atlas and encyclopaedia of the sea.* HarperCollins Publishers, 1989. 272 pp.

23 Cuny, P. J. (ed.) *Lloyd's maritime directory.* Colchester: Lloyd's of London Press.

Includes details of over 5,500 shipowners and managers, and 35,000 vessels world-wide.

24 De Kerchove, René. *International maritime dictionary.* 2nd edn. New York: French and European Publications, 1993. 1018 pp.

25 Degenhardt, Henry W. and **Gold, Edgar.** *Maritime affairs: A world handbook.* 2nd rev. edn. Harlow: Longman, 1991. 480 pp.

26 Dipl-Ling, V. and **Bensch, Erhard.** *Dictionary of shipbuilding, shipping and fisheries.* Collets, 1980. 784 pp.

27 *Directory of shipowners, shipbuilders and marine engineers.* Business Press International. Annual.

28 Emery, Chris (ed.) *Mediterranean shipping directory.* London: Lloyd's of London Press. Annual.

29 *Fairplay book of international organizations.* Coulsdon: Fairplay Publications, 1990. 62 pp.

Looks at the history, aims, and organization of 45 bodies in the shipping business arena.

30 *Fairplay world ports directory.* Coulsdon: Fairplay Information Systems.

Information is provided on over 2,400 ports across the world.

31 *Fairplay world shipping year book.* London: Fairplay Publications.

32 Goodwin, E. M. and **Kemp, J. F.** *Marine statistics: Theory and practice.* London: Stanford Maritime, 1979. 336 pp.

An introduction to statistical methods for those concerned with the operation of ships. Development of automatic, semi-automatic equipment; technological developments creating increased need for statistical approach to problems. Takes evidence contained in a numerical record of certain entities and quantifies particular features of the record for use in management context.

33 Gunton, G. Paul (ed.) *Directory of international ship registers, ship managers and manning agents.* London: Lloyd's of London Press. Annual.

34 Gunton, Paul (ed.) *Directory of international ship registers and ship management services.* London: Lloyd's of London Press. Annual.

35 Heal, David W. *Iron and steel. Shipbuilding (Reviews of United Kingdom statistical sources series vol. 16).* Oxford and New York etc: Pergamon Press, 1984. 203 pp.

Reference to sources of statistical material, both official and non official. Operating and financial statistics. Iron and steel – review to cover 1953–81. Shipbuilding – British shipbuilding since 1945. Employment; industrial disputes.

36 I.M.O. Library Information Services. *Developing a maritime library and information centre.* London: I.M.O., 1993. 90 pp.

Documentation of information sources as a guide for the development of a maritime library or information centre. Based upon information available in the I.M.O. Library; scope and content reflect mainly the Organisation's interests, although some reference is made to the commercial aspects of shipping.

37 Ivamy, E. R. Hardy. *Dictionary of shipping law.* London: Butterworths, 1984. 179 pp.

Definitions from case and statute law. Definitions drawn from statutory instruments made under the Merchant Shipping Acts. All terms fully described.

38 *Jane's containerization directory.* London: Jane's Information Group. Annual.

39 Kapoon, Peter. *The Fairplay book of shipping abbreviations. Acronyms and abbreviations used in shipping and international trade.* London: Fairplay Publications, 1980. 89 pp.

Abbreviations used in shipping.

40 *Lloyd's European Community shipping directory.* Lloyd's of London Press.

Bi-annual publication.

41 *Lloyd's nautical year book.* London: Lloyd's of London Press.

A wealth of information on the shipping world. Divided into four sections, it looks broadly at the shipping industry, Lloyd's and insurance, legal and international regulations, and general information.

42 Lloyd's of London. *Mediterranean shipping directory.* Colchester: Lloyd's of London Press.

Bi-annual publication giving reference material on companies in key areas of the shipping industry. For each country – bunker companies, cruise operators, ports, principal airports, banks, liner services. Ship classification societies, ship suppliers, shipbuilders/repairers, ship management companies, ship owners/managers, towage and salvage companies, trade associations.

43 *Lloyd's ports of the world.* London: Lloyd's of London Press.

Divided by continent, with additional sections on the UK and Scandinavia; this annual provides a comprehensive overview of ports world-wide.

44 Lloyd's Register. *Directory of services and offices worldwide.* London: Lloyd's Register of Shipping.

45 *Lloyd's shipping connections.* London: Lloyd's of London Press.

An annual reference directory for all involved in the maritime industry.

46 *Lloyd's Turkish shipping and port directory.* Colchester: Lloyd's of London Press.

47 *Lloyd's U.K. shipping contacts.* Colchester: Lloyd's of London Press.

48 Lopez, Norman J. *Bes' chartering and shipping terms.* 11th rev. edn. London: Barker & Howard, 1992. 641 pp.

Classic information and reference book. Practical guide for shipping professionals. Updates previous edition, includes changing circumstances in the shipping industry. Includes consideration of bills of lading legislation; 'sea waybills'; calculation of laytime with new decision; INCOTERMS 1990; intermodalism and electronic data interchange; LOF 90.

49 Mackenzie, C. H. C., Power, M. E. and **McDorman, Ted L.** *Liner* *shipping conferences: An annotated bibliography.* Indianapolis (Indiana): Lexington Books, 1984. 149 pp.

50 **Mason, Herbert B.** (ed.) *Encyclopedia of ships and shipping.* Gordon Press Publishers, 1977.

51 **Meanley, John A.** (ed.) *Maritime services directory.* Aegis Publications. Annual.

52 **Miller, E. Willard** and **Miller, Ruby M.** *Transportation – water: Waterways, shipping and ports, a bibliography.* USA: Vance Bibliographies, 1987. 42 pp.

Provides 500 references, including articles, government documents and other publications. Shipping – as to operations, economic aspects, regulations and docks. Commodities transported by water (e.g. coal, petroleum, minerals, grain). Ports – major geographical areas of US.

53 **Olev, Kulno.** *Maritime dictionary: English–Estonian–Russian.* French and European Publications, 1981. 560 pp.

54 **Parmer, Jean M.** and **Parmer, Jerome F.** *Maritime books bibliography.* Parmer Books, 1992.

55 **Siemss, John** (ed.) *Dictionary of maritime terms: Mercantile and naval shipping.* I.R. Publications, 1984. 838 pp.

56 **Stevens, Edward Frank.** *Dictionary of shipping terms and phrases.* 2nd edn. London: Pitman, 1971.

57 **Sullivan, Eric.** *The marine encyclopaedic dictionary.* London: Lloyd's of London Press, 1992. 497 pp.

Explanation of many thousands of terms, expressions, acronyms and abbreviations in the worlds of shipping, transport, freighting, insurance, finance, offshore oil and gas. Includes terms from wider fields of academic and technological interest.

58 **Tuin, J. D. Van Der** and **Newman, D. L.** *Elsevier's dictionary of ports and shipping.* New York: Elsevier, 1993. 368 pp.

59 **Valencia, Mark J.** and **Shepherd, Diana C.** *Marine policy and*

management in southeast Asian seas: A selected bibliography east–west center. 1981. 58 pp.

60 Vandenberghe, J. P. and **Chaballe, L. Y.** *Elsevier's nautical dictionary in six languages.* 2nd rev. edn. Amsterdam and New York: Elsevier Scientific Pub. Co., 1978. 949 pp.

Lists 18,782 entries which represents about 24,000 English terms or expressions. All English entries listed with their equivalents in French, Spanish, Italian, Dutch and German. Words also listed in separate indexes. Entire shipping issues, law and historical phrases.

61 Warsash Nautical Bookshop. *Bibliography of nautical books.* Southampton: Warsash Nautical Bookshop.

A non-annotated bibliography of nautical books. Covers books in English which are published or distributed in the UK and available or soon to be published. Updated annually with cumulative supplements. Includes books, journals, videocassettes, conference proceedings, computer programs, statutory instruments, M Notices, British Standards. Some books distributed in the USA only are included. Recently out-of-print books.

62 Waters, W. G. *A bibliography of articles relevant to transportation appearing in major economic journals, 1960–87.* Vancouver: Centre for Transportation Studies, 1987.

63 Wiktor, Christian L. and **Foster, Leslie A.** *Marine affairs bibliography: A comprehensive index to marine law and policy literature cumulation, 1980–1985.* The Hague: Martinus Nijhoff, 1987. 685 pp.

64 Wiktor, Christian L. and **Wenger, Larry B.** *Marine affairs bibliography: A comprehensive index to marine law and policy literature.* Charlottesville (Virginia): University of Virginia Law Library, 1991. 141 pp.

Comprehensive index to marine law and policy literature. Published quarterly with an annual cumulation. Reference works. Primary sources. Oceans and ocean policy. Law of the sea. Maritime transportation and communication. Marine transportation industry.

65 Zerby, J. A. and **Conlon, R. M.** *A classified and annotated select bibliography of ocean transport economics.* Australia: Centre for Applied Economic Research, University of New South Wales, 1978.

Research

66 Charles, Alison. Information source for thesis and dissertation research. *Maritime Policy and Management* 17(2), April–June 1980, 79–85.

Discusses possible process for research. Essential aspects of strategy to ensure simultaneously most effective use of time and resources.

67 La Saponara, Francesco. The shipping industry and statistical information in Italy: A survey. *Maritime Policy and Management* 13(4), October–December 1986, 307–14.

Paper presents a survey of statistical reports and analysis of value, issued periodically in Italy, in respect of technical and economic literature relating to maritime industry.

68 Webster, G. H. and **Metaxas, B. N.** A report on current research in maritime economics at the Ealing College of Higher Education. *Maritime Policy and Management* 9(1), 1982, 77.

69 Wojewokda, Czeslaw. The Maritime Institute in Gdansk, Poland. *Maritime Policy and Management* 10(2), April–June 1983, 125.

A report of current research and range of technical and economic assistant services offered in the field of maritime economics.

2 General

Textbooks

70 Alderton, Patrick M. *Sea transport: Operations and economics.* 3rd edn. London: Thomas Reed, 1984. 276 pp.

Overall picture of sea transport industry. Types of ships; propulsion and navigational equipment; crew; shipping trends; freight; operating ships; costs; chartering; cargoes; ports; documentation; safety. Includes self-test questions.

71 Branch, Alan E. *Elements of shipping.* 6th rev. edn. London and New York: Chapman and Hall, 1989. 484 pp.

Elements of shipping industry. Chartering; ship types; types of cargo and their stowage in cargo-handling equipment; international organizations; processing export consignment; containerization; ship operation; crewing. Economic, operating, commercial, political and technical aspects.

72 Bross, Steward R. *Ocean shipping.* Cambridge (Maryland): Cornell Maritime Press, 1956. 333 pp.

Provides an overall general knowledge of ocean shipping. A practical guide to the various problems involved in the carriage of goods by water. Impact of improvements in design and operating efficiency of ships. Trade routes. Marine insurance. Claims. Accounting. International relations. Military sea transportation service.

73 Couper, A. D. *The geography of sea transport.* London: Hutchinson & Co., 1972. 208 pp.

Impact of technological change in sea transport on trade links, shipping routes, economic activities. Role of sea transport in ancient and medieval worlds and influence of merchant shipping on British economic growth in the nineteenth century. Trends in world shipowning, shipbuilding, ship types against background of supply and demand.

74 Goss, R. O. *Studies in maritime economics.* Cambridge: Cambridge University Press, 1968. 194 pp.

A collection of studies covering many important aspects of the efficiency of sea transport – ships, seaports and the interface between them. A particular emphasis on quantification and on methods which are susceptible to numerical treatment.

75 **Great Britain. Central Office of Information.** *British industry today: Shipping. (Reference Pamphlet 125).* London: HMSO, 1974. 23 pp.

Pamphlet, one of a series concerned with shipping industry. Describes the merchant fleet, shipping services, role of the Government, employment, safety at sea and measures to control marine pollution.

76 **Gubbins, Edmund J.** *The shipping industry: The technology and economics of specialization.* (Volume 5 in Transportation Studies Series.) London: Gordon and Breach Science Publishers, 1986. 140 pp.

Comprehensive introduction to shipping industry. Shows how technological advances in shipping design have led to improved efficiency and transportation of more sophisticated cargoes, promoting international trade. Analyses complex interplay between technological developments, economic pressures and political focus.

77 **Kendall, Lane C.** *The business of shipping.* 5th edn. London: Chapman and Hall, 1986. 513 pp.

Business of commercial shipping. Various divisions of a shipowning and operating organization. Underlying principles of management. Developments in the industry.

78 **Nersesian, Roy L.** *Ships and shipping: A comprehensive guide.* Oklahoma: Pennwell Books, 1981. 252 pp.

View of all facets of shipping industry with aim of exploring all areas of shipping rather than narrow specialities. Goes step by step through an entire commercial transaction, from inception to signing the contract. Presents solid background of shipping business.

79 **Packard, William V.** *Sea trading. Vol. 1: The ships.* London: Fairplay Publications, 1984. 136 pp.

Description of shipping industry. A detailed description of the vessels used in all modern sea trading, their terminology, designs and operational constraints, speed consumption and operation loadlines, drafts and deadweights, machinery, capacities, classification, flag registration.

80 **Stevens, Edward F.** and **Butterfield, C. S. J.** *Shipping practice:*

With a consideration of the relevant law. 11th edn. London: Pitman, 1981. 173 pp.

Examines the business of shipping and the issues involved in the running of ships under modern conditions. Legislation – legal aspects. Hazardous cargo and pollution. Approved by many institutes and organizations as a recommended book for study.

81 Stopford, Martin. *Maritime economics.* London: HarperCollins Academic, 1988. 413 pp.

An introduction to the economics of the global shipping industry. Describes the cyclical mechanism of supply and demand for sea transport. Explains how shipping market is organized. Economics of liner and bulk shipping – conferences, unitization, bulk transport. Shipbuilding and scrapping. Maritime forecasting.

Economics

82 Chrzanowski, Ignacy H. *Shipping economics and challenges for the future.* Aldershot: Saxon House, 1975.

83 Chrzanowski, Ignacy H. *An introduction to shipping economics.* London: Fairplay Publications, 1985. 131 pp.

Concise study in which all sectors of shipping industry are covered, being put into their historical perspective and analysed. Influence of freight markets, cost structures, pricing, role of shipping industry in national economic context. Shipping policy. Gradations from protectionism to liberalism.

84 Chrzanowski, Ignacy H., Krzyzanowski, Maciej and **Luks, Krzysztof.** *Shipping economics and policy: A socialist view.* London: Fairplay Publications, 1979. 297 pp.

Economics and organization of sea transport. Activity of fleets in socialist countries. International division of labour. Competition. Monopolistic agreements. Contradictions in contemporary shipping policies.

85 Drewry Shipping Consultants. *Shipping statistics and economics.* London: Drewry Shipping Consultants, 1988. Annual.

86 Evans, John and **Marlow, Peter.** *Quantitative methods in maritime economics.* 2nd edn. Coulsdon: Fairplay Publications, 1990. 282 pp.

Principles, models, methodologies. Mathematical developments. Covers important issues of maritime economics and others concerned with shipping. Self-contained and caters for general needs of students.

87 Goss, R. O. (ed.) *Advances in maritime economics.* Cambridge: Cambridge University Press, 1977. 294 pp.

Eight studies of different aspects of maritime economics. Emphasis on economic efficiency of sea transport system. Flag discrimination; taxation; ship size/delays – effect on costs; economics of congestion; application of cost–benefit analysis to safety of life at sea. Substantial introduction – compares and extends some of the arguments.

88 Gwilliam, K. M. *Current issues in maritime economics.* Dordrecht: Kluwer Academic Publishers, 1993. 177 pp.

Contains a selection of papers presented at an international conference in Rotterdam, June 1991. Contains 11 papers and addresses three major areas of interest: (i) the changing international context; (ii) relationship between market structure and the workability of competition; (iii) decision processes of firms in the shipping world.

89 Gwilliam, K. M. Maritime economics in transition. In: K. M. Gwilliam (ed.) *Current issues in maritime economics.* Dordrecht: Kluwer Academic Publishers, 1993, 1–8.

An editorial introduction of collection of papers from a conference at Erasmus University in Rotterdam, June 1991 for the retirement of Professor H. J. Molenaar. Introductory paper juxtaposes the issues as identified by Molenaar in his valedictory observations and the opinions on them offered by the conference contributors.

90 Heaver, T. D. The many facets of maritime economics. *Maritime Policy and Management* 20(2), April–June 1993, 121–32.

Paper for inaugural address to founding meeting of the International Association of Maritime Economists. Component parts of structure of maritime economics. Potential contribution of IAME to study of maritime economics. Directions for future research.

91 Hope, Einar. The contributions of Arnljot Stromme Svendsen to shipping economics. In: Einar Hope (ed.) *Studies in shipping economics: In honour of Professor Arnljot Stromme Svendsen.* Oslo: Bedriftsokonomens Forlag A/S, 1981. 8 pp.

Concentrates on Arnljot Stromme Svendsen's published scientific works, i.e. shipping economics.

92 Hope, Einar (ed.) *Studies in shipping economics: In honour of*

Professor Arnljot Stromme Svendsen. Oslo: Bedriftsokonomens Forlag A/S, 1981. 103 pp.

Collection of studies *re* shipping economics. Present-day empirical research into various economic aspects of shipping.

93 Hughes, C. N. *Shipping: A technoeconomic approach.* London: Lloyd's of London Press, 1989. 135 pp.

Considers both purely technical and commercial viewpoints of decision-making in shipping. Shipbuilding considerations. Ship types – bulkers, passenger vessels, oil and chemical tankers, gas carriers. Main propulsion systems. Auxiliary power generation. Registry, manning, classification. Operational technoeconomics. Onboard systems.

94 Lawrence, S. A. *International sea transport: The years ahead.* USA, Canada and London: Lexington Books, 1972. 316 pp.

Firm background for understanding the issues confronting shipping industry. Impact of political pressure, technological change. Analyses world-wide shipping within framework of economic interdependence and global technologies.

95 Metaxas, B. N. Maritime economics: Problems and challenges for the future. *Maritime Policy and Management* 10(3), July–September 1983, 145–64.

Paper suggests that subjects of the trade cycle, maritime evolution of the developing countries and the increase in real cost of maritime transport to the world community are the three main areas requiring thought, research and analysis.

96 OECD. *Maritime transport.* Paris: OECD, 1990.

Annual report. Provides comprehensive coverage of international developments in maritime transport, with particular reference to national shipping policies and longer-term trends in international shipping and trade. Statistical annex of up-to-date data on seaborne trade, the movement of bulk commodities, world fleet, freight markets.

97 Sharp, C. H. *Transport economics. Macmillan studies in economics.* London: Macmillan, 1973. 80 pp.

Identifies basic economic problems of transport in Britain. Introductory coverage of wide variety of topics. Methods of measuring output of transport services. Importance of different modes. Economic theory of investment. Pricing theory. Evaluating in monetary terms 'intangible' benefits. Government taxation regulation policies.

98 Shimojo, T. *Economic analysis of shipping freights.* Kobe: Kobe University, 1979. 304 pp.

99 Sturmey, S. G. *Shipping economics: Collected papers.* London: Macmillan Press, 1975. 256 pp.

Collection of papers and specially written introduction where author pays attention to problem of protecting shippers from power of liner conferences. Extensive quotation from first public lecture on Code of Conduct for Liner Conferences adopted by the United Nations.

100 Thorburn, Thomas. *Supply and demand of water transport: Studies in cost and revenue structures of ships, ports and transport buyers with respect to their effects on supply and demand of water transport of goods.* Stockholm: Business Research Institute, Stockholm School of Economics, 1960. 232 pp.

Terminal part of water transport. Carrying part of water transport and price-making.

101 United Nations Conference on Trade and Development. *Review of maritime transport.* New York: United Nations.

Annual publication prepared by UNCTAD Secretariat in accordance with Section VI of the programme of work of the Committee of Shipping. Identifies main developments in world maritime transport and provides relevant statistical data.

102 Walters, A. A. A method for collecting economic statistics of shipping. *Journal of the Royal Statistical Society* Series A 134(3), 1971, 440–48.

103 Westinform Service. *Increasing vessel size: Shoreward costs, seaward savings?* London: Westinform Service, 1976. 64 pp.

Considers the implications of the building for and employment of successively larger and larger vessels on long-haul ocean routes. Examines the consequent comparable changes to port and berth accommodation. Economies of sea transport cost and costs of terminal adaptation.

3 Maritime Development (Post-1950)

Shipping Development

104 Apostolidis, A. V. The Greek shipping industry and the Greek economy. Thesis submitted for MSc Econ, Wales, UWIST, Cardiff, 1988.

This analyses the growth of the Greek shipping industry and the Greek economy and tries to identify the relationship that exists between the two over the time period 1960–84.

105 Bess, H. David and **Farris, Martin T.** *U.S. maritime policy: History and prospects.* New York: Praeger Publishers, 1981. 218 pp.

Maritime transportation. History of maritime policy. Early policy decisions and events at beginning of the twentieth century. Comprehensive analysis of Merchant Marine Act 1936. Maritime policy in 1970s. Technological changes; pricing problems; antitrust. Federal Maritime Commission.

106 Beth, H. L. Hader, A. and **Kappel, R.** *25 years of world shipping.* London: Fairplay Publications, 1984. 205 pp.

Development of shipping late 1950s to mid-1980s, against background of world economy. Changes in passenger traffic; offshore markets; technology and safety; ports; shipbuilding industry; cyclical developments. Future prospects?

107 *The city and the sea.* Colchester: Seatrade Publications, 1978. 32 pp.

Study carried out on the basis of what the shipping industry sees in the City. How the City serves the shipping industry. Lloyd's and insurance; P & I; the Baltic Exchange; arbitration; sale and purchase; alternative markets.

108 Clegg, Paul. *British shipping.* Weybridge: Allan (Ian) Ltd, 1988.

109 Corlett, Ewan. *The revolution in merchant shipping.* New York: Sheridan House, 1981. 60 pp.

110 Couper, Alastair D. (ed.) (Series editor: Robert Gardiner.) *The shipping revolution; The modern merchant ship.* London: Conway Maritime Press, 1992. 192 pp.

Third volume in Conway's 'History of the Ship' series. Starts with an overview of world shipping and shipping revolution against background of changing operational environment in period from 1960s. Developments in design and functions of modern ships. New offshore industries support vessels. Shipbuilding. Propulsion. Navigation. Equipment. Ownership of vessels.

111 Craig, R. S. The business historian and the shipping industry: A review and prospect. *Business Archives,* no. 33, December 1970, 7–15.

112 Davis, J. G. *World shipping today and its prospects. Wakeford Memorial Lecture.* University of Southampton, 1988. 16 pp.

113 De La Pedraja Toman, René. *The rise and decline of U.S. merchant shipping in the twentieth century.* New York: Macmillan Publishing Company, 1992. 200 pp.

114 Duff, Peter. *British ships and shipping.* Gordon Press Publishers, 1977.

115 No entry.

116 Frankel, Ernst G. *Structural adjustment and integration of ocean shipping: 2nd Chua Chor Teck Annual Memorial Lecture.* London: Chua Chor Teck Memorial Fund Trustee Committee, 1988. 14 pp.

Memorial lecture given Wednesday, 13 January 1988. Reviews the changing conditions in ocean shipping. Technological developments. Economic impact changes in shipping. Structural changes in shipping. Future prospects and developments. The new environment and impact on shipping.

117 Frankel, Ernst Gabriel. *The world shipping industry.* London and New York: Croom Helm, 1987. 261 pp.

Reviews expanding role of developing countries in shipping and evaluates contribution of shipping to development. Changes in institutional and environmental framework of shipping. Social impact. Challenges, opportunities, problems of shipping industry.

118 Gandhi, D. S. *Shipowners, mariners and the new law of the sea: The effects of technical, legal and political constraints upon the traditional users of the world's oceans.* London: Fairplay Publications, 1977. 83 pp.

119 Graham, M. G. and **Hughes, D. O.** *Containerisation in the eighties.* London: Lloyd's of London Press, 1985. 237 pp.

Historical account of the industry since advent of containerization. Problems affecting future. Recent developments in 1980s. What is important to container services. Future impact of UN Liner Code and UN Multimodal Convention.

120 Hayter, D. M. *Changing patterns in U.K. trade and unitized freight transportation, 1970–1978.* Public Sector Economics Research Centre, University of Leicester, 1980. 44 pp.

121 Hope, Ronald. *The Merchant Navy.* London: Stanford Maritime, 1980. 132 pp.

The British shipping industry, the companies of which it is comprised, and its structure. Organization of the merchant ship and the law as it applies to merchant seamen. Merchant Navy terms and phrases. Merchant Navy institutions. Place of UK fleet in the world.

121a Hope, Ronald. *A new history of British shipping.* London: J. Murray, 1990. 533 pp.

Influence of ships. Changing cargoes; cargo destinations; hull improvements; propulsion; navigation. Living conditions of crew; crew wages and safety. Government policy; economic theory.

122 Ingpen, B. D. *South African merchant ships: An illustrated recent history of coasters, colliers, containerships, tugs and other vessels.* Cape Town: A. A. Balkema, 1979. 149 pp.

A history of South African merchant shipping, with special reference to the post-1910 period. Anecdotes about ships, owners, shipwrecks, cargoes, commercial problems, financing and the personalities involved. 108 photographs of ships presented to amplify the text, and maps and diagrams which illustrate development.

123 Kearsley, Harold. *Maritime power and the twenty first century.* Aldershot: Dartmouth Publication Co., 1992. 150 pp.

124 Lauriat, George. *China shipping: The great leap forward.* (A

Lloyd's Shipping Economist study.) Colchester: Lloyd's of London Press, 1983. 196 pp.

Reviews political situation from 1949 to 1976. The 'Great Leap' – the mass assault on economic problems. Impact of inefficiency of this assault on economy. Maritime industry. Ways used to acquire merchant tonnage. Building a national fleet, China and the sea. Ports and waterways. Marine insurance. Shipbuilding and repairs. Future prospects.

125 Lee, M. Shipping policies of the Republic of China on Taiwan, 1950–1985: A study of the national fleet development. Thesis submitted for PhD, University of Liverpool, 1988.

This describes the shipping policies on Taiwan and their effects on the national economy and the national fleet over the period 1950–85. It comments on the strategy in formulating future policy.

126 Rinman, Thorsten and **Brodefors, Rigmor.** *The commercial history of shipping.* Gothenburg, Sweden: Rinman & Linden, 1983. 191 pp.

The commercial development of shipping during the last 200 years. Shows how economic and technical developments have affected the international shipping industry.

127 Rinman, Thorsten and **Linden, Rigmor.** *Shipping: How it works.* Gothenburg, Sweden: Rinman & Linden, 1978. 191 pp.

Historical roots of modern shipping industry, from circa 1818. World seaborne trade. Industrial and economic developments – impact on shipping. Structure of shipping market.

128 Schönknecht, Rolf and others. *Ships and shipping of tomorrow.* Hounslow: Macgregor Publications, 1983. 240 pp.

Developments in ships and shipping – scientifically based examination of development possibilities. Advantages and drawbacks of solutions that are technologically and economically practicable. Developments described are illustrated in multi-colour drawings.

129 *Seatrade: 20th anniversary study: 1970–1990.* Colchester: Seatrade Publications, 1990. 44 pp.

Considers the shipping industry's triumphs and disasters of past two decades, and what's in store for world shipping. Personal opinions given by some of shipping's outstanding personalities.

130 Sjostrom, William. On the origin of shipping conferences: Excess capacity and the opening of the Suez Canal. *International Journal of Transport Economics* 16(3), October 1989, 329–35.

131 Sturmey, S. G. *British shipping and world competition.* London: Athlone Press, University of London, 1962. 436 pp.

Why has UK tonnage declined? Critical analysis. Traces history of shipping industry in its international context. Examines various explanations for decline. Prospects for future.

132 United Nations Conference on Trade and Development Secretariat. *Major issues in world shipping: a) Merchant fleet development; b) Structure of world shipping.* Geneva: United Nations, 1986. 43 pp.

Consolidated report as response to requests of Conference and Committee at 10th and 11th sessions. Considers imbalance between supply and demand in world shipping. Developments and issues in world bulk and liner shipping. Recommendations.

133 United Nations Conference on Trade and Development. *Shipping in the 70's.* New York: United Nations, 1972.

Shipowners; Entrepreneurs; Companies

134 Davies, P. N. *Sir Alfred Jones: Shipping entrepreneur par excellence.* London: Europa Publications, 1978. 162 pp.

Biography of life and career of Alfred Jones (1845–1909), dominant figure in development of West African trade and contributor to establishment of British West African colonies.

135 Davies, Peter N. *Henry Tyrer: A Liverpool shipping agent and his enterprise, 1879–1979.* London: Croom Helm, 1979. 159 pp.

Investigates the development of the businessman Henry Tyrer and his background within Liverpool commerce in the nineteenth and twentieth centuries, within various aspects of shipping in Liverpool, Preston and London.

136 Garrett, K. S. *Everard of Greenhithe.* Kendal: World Ship Society, 1991. 240 pp.

137 Hodgson, Godfrey. *Lloyd's of London: A reputation at risk.* Harmondsworth: Penguin Books, 1986. 414 pp.

Traces the company's past and present and monitors the ebb and flow of Lloyd's and its work. The author chronicles a series of scandals from the Savonita affair, through the computer-leasing debacle and the Sasse Syndicate to Ian Posgate.

138 **Howarth, Stephen.** *Sea shell: Story of Shell's British tanker fleets, 1892–1992.* New Malden: T. Reed, 1992.

139 **Lloyd's List International.** *Cunard Line.* Colchester: Lloyd's of London Press. 48 pp.

A special supplement to commemorate the 150th anniversary of Cunard. The philosophy of modern expansion. The beginnings of Cunard. Development and expansion. Liverpool to fete the QE2. 150 years at sea.

140 **Mitchell, William Harry** and **Sawyer, L. A.** *Sailing ships to supertanker: Hundred year story of British Esso and its ships.* Sudbury: T. Dalton, 1987. 264 pp.

141 **Naess, Erling D.** *Autobiography of a shipping man.* Colchester: Seatrade Publications, 1977. 266 pp.

142 **Nakagawa, Keiichiro** and **Yui, Tsunehiko** (eds) *Business history of shipping.* Irvington (New York): Columbia University Press, 1985. 300 pp.

143 **Padfield, Peter.** *Beneath the house flag of the P & O.* London: Hutchinson & Co., 1981. 143 pp.

A nostalgic evocation of the great days of the Peninsular and Oriental Line, which plied the steamship route to India and the East.

144 **Rees, D. J.** A study of management and organisation development in a U.K. shipping transport and trading company, 1972–80. Thesis submitted for MA, Durham, 1987.

145 **Sedgwick, Stanley** and **Sprake, Raymond F.** *London and Overseas Freighters Ltd., 1949–77.* Kendal: World Shipping Society, 1977. 60 pp.

146 **Sivewright, Sandy.** *One man's mission: 20,000 ships.* London: Lloyd's of London Press, 1989. 272 pp.

147 Somner, Graeme H. *Ben Line: Fleet list and short history.* Kendal: World Shipping Society, 1980. 120 pp.

Describes the Ben Line Fleet. A short history of the Ben Line. The Board of The Ben Line Steamers Ltd. Fleet list – sailing vessels; Ben Line; Bamburg Shipping Co. Ltd; Ben Odeco Ltd; Atlantic Drilling Co. Ltd. Vessels managed for the British Government; Ben Line Ship Management Ltd.

148 Yui, Tsunehiko and **Nakagawa, Keiichiro** (eds) *Business history of shipping: Strategy and structure – conference proceedings.* Tokyo: University of Tokyo Press, 1985. 330 pp.

4 Demand

General

149 Drewry Shipping Consultants. *Asian oil trades and tanker demand.* London: Drewry Shipping Consultants, 1983.

150 Winston, Clifford. A multinominal probit prediction of the demand for domestic ocean container service. *Journal of Transport Economics and Policy* 15(3), September 1981, 243–52.

This attempts to demonstrate the usefulness and limitations of the prediction capability of statistical choice models in analysing actual instances of a new alternative – the ocean container service.

General

151 Chizea, B. I. Ship fleet size in relation to available demand for cargo shipments and to route characteristics. Thesis submitted for PhD, University of Manchester, 1981.

152 Market Development Dept. B & W Diesel A/S. *Balance in maritime sectors: Transport demand, scrapping and contracting in the 80's.* Denmark: Market Development Department, B & W Diesel AS., 1980. 84 pp.

Report prepared for the B & W Diesel A/S licence meeting in Copenhagen, August 1980. Updates and elaborates earlier prospective report. Summary of shipbuilding market. Seaborne oil and dry bulk transports. General cargo transports. Methodologies.

153 Oum, Tae Hoon. Derived demand for freight transport and intermodal competition in Canada. *Journal of Transport Economics and Policy* XIII (2), May 1979, 149–68.

Formulates a demand model for intercity freight transport as intermediate input to production and distribution sectors of economy. Estimates price elasticities, elasticities of substitution between, *inter alia*, waterway freight transport, excluding ocean shipping.

154 Winston, C. The demand for freight transportation: Models and applications. *Transportation Research* Series A 17(6), 1983, 419–27.

Forecasting

155 Binkley, James K. and **Bessler, David A.** Expectations in bulk ocean shipping: An application of autoregressive modeling. *Review of Economics and Statistics* 65(3), August 1983, 516–20.

This is an analysis of the behaviour in a market with a high degree of instability, ocean shipping, of dry bulk commodities. It argues that the combination of a fixed supply and a variable demand results in highly fluctuating rates.

156 Drewry Shipping Consultants. *Tanker supply, demand and profitability to 1996.* London: Drewry Shipping Consultants, 1991. 107 pp.

Part of the Seaborne Trade and Transport Reports. Historical trading performance. Forecast profitability to 1996. The nature of shipping costs. Development of the tanker fleet.

157 Fairplay Publications. *Review and outlook of the shipping market.* London: Fairplay Publications. Annual.

158 Grafton, R. G. T. The forecasting of shipping charter rates. Thesis submitted for PhD, University of Liverpool, 1981.

159 Hampton, Michael J. *Long and short shipping cycles: The rhythms and psychology of shipping markets.* 3rd edn. Cambridge: Cambridge Academy of Transport, 1991. 71 pp.

Predictions for the performance of the 1990s tanker freight market. Psychological basis for forecasts. Short cycles of three to four years. Long cycles of about twenty years. Long cycle peaks and lows. What may lie ahead. Collapse in general economy? After 1990: tanker feast and dry-bulk famine.

160 Moore, C. G. and **Pomrehn, H. P.** Technological forecast of marine transportation systems 1970–2000. *Technological Forecasting and Social Change* 3, 1971, 99–135.

161 **Nagatsuka, Seiji.** *Trends of the world shipping and shipbuilding in 1991 and prospects for the same in the near future.* Tokyo: Japan Maritime Research Institute, 1992. 68 pp.

Booklet, part of series related to maritime affairs. Trends and future prospects of major industries demanding shipping services. Trends and prospects of shipping. Trends and prospects of shipbuilding. Trends in Japanese shipping and shipbuilding. Updates previous titles.

162 **Sychrava, Lev** and **Bush, Marian.** *Forecasting ship demand.* London: Seatrade, 1971. 20 pp.

Examines the EEC global forecasts for demand for ships. Discussion of forecasting approach, plus modifications. Illustrates effect of some of modifications on revised forecast for 1975.

163 **Wright, G.** Expectations in the shipping sector. *International Journal of Transport Economics* 20(1), February 1993, 67–76.

What is the basis for expectations formed about the shipping sector?: rational and non-rational expectations are considered.

Economic Theory

General

164 **Hattori, Katsuhiko.** *Tonnage demand: Supply trends and outlook of scrapped tonnage.* Tokyo: Japan Maritime Research Institute, 1984. 38 pp.

Outlines trends at the time of tonnage demand and supply and the scrapping situation in that connection, to be followed by attempt to predict future tendencies in long-term perspective.

165 **Jansson, J. O.** and **Shneerson, D.** The effect of capacity costs and demand elasticities on the structure of liner freight rates. *Logistics and Transportation Review* 22(1), March 1986, 3–25.

5 Supply

General

General

166 **Abrahamsson, B. J.** The marine environment and ocean shipping. *International Organization*, Spring 1978, 291–311.

167 **Azzee, B. B. H.** Shipping and the transfer of technology. *Maritime Policy and Management* 9(3), 1982, 229–34.

Contends that shipping qualifies as a decisive and effective supplementary element in the total process of technology transfer.

168 **Cameron, R. M.** Economic ship design. Thesis submitted for PhD, University of Glasgow, 1971.

169 **Casson, M., Barry, D.** and **Horner, D.** The shipping industry. In: M. Casson and others. *Multinationals and world trade: Vertical integration and the division of labour in world industries.* London: Allen & Unwin, 1986, pp. 343–73.

170 **Corkhill, Michael.** *Tonnage measurement of ships: Towards a universal system.* London: Fairplay Publications, 1977. 110 pp.

History of tonnage measurement and the build-up to the acceptance of the IMCO's International Convention for the Tonnage Measurement of Ships, 1969. Effect of Convention on tonnage users. Assessment of IMCO's role in helping to implement Convention.

171 **Corkhill, Michael** (rev. by Andrew Moyse). *Chemical tankers: The ships and the markets.* 2nd edn. London: Fairplay Publications, 1981. 196 pp.

Comprehensive technical and commercial study of the chemical fleet and its market, through post-oil-crisis years to subsequent boom. Progress and acceptance of IMCO Code for the Construction and Equipment of Chemical Tankers.

172 Drewry Shipping Consultants. *World bulk fleets.* London: Drewry Shipping Consultants, 1988. 78 pp.

Report – part of Seaborne Trade and Transport series covering world's bulk shipping. The secondhand ship market. Fleet profile, 1983–8. Removals from the fleet. Additions to the bulk fleet. Current outlook.

173 Madigan, Richard E. *Taxation of the shipping industry.* 2nd edn. Cornell Maritime Press, 1982. 108 pp.

174 Mayer, H. Some geographic aspects of technological changes in maritime transportation. *Economic geography* 49, 1973, 145–55.

175 Shimojo, T. A proposal of a measure for the shipping services. *Review of Kobe University of Mercantile Marine,* 23, 1975, 310–89.

176 Spencer, C. F. *Spencer on ships: Modern commercial merchant vessels and operational techniques.* Worcester: Square One Publications, 1992. 346 pp.

Introduction to a wide variety of vessels, their typical cargoes and operations. Ships, their features, problems and operations. Merchant vessels – commercial; oil industry specials. Civil engineering – marine operations. Operational techniques. Surveys, notes.

177 Sturmey, S. G. Social costs and pricing in shipping. In: K. M. Gwilliam (ed.) *Current issues in maritime economics.* Dordrecht: Kluwer Academic Publishers, 1993, pp. 101–11.

Considers the impact of social costs upon shipping. Adopts a two-part classification of costs – costs for which enterprises make payment; costs which fall on society without compensation.

178 Taylor, A. J. The modelling of supply in the shipping industry. *Omega* 4(2), 1976.

179 United Nations Conference on Trade and Development. *Technical and financial assistance in shipping and ports.* Geneva: 1977.

Productivity

180 Claessens, E. Basic cost–size relations in general cargo ships. *Maritime Policy and Management* 14(2), 1987, 163–70.

Discusses basic cost relationship applied to purchase costs of general cargo ships, sold in yen, in period 1975–84 and delivered by Japanese shipyards.

181 Frankel, E. G. The economics of technological change in shipping. *Maritime Policy and Management* 18(1), 1991, 43–53.

How to use technology productivity analysis as method for effective management of technological change in shipping, with models to represent economic impact of technological change.

182 Goss, Richard O. The measure of productivity in shipping. In: Einar Hope (ed.) *Studies in shipping economics: In honour of Professor Arnljot Stromme Svendsen.* Oslo: Bedriftsokonomens Forlag A/S, 1981. 5 pp.

Physical measures of productivity. Economic measure of productivity.

183 Ha, Yeong Seok. Estimations of translog cost function and productivity in the shipping industry in Korea. Thesis submitted for PhD, City University of New York, 1991.

Congestion

184 Vanags, A. H. The application of the economics of congestion to marine problems. *Journal of Navigation* 25(3), July 1972, 392–4.

185 Vanags, A. H. Maritime congestion: An economic analysis. In: R. O. Goss (ed.) *Advances in maritime economics.* Cambridge: Cambridge University Press, 1977, 192–212.

Economic aspects of congestion in maritime context. A deterministic model of congestion. A queuing model of congestion. Practical implications. Appendix to elaborate technical points of argument.

Safety

186 Barker, Simon. Hazardous goods at sea: Are safe ships and clean seas mutually exclusive? A Canadian perspective. *Marine Policy* 16(4), July 1992, 306–32.

Examines how the international community and Canada have dealt with the carriage of dangerous cargo at sea and considers the shift in the focus on safety to focus on the protection of the marine environment.

187 **Burger, W.** The use of marine space off north-west Europe: Including associated problems, with particular emphasis on shipping accidents together with some proposals. Thesis submitted for MSc, UWIST, Cardiff, 1980.

188 **Butcher, John** and **De Boo, Arie.** *Safety at sea.* London: Conservative Political Centre, 1973. 36 pp.

Examines attitudes towards safety at sea and improving standards of safety at sea.

189 **Gardner, B. M.** and **Goss, R. O.** Lifeboats versus inflatable liferafts: A comparison of costs and benefits. In: R. O. Goss (ed.) *Advances in maritime economics.* Cambridge: Cambridge University Press, 1977, 257–87.

Reports on economic study: (i) to test practicability of applying cost–benefit analysis to problems in marine safety field; (ii) to investigate extent to which costs and benefits of substituting inflatable liferafts for lifeboats could be assessed.

190 **Gates, Edward T.** *Maritime accidents: What went wrong.* Houston (Texas): Gulf Publishing Co., 1990. 144 pp.

191 **Giziakis, K.** The cost of marine traffic accidents world wide with particular reference to North-West European waters. Thesis submitted for PhD, CNAA, City of London Polytechnic, 1987.

This introduces an economic concept to maritime traffic accidents and argues it is useful for shipowners, marine insurers, state and international organizations to understand the importance of the problems according to economic ordering and to do cost–benefit analysis for proposed policies or to evaluate the existing ones.

192 **Gold, Edgar.** Legal aspects of the transportation of dangerous goods at sea. *Marine Policy* 10(3), July 1986, 185–90.

Traces the background to the regulation of and liability for the carriage of dangerous, noxious and hazardous substances at sea.

193 **Goss, R. O.** Rational approaches to maritime safety. *Transaction*

of N.E. Coast Institute of Engineers and Shipbuilders 105(3), June 1989.

194 Goss, R. O. and **Vanags, A. H.** The costs and benefits of navigational aids in port approaches. In: R. O. Goss (ed.) *Advances in maritime economics.* Cambridge: Cambridge University Press, 1977, 213–54.

Cost–benefit analysis in principle. Calculation of costs and benefits. Conclusions and practical problems.

195 Lloyd's Shipping Economist. *Shipping and the environment.* London: Lloyd's of London Press, 1991. 28 pp.

196 MacAlister, Terry. Changes at sea. *International Management* 47(3), March 1992, 58–61.

European shipowners are having to face the fact that ignoring safety and environmental concerns is going to be costlier than prevention in the first place.

197 Nautical Institute. *Management of safety in shipping: Operations and quality assurance.* London: Nautical Institute, 1991. 238 pp.

A collection of papers which collectively consider the question of how ship safety can be managed effectively in the current competitive operational climate.

198 Ramwell, Dave and **Madge, Tim.** *Ship too far: Truth about the loss of the 'Derbyshire'.* Hodder, 1992. 256 pp.

Routeing

199 Caillol, R. Rising fuel costs and the geographical advantages of Mediterranean ports. *Maritime Policy and Management* 8(1), 1981, 41–7.

A methodological framework for computer simulation analysis of vessel routeing. Shows that Mediterranean ports have a competitive edge, between Far East and Europe, over North-West Europe.

200 Davis, P. V. Computer modelling of marine traffic behaviour. Thesis submitted for PhD, CNAA, 1981.

201 Fleming, D. K. Imagination of movement. *Maritime Studies and Management* 3(4), April 1976, 193–4.

Editorial beginning which suggests that the less frequent and more perfunctory use of maps by transport specialists results in lost opportunities in analysis of the maritime scene.

202 Gilman, S. Optimal shipping technologies for routes to developing countries. *Journal of Transport Economics and Policy* XI(1), January 1977, 24–44.

Explores relationship between route conditions and the choice of handling technology, with particular reference to routes between developed and developing nations. Comparison of cargo handling rates; costs at sea, in ports.

203 Higham, J. R. Optimal ship routeing for the North Atlantic Ocean using a dynamic programming model. Thesis submitted for M.Phil, CNAA, Plymouth Polytechnic, 1988.

A survey of the development of optimal ship routeing has been carried out with particular attention being given to optimization techniques; the advantage of optimal routeing has also been documented.

204 Inter-Governmental Maritime Consultative Organisation. *Ships' routeing.* London: I.M.O., 1978. 130 pp.

Describes the general provisions of routeing, traffic separation schemes, deep water routes and areas to be avoided, which have been adopted by IMCO. (Supersedes all previous editions.)

205 McKenzie, J. S. The routeing of ships. *Journal of Transport Economics and Policy* V(2), May 1971, 201–15.

Probable costs and benefits of area and ultimately world-wide control over route and speed of each ship. Suggest this would eliminate: (i) queuing for berths; (ii) inadequate use of berths. Would prevent collisions.

206 Motte, R. *Weather routeing of ships.* London: The Maritime Press, 1972. 149 pp.

Outlines methods for an effective system of ship routeing and analyses routeing techniques in use. Includes an economic summary to provide an indication of the potentialities of this technique.

207 Norton, Roger D. Policy issues in the routeing of radioactive materials shipments. *Natural Resources Journal* 21(4), October 1981, 735–55.

208 Plant, G. International legal aspects of vessel traffic services. *Marine Policy* 14(1), January 1990, 71–80.

Examines the legality of establishing Vessel Traffic Services in all sea areas. Considers suitability in different waters for imposing (a) mandatory systems or (b) voluntary compliance.

209 Quon, T. K. S., Bushell, G. E. and **Laube, J. A.** Risk analysis of vessel traffic systems. *Maritime Policy and Management* 19(4), October–December 1992, 319–36.

Risk analysis approach to capital planning of vessel traffic services (embedded within benefit–cost methodology) used in Canadian context.

210 Ronen, D. Cargo ships routeing and scheduling: Survey of models and problems. *European Journal of Operational Research* 12, 1983, 119–26.

211 Schechter, Joel. A minimum cost vessel routeing model applied to the U.S. trust territory of the Pacific Islands. *Maritime Studies and Management* 3(4), April 1976, 195–204.

Paper deals with the determination of minimum-cost vessel routeing as an aspect of the economic, social and political relationships that impinge upon shipping problems of US Trust Territory.

212 Veldman, Simme J. H. *Route planning in a centrally assisted liner system.* Delft: Eburon, 1990. 233 pp.

213 Willingale, M. C. The port-routeing behaviour of short-sea ship operators: Theory and practice. *Maritime Policy and Management* 8(2), 1981, 109–20.

A macro theory of port routeing is presented, and an examination of the business environment of operation. In-depth interviews reveal operating policies of individual shipping companies.

214 Willingale, M. C. Port-routeing decision making. Thesis submitted for PhD, University of Southampton, 1982.

215 Willingale, M. C. Ship-operator port – routeing behaviour and the development process. In: B. S. Hoyle and D. Hilling (eds) *Seaport systems and spatial development: Technology, industry and development strategies.* Chichester/New York (*et al.*): John Wiley & Sons, 1984, pp. 43–59.

Examines port planning in Britain through the eyes of the primary customers of ports – the shipping lines. Port planning in Britain 1965 to 1980. Circumstances underlying new-service establishment. Service establishment's decision-making process. Port selection.

Economic theory

General; shipping problems

216 **Charemza, W.** and **Gronicki, M.** An econometric model of world shipping and shipbuilding. *Maritime Policy and Management* 8(1), 1981, 21–30.

Using the SHACDEM series of econometric models for analysing, quantitatively, supply-and-demand balances in world shipping and shipbuilding markets.

217 **Drewry Shipping Consultants.** *Bulk fleet growth and tonnage supply in the 1990's.* London: Drewry Shipping Consultants, 1990. 86 pp.

Report – part of series covering world's bulk shipping. The bulk fleets – mid-1990 profile. The new building market and vessel deliveries. Recent trends in scrapping and loss. The outlook for bulk fleet supply in the 1990s. The second-hand ship market.

218 **Frankel, E. G.** Shipping: Choice of technology. *Maritime Policy and Management* 10(1), 1983, 1–15.

Technology choice in shipping is a complex issue due to the wide variety of alternatives and to demands for more specialization in cargo form, containment, types of handling, methods of transport and efficient intermodal interface.

219 **Goss, R. O.** and **Jones, C. D.** *Economics of size in dry bulk carriers.* London: HMSO, 1971. 49 pp.

220 **Goulielmos, A.** A vintage model approach to some problems of shipping economics. Thesis submitted for PhD, Brunel University, 1974.

221 **Hope, Ronald.** Solving shipping's problems: Johnson's law. *Maritime Studies and Management* 2(1), July 1974, 1–4.

Editorial introduction which suggests that many problems are tackled in an amateur fashion. Absence of serious and critical examination gives cause for concern.

222 **Otimong, S.** A quantitative approach to model port shipping operations, with reference to Dar-es-Salaam Port. Thesis submitted for MSc, UWIST, Cardiff, 1983.

Ship price; second-hand

223 **Beenstock, Michael.** *A theory of ship prices.* London: City University Business School, 1984. 24 pp.

Working Paper No. 68 of the Working Paper Series. Examines the theoretical determinants of ship prices. Examines capital theory for appropriate underpinning in determining ship prices.

224 **Drewry Shipping Consultants.** *Investment opportunities in second hand ships.* London: Drewry Shipping Consultants, 1987. 87 pp.

225 **Drewry Shipping Consultants.** *Investment opportunities in second-hand tonnage: A review of profitability, 1971–76 and prospects, 1976–81.* London: H. P. Drewry, 1976. 87 pp.

Part of series covering world shipping. The economic structure of the shipping market. The economics of shipowning. Shipping revenues. Shipowners' profitability 1971–6. Investment opportunities in second-hand tonnage 1976.

226 **Drewry Shipping Consultants.** *The market for secondhand tankers and bulkers.* London: Drewry Shipping Consultants, 1985. 97 pp.

A study which examines the tanker sale and purchase market. Evaluates post-1979 sale activity by volume and vessel price. Examines the documentation used and channels followed in representative sale and purchase transactions. Role of agents involved. Ships most likely to be bought and sold in future years.

227 **Drewry Shipping Consultants.** *Secondhand dry cargo ships: A review of trends in the sale and purchase market.* London: Drewry Shipping Consultants, 1981. 81 pp.

Report – part of series covering world's bulk shipping. Factors affecting the sale and purchase market. A historical review 1975–80. National sale and purchase activity. Future prospects for second-hand dry cargo ship values.

228 Drewry Shipping Consultants. *Secondhand ships: Market sector prospects and investment options.* London: Drewry Shipping Consultants, 1992. 163 pp.

229 Hale, C. and **Vanags, A.** The market for second-hand ships: Some results on efficiency using co-integration. *Maritime Policy and Management* 19(1), 1992, 31–9.

Looks again at structure of shipping markets and asks whether they are efficient. Concept of co-integration used to analyse world-wide markets for second-hand dry bulk carriers.

230 Kappel, R. *Developments in the second-hand market.* Bremen: Institute of Shipping Economics, 1984. 46 pp.

Lay-up

231 Alderton, Patrick M. *When to lay up: The theory and practice.* London: City of London Polytechnic, 1988. 12 pp.

Considers the general theory as to when a shipowner should lay-up his vessel. Observes how shipowners have reacted to low freight rates over last thirty years.

232 Goss, R. O. All deadweight tonnes are not the same. (Editorial.) *Maritime Policy and Management* 11(4), 1984, 223–6.

Gives reasons why continued existence of large volumes of laid-up tonnage does not mean entrepreneurs are wrong to enter shipping or to build new ships.

233 Haralambides, Hercules E. Estimation of laid-up tonnage in competitive shipping markets. *Logistics and Transportation Review* 22(2), June 1986, 184–92.

234 Kanjilal, S. M. The economics of ship lay-up and restoration. Thesis submitted for MSc, University of Strathclyde, 1979.

Ship cost

235 *Characteristics and cost structure of the bulk shipping fleet.* Australia: Bureau of Transport and Communications Economics, 1988. 244 pp.

236 Douglas, P. S. The effects of ship financing and leasing on the measurement of shipping costs. *Maritime Policy and Management* 12(1), 1985, 27–34.

Ships' capital costs, largest single element of total cost of providing shipping services, are essentially unquantifiable. High-percentage financing is a common feature of ship purchase.

237 Drewry Shipping Consultants. *Ship costs: Their structure and significance.* London: Drewry Shipping Consultants, 1990. 177 pp.

Report – part of Seaborne Trade and Transport series covering world's bulk shipping. The cost of shipping. Ship acquisition costs. Operating costs. Voyage costs. External cost influences. Future prospects.

238 Gilman, S. A review of ships' costs. *Maritime Policy and Management* 12(1), 1985, 91–8.

Examines methodology, definitions and practices regarding ships' costs. Issues arising from cost relationships, parametric studies and network costs.

239 Goss, R. O. Ships' costs: A review article. *Maritime Policy and Management* 10(2), April–June 1983, 127–31.

Reviews situation of available cost data on ships' capital and operating costs. Suggests need for agreement on methods, sources of raw data to be used. Reviews progress in this area.

240 Goss, R. O. Ships' costs: The overall problem and some solutions. *Maritime Policy and Management* 12(1), 1985, 1–8.

Opening paper at Ships' Costs Conference, September 1984, the object of which was 'to discuss and seek agreement on definitions, practices and procedures involved in evaluating costs of sea transport'.

241 Goss, R. O. and others. *The cost of ships' time.* London: HMSO, 1974. 45 pp.

Presents a method of evaluating the cost of ships' time that could be used to evaluate predictable delays or time savings to ships. Methods; results; delays to cargo; delays to containers.

242 Harding, A. S. and **Ryder, S. C.** *The impact of irregular ship arrivals on liner terminals with high fixed costs.* Liverpool: Marine Transport Centre, University of Liverpool, 1978. 78 pp.

Examines the effect and port costs of irregular arrivals; identifies circumstances under which shipping companies have some room for manoeuvre and where some system benefit may be possible; suggests possible approaches to improving the ship–port match.

243 United Nations Conference on Trade and Development. *Technological progress in shipping: The impact on world seaborne trade of changes in shipping costs.* 1970.

Operational research

244 Fleming, D. K. The U.S. maritime dilemma: Skaarup's view. *Maritime Policy and Management* 15(4), 1988, 265–7.

O. Skaarup says that the US maritime industry is in a deplorable condition despite direct and indirect subsidies. Without subsidy it is totally uncompetitive; with subsidy, incentive to improve vanishes.

245 Moreby, D. H. Organizational analysis in shipping. Thesis submitted for PhD, University of Bath, 1974.

246 Tomlinson, D. S. Developing a data base for planning: A shipping industry case. *Long Range Planning* 13(6), 1980, 42–5.

Speed

247 Avi-Itzhak, B. *Speed, fuel consumption and output of ships: Some quantitative economical and national implications of the oil crisis.* Haifa: Israel Shipping Research Institute, 1974. 35 pp.

Deals with some aspects of both price hikes and possible shortages arising from the oil crisis. Mathematical relations between fuel consumption, steaming speed, turnaround time and output. Development of quantitative models for determining optimal economic speed as function of bunker costs. Allocating fuel to ships in cases of shortage.

248 Buxton, I. L., Zachariades, A. and **Svensen, T. E.** *Ship speed and economics.* Newcastle: School of Marine Technology, University of Newcastle, 1983. 123 pp.

Short course notes from 11–13 January 1983. Fleets and freight market; ship types and machinery; calculating optimal speed; economic factors affecting optimal speed; technical and other factors affecting optimal speed; estimating optimal speed; speed and performance in service; liners and new ships.

Elasticity

249 Evans, J. J. The elasticity of supply of sea transport. *Maritime Policy and Management* 15(4), 1988, 309–13.

Looks at theory of short-run supply function of tramp shipping and concludes that elasticity of supply is constant at all levels of output from lay-up point to maximum design speed.

Shipowning; Management

General

250 Chapman, Stephen E. *A global analysis of ship ownership and ship management.* Colchester: Lloyd's of London Press, 1989. 36 pp.

An industry survey, summarizing an analysis of 340 responses to a Lloyd's *Ship Manager and Shipping News International* questionnaire sent to senior executives in the shipping industry in 1989. Characterizes the size, location, trades and markets in which ship owning and ship management companies operate.

251 Chrzanowski, Ignacy H. *Concentration and centralisation of capital in shipping.* Farnborough, Hants: Saxon House, 1974. 167 pp.

Differing attitudes of Marxist methods of economic analysis from analytical methods used in the West, with regard to shipping. Present structure of world's maritime transport system from concentration and centralization of capital. Can East and West make full use of merchant fleets on a mutually acceptable economic basis?

252 Chrzanowski, Ignacy H. Concentration in shipping. *Journal of Transport Economics and Policy* VIII (2), May 1974, 174–9.

Considers quantitative and qualitative aspects of concentration in shipping. Degrees of concentration of fleets in Japan and Great Britain, USA, West Germany, Sweden.

253 Cohen, D. and **Shneerson, D.** The domestic resource costs of establishing/expanding a national fleet. *Maritime Studies and Management* 3(4), April 1976, 221–36.

Paper examines economic criteria for a policy tool that would aid decision-making regarding establishing/expanding of national fleets. Considers use of domestic resource costs of net foreign exchange earned as investment criterion for export industries in developing countries.

254 Cullinane, K. The utility analysis of risk attitudes in shipping. *Maritime Policy and Management* 18(3), 1991, 157–69.

Presents and justifies method for assessing and measuring attitudes towards shipping market risk, and applies method to sample of shipowners in UK, Norway, Hong Kong and Greece.

255 Datz, I. M. *Planning tools for ocean transportation.* Cambridge (Maryland): Cornell Maritime Press, 1971. 168 pp.

256 European Shippers Councils, Council of European and Japanese National Shipowners Associations. *Changing maritime environment. 2nd Symposium.* Montreux, Switzerland, 1980.

257 I.C.C. Business Ratios. *Shipowners: An industry sector analysis.* 2nd edn. London: I.C.C. Business Ratios, 1979.

258 Kindleberger, Charles P. Multinational ownership of shipping activities. *World Economy* 8(3), September 1985, 249–65.

This deals with shipping companies whose owners are nationals of several countries but concentrates primarily on the role of multinational corporations owning their own vessels.

259 Kruse, H. J. Corporate planning in shipping. *Chartered Institute of Transport Journal*, May 1975, 244–8.

260 Low, K. M. S. Development of ASEAN shipping and the role of shipowners' organisations. Thesis submitted for MSc, UWIST, Cardiff, 1988.

This describes the role and development of the ASEAN shipping and shipowners' association in each of the member states and also the part played by the Federation of ASEAN shipowners' associations. It attempts to assess the effectiveness of these associations in achieving the objectives of their members.

261 Lowry, Nigel. World shipping: New captains at the helm. *International Management* 44(7), July/August 1989, 24–7.

The shipping industry plays a strategic and highly efficient role in world trade. As the industry emerges from serious recession a new type of shipowner has arisen.

262 *Shipowners. Seatrade – Papers from a conference on shipping in Hong Kong.* Hong Kong, 1975.

263 Söhmen, H. The concepts of shipowners in today's market and regulatory environment. *Maritime Policy and Management* 13(1), January–March 1986, 1–15.

Market and regulatory developments: commercial and legal background; shipowners' reaction. Emphasizes need for training in shipping industry and for effective administration founded in legislation.

264 Soper, R. T. Corporate planning in liner shipping. In: Hans Ludwig Beth (ed.) *Liner shipping in the eighties: Report on the International Symposium, Bremen, October 24th–26th 1979.* Bremen: Institute of Shipping Economics, 1980, pp. 63–78.

Concentrates on containership field. Emphasizes need to set out an index of assets available to establish criteria of planning e.g. financial assets; physical assets; human assets. Considers the 'Energy Crisis' and the changing philosophy of available human assets – the 'Work Ethic'.

265 Tourkolias, A. The future of Greek shipping (from the shipowner's perspective). Thesis submitted for MSc, UWIST, Cardiff, 1976.

266 Vidali, M. W. Chiot shipowners in London and the persistence of ethnicity. Thesis submitted for M.Phil, CNAA, The Polytechnic of Central London, 1988.

267 Warner, David W. Merchant fleet feasibility study: Guidelines for developing countries. *Maritime Policy and Management* 10(4), October–December 1983, 251–64.

Defines the purpose of a feasibility study and sets out representative guidelines for incorporating the scope, work plan and type of information required by consumers of a feasibility study.

268 Wergeland, Tor. Shipping companies and market research. In: Einar Hope (ed.) *Studies in shipping economics: In honour of Professor Arnljot Stromme Svendsen.* Oslo: Bedriftsokonomens Forlag A/S, 1981.

Seeking information *re* shipping market. Structuring the market scene. Market models. Formal market models – pros and cons.

Enterprises

269 Nautical Institute. *New ways of stimulating enterprise.* London: Nautical Institute, 1987.

270 Pollack, Gerald A. and **Nicolich, Lillian.** The business economist at work: Overseas shipholding group. *Business Economics* 24(3), July 1989, 48–51.

271 Svendsen, A. S. The role of the entrepreneur in the shipping industry. (Editorial.) *Maritime Policy and Management* 8(3), 1981, 137–40.

Explaining a country's position in international shipping by referring to the enterprising shipowner or the entrepreneur in shipping.

272 Swayne, Sir Ronald. Shipping: Some political and economic consequences of changes in technology and ownership. *Royal Society of Arts Journal* 130, October 1982, 686–703.

The paper given and discussion following at the Thomas Gray Memorial Lecture, April 1982, to the Society. Examines the impact that changes in technology and ownership have had on merchant shipping and the economic and political consequences to be considered.

273 United Nations Conference on Trade and Development. *Multinational shipping enterprises.* New York: United Nations, 1972. 24 pp.

274 Wepster, A. Shipping and maritime studies and management. *Maritime Studies and Management* 1(4), April 1974, 195–6.

Editorial introduction. Aim of journal to provide information service to shipping industry.

275 Zurek, Janusz. The main elements of the economic analysis of the Polish shipping enterprise. *Maritime Policy and Management* 10(2), April–June 1983, 113–20.

Commentary which considers the aim of economic analysis in respect of the question as to how far economic activity of an enterprise conforms to the rational economic rule.

Ship management

276 Branch, Alan E. *Economics of shipping practice and management.* 2nd edn. London and New York: Chapman and Hall, 1988. 360 pp.

Shipping practice and management. Economic considerations. Role of management, especially *vis-à-vis* budgets, finance, personnel and marketing. Ship design; ship investment criteria; ro-ro vessels; freight market; counter trade; computerization; combined transport operations.

277 Downard, John M. *Managing ships.* London: Fairplay Publications, 1984. 154 pp.

Different ways in which ships are managed, plus common elements of ship management. Objectives, functions, restraints. Regulations, policies, control, communication, motivation, research and development. General guidelines.

278 Packard, William V. *Voyage estimating.* 2nd rev. edn. London: Fairplay Publications, 1981. 80 pp.

Practical introduction to the art of operating and broking ships. Demonstrates how a realistic estimate is built up from its elements, with work examples for real voyages under modern conditions.

Tankers

279 Drewry Shipping Consultants. *Independent tanker owners.* London: H. P. Drewry, 1979. 61 pp.

Report – part of series covering world's bulk shipping. Leading independent tanker owners of the 1970s. Comparison of fleet and charter positions of leading independent tanker owners and the general outlook for the independent sector.

280 Drewry Shipping Consultants. *Tanker building shipyards.* London: H. P. Drewry Shipping Consultants, 1981. 62 pp.

Report – part of series covering world's bulk shipping. Economic trends and tanker industry conditions in the 1970s. Shipyard developments in the 1970s. Estimated levels of world tanker building capacity.

Finance, Cost and Investment

General

281 Cheng, Phillip C. *Financial management in the shipping industry.* Centreville (Maryland): Cornell Maritime Press, 1979. 360 pp.

Deals with financial management of shipping companies. Accounting rules; financial requirements; management controls on shipping industry. Financial executives. Empirical approach reveals information concerning practical side of problems.

282 Goss, R. O. How to make money in shipping. (Editorial.) *Maritime Policy and Management* 14(1), 1987, 1–3.

Shipping markets depressed at present. Suggests three ways to make money: buy ships cheaply and sell them dearly; do what others do but better; adopt technical advances that are economic.

283 Laing, E. T. *Costs of deep sea general cargo shipping.* Liverpool: Marine Transport Centre, University of Liverpool, 1974.

283a Lorange, Peter and **Norman, Victor D.** *Risk preference and strategic decision making in large Scandinavian shipping enterprises.* Bergen: Institute of Shipping Research, 1971. 41 pp.

284 Lorange, Peter and **Norman, Victor D.** Risk preference in Scandinavian shipping. *Applied Economics* 5(1), March 1973, 49–59.

This concluded that most shipowners were not neutral in their attitude towards risk and this was reflected in their choice of business policy.

285 Rossi, James A. The Title XI ship financing guarantee program: An historical overview. *Journal of Maritime Law and Commerce* 12(2), January 1981, 183–95.

Long-term financing aims for shipowners under Title XI program. Development of the program and application. The Federal Ship Financing Act 1972. Impact on Title XI program. Growth of Title XI by vessel type. New regulations, 1979.

286 Seatrade Conference Proceedings. *Money and ships.* London, 1973.

287 Seatrade Conference Proceedings. *Money and ships.* London, 1974.

288 Seatrade Conference Proceedings. *Money and ships 75.* London, 1975.

289 Seatrade Conference Proceedings. *Money and ships 77.* London, 1977.

290 Seatrade Conference Proceedings. *Money and ships 78.* London, 1978.

291 Seatrade Conference Proceedings. *Money and ships 79: 7th Conference.* London, 1979.

292 Seatrade Conference Proceedings. *Money and ships: 9th Conference.* London, 1981.

293 Seatrade Conference Proceedings. *Money and ships: 10th Conference.* London, 1982.

294 Seatrade Conference Proceedings. *Money and ships: 16th Conference.* London, 1989.

295 *Ship loans in default.* Drewry–Whitehall Ltd, 1983.

Ship finance

296 Buchan. *Mortgages of ships: Marine security in Canada.* Butterworth US, Legal Publishers, 1986. 364 pp.

297 Drewry Shipping Consultants. *Finance for ships: Problems and options in shipping investment.* London: Drewry Shipping Consultants.

298 Drewry Shipping Consultants. *Financing ships: The challenge of the 1990's.* London: Drewry Shipping Consultants, 1989. 134 pp.

299 Drewry Shipping Consultants. *Shipping finance and investment.* London: Drewry Shipping Consultants, 1983.

300 Drury, Charles and **Stokes, Peter.** *Ship finance: The credit crisis:*

Can the debt/equity balance be restored? Colchester: Lloyd's of London Press, 1983. 177 pp.

Examines the ways in which the shipping industry has chosen/been persuaded to fund its operations and investments over the two decades prior to writing. Where and how errors are made, resulting in shipping arriving at insufficient equity and poor profitability. Suggestions for improvement.

301 Fairplay Publications. *Ship mortgages.* London: Fairplay Publications, 1987.

301a Gallagher, Thomas J. Maritime tax: The capital construction fund of the Merchant Marine Act 1970. *Journal of Maritime Law and Commerce* 6(1), October 1974, 105–18.

Tax deferral provisions under US Merchant Marine Act 1970. Application of Act. Impact on owner or lessee of eligible vessel.

302 Gardner, B. M., Goss, R. O. and **Marlow, P. B.** *Ship finance and fiscal policy.* Cardiff: UWIST, 1982. 58 pp.

303 Grammenos, C. Th. *Bank finance for ship purchase.* Bangor Occasional Paper in Economics, No. 16. Cardiff: University of Wales Press, 1979. 58 pp.

304 Gray, James. *Shipping futures.* 2nd rev. edn. London: Lloyd's of London Press, 1990. 192 pp.

Use of freight futures. Making effective use of futures market in freight, oil and finance. Addresses question of how shipping industry manages risk. How risk controlled and limited.

305 Gray, James W. *Financial risk management in the shipping industry.* London: Fairplay Publications, 1986. 88 pp.

Discusses the management of marine commercial risk, comparing traditional period timechartering, contracts of affreightment against new buildings and joint ventures with freight futures. How shipping industry can use financial futures market to control both currency and interest rate risks.

306 Haralambides, Hercules E. Sensitivity analysis of risk in shipping finance. In: K. M. Gwilliam (ed.) *Current issues in maritime economics.* Dordrecht: Kluwer Academic Publishers, 1993, pp. 168–77.

Examines the advantages of the risk-analysis approach to shipping finance project appraisal over the traditional single-estimate approach.

307 *International ship finance – the margin squeeze – problems and opportunities.* London: Lloyd's of London Press, 1993. 100 pp.

Leading developments in ship finance based on selected papers from the 5th International Lloyd's Shipping Economist Shipping Finance Conference 1992.

308 Marlow, P. B. Effect of the 1984 Budget proposals on the U.K. shipping industry. (Editorial.) *Maritime Policy and Management* 11(2), 1984, 75–7.

Reduction in the rate of corporation tax and changes in capital allowances over several years proposed. Result may be to increase number of vessels using flags of convenience, thus reducing size of UK shipping fleet.

309 Maruno, Y. Practice and law of international ship finance. Thesis submitted for PhD, University of Southampton, 1979.

310 Paine, Frank. *The financing of ship acquisitions.* London: Fairplay Publications, 1989. 148 pp.

311 Saaban, Hairani. *Ship financing in Singapore.* Butterworth, US, 1983. 305 pp.

312 *Shipping finance: Options and opportunities for the 1990's.* Lloyd's Shipping Economist, Maritime Consultants Ltd, 1988.

313 Sloggett, J.E. *Shipping finance: Financing ships and mobile offshore installations.* London: Fairplay Publications, 1984. 116 pp.

Procedures and terminology of shipping finance. New and second-hand craft. Financial appraisal of new projects, use of foreign currency, taxation and leasing. Investment in ships.

314 Stephenson Harwood. *Shipping finance.* London: Euromoney Publications PLC, 1990. 365 pp.

Examines some of the whys and wherefores of the financing of both new and second-hand ships. Considers the documents involved in everyday shipping finance transactions. English and foreign law. Marine insurance.

315 Stokes, Peter. *Ship finance: Credit expansions and the boom–bust cycle.* London: Lloyd's of London Press, 1992. 236 pp.

Examines the techniques of financing the shipping industry in relation to historical development of shipping finance in the past three decades. Demonstrates how shipping industry's increasingly highly geared capital structure in 1960s and 1970s led to financial crisis.

316 Stopford, Martin. Shipping cycles and ship finance: A special overview. *International Financial Law Review* August 1992, 3–5.

317 Symeonides, N. An investigation into shipping and the way it is financed. Thesis submitted for MSc, Heriot-Watt University, Edinburgh, 1980.

318 Wistner, E. J. Banks and shipping finance. Thesis submitted for MSc, Heriot-Watt University, Edinburgh, 1990.

Banks have traditionally played a role in the financing of the shipping industry. Many banks, however, have sustained large losses in their involvement with shipping. To control and manage the risk involved with financing the shipping industry, the banks must have a clear understanding of the factors governing the market. They must also critically evaluate each project to find those projects that have an acceptable return profile, it is argued.

319 Ysern, M. Gorina. *Ship leasing.* London: Lloyd's of London Press, 1988.

Bunker costs

320 Buxton, I. L. Fuel costs and their relationship with capital and operating costs. *Maritime Policy and Management* 12(1), 1985, 47–54.

Fuel costs are said to account for over 50 per cent of total operating costs, but what do we mean by operating costs?

321 Ewart, W. D. *Bunkers: A guide for the ship operator.* London: Fairplay Publications, 1982. 72 pp.

Written with manager and operator in mind. Current problems examined. Considers most practical method of obtaining overall increase in operating efficiency. Marine fuel grades. Legal problems. Shipboard fuel systems. Bunker adjustment factors and freight rates. Marine fuel specification. Bunker costs.

321a Lavon, B. and **Shneerson, D.** Optimal speed of operating ships. *Maritime Policy and Management* 8(1), 1981, 31–9.

Fuel costs a dominant item. Paper shows how optimal speed is set – i.e. by minimizing fuel consumption per mile and minimizing total operating costs of the ship.

322 Seatrade. *Fuel economy: Shipping's vital technology challenge.* Colchester: Seatrade Publications, 1982. 64 pp.

Collection of papers, part of series covering world shipping. Technical progress. Owners' options. Technology. Bunkers. Alternative power.

323 Seatrade. *Fuel economy in shipping.* Colchester: Seatrade Publications, 1981. 68 pp.

Collection of papers, part of series covering world shipping. The drive to change. The market – organization: bunker supplies. The operators – bunkering; re-engining; economical steaming. Technology – propulsion; hulls and props; coatings. Alternative fuels.

Operating costs

324 Alderton, P. M. An appraisal of the costs of operating British general cargo ships from 1870–1970, with specific reference to the trade between London and Sydney. Thesis submitted for M.Phil, CNAA, 1978.

325 Casas, J. *Shipping costs and revenues.* Bremen: Institute of Shipping Economics, 1981. 38 pp.

Analysis of economics of ship operation. Investigates revenues and costs of a single vessel being engaged in a round voyage.

326 Downard, John M. *Running costs.* London: Fairplay Publications, 1981. 142 pp.

Examines the costs of all activities associated with management of the ship. Owner's policies, condition of the ship, mode of operation. Tasks and responsibilities associated with ship management.

327 Drewry Shipping Consultants. *The impact of inflationary costs tendencies on dry cargo shipping operations.* London: H. P. Drewry Shipping Consultants, 1982. 124 pp.

Part of series covering world's bulk shipping. Vessel acquisition cost. Development of operating costs 1973–81. Voyage costs. Fully built-up costs. Impact of currency fluctuations on shipping costs. Impact of inflationary cost pressures. The dry cargo shipping market of the 1980s.

328 Gentle, N. F. and **Perkins, R. J.** *An estimate of operating costs for bulk, ro-ro and container ships.* Canberra: Bureau of Transport Economics, 1982.

329 Heaver, T. D. The treatment of ships' operating costs. *Maritime Policy and Management* 12(1), 1985, 35–46.

Growth of world trade has seen great increase in specialization of ships by type and size. Effects of this development have been diverse and fundamental.

330 Perakis, Anastassios N. A second look at fleet development. *Maritime Policy and Management* 12(3), July–September 1985, 209–14.

Paper considers problem of fleet operating costs and annual cargo capacity characteristics. Finds artificial constraint on dealing with this problem imposed by 'A simple approach to fleet deployment', H. Benford, *Maritime Policy and Management*, 8(2), 223–8. Seeks to formulate problem correctly.

Investment

331 Andreassen, J. A. B. Risk and investment decisions in non-liner shipping. *Maritime Policy and Management* 17(1), 1990, 23–30.

Investment decisions of non-liner shipping firms are discussed. A stochastic model was formed and used to examine and contrast two different strategic groups: risk averters and risk lovers.

332 Ariel, A. Evaluation of political risk in a shipping project. *Maritime Policy and Management* 17(1), 1990, 15–21.

Shipping industry must take account of risks outside its control, such as expropriation of shore facilities, when selecting investment projects. Such risks have a cash value to be levied.

333 *Capital for shipping: A comprehensive directory of financial services for shipping.* Colchester: Lloyd's Shipping Economist, 1992. 52 pp.

A financial and legal overview of the shipping industry is given followed by an A–Z listing of shipping financial and legal services.

334 Drewry Shipping Consultants. *The rising costs of ship construction.* London: Drewry Shipping Consultants, 1974. 54 pp.

Report – part of series covering world's bulk shipping. New building prices and inflation. The basic construction cost. Non-shipyard cost influences. Cost reducing developments.

335 Evans, J. J. Some practical aspects of investment appraisal in shipping. *Maritime Policy and Management* 11(3), 1984, 197–222.

Application of techniques of investment appraisal to shipping finance, using discounted cash flow methods.

336 Gardner, B. and **Richardson, P.** Fiscal treatment of shipping. *Journal of Industrial Economics* 22(2), 1973/74, 95–117.

This article looks at the effects of fiscal and financial arrangements of the incentive to invest in shipping and how it can be encapsulated to compare the different arrangements.

337 Gardner, Bernard and **Marlow, Peter.** An international comparison of the fiscal treatment of shipping. *Journal of Industrial Economics* 31(4), June 1983, 397–415.

This attempts to compare the subsidies and aids offered by different countries which influence the initiative to invest in shipping within those countries, the focus being mainly on Europe. This looks at the way incentives to invest in shipping can be calculated and also compares different arrangements. It deals with such fiscal measures as corporate tax, investment grants and free depreciation allowances, and financing by retaining profits and loan capital.

338 Goss, R. Assessing investments in shipping: A modular approach. *Maritime Policy and Management* 14(3), 1987, 197–225.

Offers modules for the assessment of investments in shipping, allowing for variations in profitability, finance, inflation, taxes and tax allowances.

339 Harvey, D. A. The analysis and financing of capital projects in shipping. Thesis submitted for PhD, University of Bath, 1987.

A descriptive model is developed of investment and financial decisions in the shipping industry, which seems likely to provide insights into behaviour in other industries.

340 Hope, Einar and **Boe-Oystein.** Investment behaviour in Norwegian bulk shipping. In: Einar Hope (ed.) *Studies in shipping economics: In honour of Professor Arnljot Stromme Svendsen.* Oslo: Bedriftsokonomens Forlag A/S, 1981. 12 pp.

Theory and practice concerning investment decisions in Norwegian bulk shipping.

341 *Investing in shipping's revival.* London: Lloyd's of London Press, 1989. 135 pp.

342 **Klausner, R.** The evaluation of risk in marine capital investment. *Marine Technology* 7(4), 1970.

343 **Lorange, P.** and **Norman, V. D.** *Risk preference patterns among Scandinavian tankship operators.* Bergen: Institute for Shipping Research, 1970.

344 **Miyashita, K.** A quarterly econometric analysis of ship investment planning behaviour in the world bulk carrier industry. *Maritime Policy and Management* 9(2), 1982, 115–33.

As bulk carriers grow rapidly in share of total world seaborne dry bulk and oil trades, what is the effect on ship-ordering behaviour? Attempts to substantiate this effect by econometric model analysis.

345 **Pineus, Kaj.** *Ship's value.* 2nd edn. London: Lloyd's of London Press, 1986. 126 pp.

Considers general issues relating to value of ship – the market value; dividend value; expectation value; utility value; replacement value; liquidation value. Moves on to particular issues where ship's value is of importance. The English method of valuation.

346 **Sparshott, K. J.** An operational research approach to shipping. *Maritime Studies and Management* 1(3), January 1974, 147–54.

How operational research approach can be used in shipping industry to obtain effective savings. Optimal dry dock policy where principal criterion is to minimize total cost of operation.

347 **Volk, B.** *Shipping investment in recession.* Bremen: Institute of Shipping Economics, 1984.

348 **Walsh, Robert G.** Estimated return on investment of oil tankships. *Marine Technology* 15(1), 1978.

Shipbuilding and repair

General

349 **Bauchet, P.** Is the West doomed to lose its maritime-related industries? *Maritime Policy and Management* 17(1), 1990, 3–13.

Decline in shipping, shipbuilding and metallurgy in Europe has occurred at same time as their growth in Asia. Sociological factors are more relevant than economic ones in explaining this trend.

350 Benford, H. A get-well recipe for West European shipbuilders. (Editorial.) *Maritime Policy and Management* 12(2), 1985, 107–11.

The present condition and suggestions for a way forward for shipbuilders in Western Europe.

351 Benford, H. On lifting the veil. (Editorial.) *Maritime Policy and Management* 9(3), 1982, 163–4.

On so-called 'black art' of estimating shipbuilding costs and inability of shipowners and naval architects to relate design decisions to dollars or pounds.

352 Benford, H. Ships' capital costs: The approaches of economists, naval architects and business managers. *Maritime Policy and Management* 12(1), 1985, 9–26.

On methods for estimating costs of building ships, from various points of view. Sees need for better cost-estimating procedures, uniform accounting systems and sharing cost data.

353 Cashman, J. P. A worldwide view of shipping and shipbuilding. *Maritime Policy and Management* 12(2), 1985, 113–34.

Outlines world fleet situation, gives précis of events leading to current deep recession and effects of latter on world shipbuilding industry, and reviews state of order books and future trends.

354 Clarke, Malcolm A. (ed.) *Shipbuilding contracts: A comparative analysis of contracts in the major maritime jurisdictions.* 2nd rev. edn. London: Lloyd's of London Press, 1992. 265 pp.

A general study of shipbuilding contracts as might be drawn and interpreted in various countries, notably Argentina, Denmark, France, Germany, Italy, Japan, Korea, Norway, Sweden, the UK, the US, and Yugoslavia. Particular attention to certain standard forms where rules governing disputes are found.

355 Commission of the European Communities. *Report on the long and medium term development of the shipbuilding market.* Luxembourg: Office for Official Publications of the European Communities, 1972. 166 pp.

A report drawn up at the request of the European Economic Community Shipbuilding Liaison Committee, in accordance with suggestions from the Council of European Community Ministers' Committee for medium-term economic policy. (i) Development of the new building demand; (ii) estimated future trend of the world shipbuilding supply.

356 Curtis, Simon. *Law of shipbuilding contracts.* London: Lloyd's of London Press, 1991. 330 pp.

Explains principles of English law of contract as these relate to shipbuilding and are currently applied by courts and arbitration tribunals in London. Particular reference to principal standard form published 1974 by Shipbuilders' Association of Japan.

357 Dobler, Jean-Pierre. Is the balancing of the world newbuilding market on the long term a goal out of reach? In: K. M. Gwilliam (ed.) *Current issues in maritime economics.* Dordrecht: Kluwer Academic Publishers, 1993, pp. 85–100.

A review of the forces and constraints at work on the world shipbuilding market which tend to prevent it reaching a state of long-lasting balance.

358 Drewry Shipping Consultants. *The outlook for world shipbuilding.* London: H. P. Drewry Shipping Consultants, 1983. 64 pp.

Part of series covering world shipping industry. This study discusses the fundamental changes which occurred in the world's shipbuilding industry in the post-1970 decade in terms of overall size, geographical distribution, product mix and degree of government assistance. Forecast for industry for end of 1980s.

359 Drewry Shipping Consultants. *Prospects for the world shipbuilding industry: A forecast of new building requirement to the late 1980's.* London: H. P. Drewry Shipping Consultants, 1979. 102 pp.

Report – part of series covering world's bulk shipping. Growth of world shipbuilding capacity during the 1970s. Government measures to assist shipbuilding. Building demand 1979–88. Distribution of world shipbuilding output 1979–88.

360 Drewry Shipping Consultants. *Shipbuilding credits and government aid.* London: Drewry Shipping Consultants, 1973. 36 pp.

Report – part of series covering world's bulk shipping. Historical background. Credits and aids in main shipbuilding countries.

361 Drewry Shipping Consultants. *Shipbuilding market: Medium to long term prospects 1988–1997.* London: Drewry Shipping Consultants.

362 Drewry Shipping Consultants. *Support for the world shipbuilding industry.* London: Drewry Shipping Consultants, 1976. 54 pp.

Report – part of series covering world's bulk shipping. Growth of world shipbuilding output 1970–76. Government support to shipbuilding. Policies of major shipbuilding countries. World shipbuilding 1980–85.

363 Drewry Shipping Consultants. *World shipbuilding: The next ten years.* London: Drewry Shipping Consultants, 1992. 120 pp.

364 Drewry Shipping Consultants. *The world shipbuilding industry.* London: Drewry Shipping Consultants, 1976.

365 Drewry Shipping Consultants. *World shipbuilding market: Prospects to 2000.* London: Drewry Shipping Consultants.

366 Harrison, Richard T. *Industrial organization and changing technology in United Kingdom shipbuilding: Historical development and future implications.* Aldershot: Avebury, 1990. 260 pp.

367 Law, Stuart C. Title XI of the Merchant Marine Act, 1936: Proposals for its amendment. *Journal of Maritime Law and Commerce* 2(1), October 1970, 63–89.

Suggested amendments to Title XI of US Merchant Marine Act 1936 which authorized government to insure private loans of up to 75 per cent of private parties' costs of constructing American flag merchant ships in American yards.

368 Nagatsuka, Seiji. *Changes in the world's shipbuilding facilities for large size vessels and future prospects thereof: Supply capacity exceeds construction demand for VLCC's.* Tokyo: Japan Maritime Research Institute, 1991. 49 pp.

Part of series covering maritime topics in Japan and abroad. What is a large-size vessel? Changes in the development of large-size vessels. World's shipbuilding facilities for large-size vessels. Volume of VLCC builds. Shipbuilding capacity for VLCC in the 1990s.

369 Nagatsuka, Seiji. *Outlook for demand–supply of world shipbuilding in the 1990's: Shipbuilding capacity as viewed from equipment and work force.* Tokyo: Japan Maritime Research Institute, 1989. 44 pp.

Part of series covering maritime topics in Japan and abroad. Changes in world shipbuilding environments. Changes in new building supply capacity.

Supply–demand of new buildings in the 1990s. Future problems and measures to cope with them.

370 Riddle, Ian. *Shipbuilding credit.* Bangor: University of Wales Press, 1983. 100 pp.

371 Schaefer, Heinz and **Speidel, Volker** (eds) *The restructured West European shipbuilding industry: A survey.* Bremen: Institute of Shipping Economics and Logistics, 1985, 95 pp.

372 Stopford, R. M. A new life for shipbuilding in the 1990's? *Maritime Policy and Management* 14(4), 1987, 301–12.

On nature of problems facing world shipbuilding – in state of crisis for ten years, an economic and political battlefield, a financial disaster area – and makes prognosis for next ten years.

373 Strath, Bo. *Politics of deindustrialization: Contraction of the West European shipbuilding industry.* Beckenham: Croom Helm, 1987. 320 pp.

374 Strath, Bo. Redundancy and solidarity: Tripartite politics and the contraction of the West European shipbuilding industry. *Cambridge Journal of Economics* 10(2), June 1986, 147–63.

This paper analyses the contraction in shipbuilding by examining the reaction of the crisis-hit shipyard managers, workers, and their unions at large yard locations in Europe and Scandinavia and considers the reaction to the crisis at national level of the industry, its union and governments.

375 Taylor, R. The shipbuilding shake-up. *Management Today,* April 1979, 51–8.

This looks at the collapse of all the ports around the world, which has left British shipbuilding, it argues, with no option save to contract. The issue is how much it will cost in public money.

376 Todd, Daniel. *Industrial dislocation: The case of global shipbuilding.* London and New York: Routledge, 1991. 292 pp.

Examines forces leading to dislocations in shipbuilding industry. Uses Japan as linchpin. Assailment of its market share – reasons, and diversification strategies. Differing attitudes of shipbuilders in advanced industrial countries from those in newly industrializing countries. Future prospects.

377 Todd, Daniel. *The world shipbuilding industry.* London: Croom Helm, 1985. 376 pp.

Presents a comprehensive overview of the world shipbuilding industry against the background of its global realignment. Contrasts the conditions which foster its development in newly industrialized countries (e.g. Japan, South Korea, Brazil) with the problems leading to its decline in Western Europe and North America.

378 Walker, Fred M. *Steel shipbuilding.* Shire Publications, 1981. 32 pp.

379 Woodhead, Gregory. Bilateral monopoly in the shipbuilding industry: A study of wage rate determination. Thesis submitted for PhD, University of New Hampshire, 1989.

380 Yoshida, Shigeru. *Background of recent excessive shipbuilding orders.* Tokyo: Japan Maritime Research Institute, 1985. 49 pp.

Part of series covering maritime topics in Japan and abroad. Relationship between shipping market and volume of shipbuilding orders: Orders placed in 1983. Background of excessive shipbuilding orders. Vessel scrapping as a means to reduce overtonnage.

Government

381 Agapos, A. M. Government legislation and the shipbuilding industry. *Journal of Business Research* 2(3), July 1974, 311–24.

382 Apostolopoulos, Y. N. The European Economic Community shipbuilding industry, with special reference to Greece. Thesis submitted for PhD, University of Strathclyde, 1976.

383 Bloom, Murray A. Repayment of construction–differential subsidy. *Journal of Maritime Law and Commerce* 11(2), January 1980, 183–94.

Financial aid programme under Merchant Marine Act 1936. The Construction–Differential Subsidy (CD) programme – the payment of subsidy to US shipyards to help offset disparity between construction costs of vessels in the US and similar vessels in foreign countries.

384 Campbell, M. R. Labour and public ownership: The nationalization of aerospace and shipbuilding, 1974–79. Thesis submitted for MA, University of Manchester, 1989.

385 Drewry Shipping Consultants. *Shipbuilding credits and government aid.* London: H.P. Drewry Shipping Consultants, 1973. 36 pp.

Part of series covering world shipping industry. This study looks at the major shipbuilding countries and examines the aids and credit facilities in each.

386 Evans, J. J. Replacement, obsolescence and modifications of ships. *Maritime Policy and Management* 16(3), 1989, 223–31.

Discusses theory of replacement in ocean-going vessels context. Shipowners must be aware of cost changes and technological improvements in case early replacement is necessary.

387 Hogwood, Brian W. *Government and shipbuilding: The politics of industrial change.* Farnborough, Hants: Saxon House, 1979. 302 pp.

Analysis of government industrial policy in Britain, in respect of shipbuilding, 1959–78. Examines evolution of shipbuilding policy; characteristics of involved government bodies; policy delivery aspects. Declared expectations. Actual outcomes. Implications for future policy.

388 No entry.

389 Lones, Trevor. Shipbuilding: Seeking a common salvation. In: *EEC shipping: A guide to the sea transport industries of the European Economic Community.* Colchester: Seatrade Publications, pp. 51–3.

Examines attempts to establish cohesive EEC shipbuilding policy. Future prospects.

390 Stopford, R. M. and **Barton, J. R.** Economic problems of shipbuilding and the state. *Maritime Policy and Management* 13(1), January–March 1986, 17–44.

Examines nature of international problems presented to governments by shipbuilding industry, mainly from European viewpoint.

National shipbuilding

391 Basilico, Rocco. The Japanese and the Italian shipbuilding industries: A different role in the period of growth and during the present crisis. *Rivista Internazionale di Scienze Economiche e Commerciali* 27(7–8), July–August 1980, 655–64.

392 Chida, Tomohei and **Davies, Peter N.** *The Japanese shipping and shipbuilding industries: A history of their modern growth.* London: Athlone Press, 1990. 220 pp.

Considers both shipping lines and shipbuilding industry. Focuses principally upon the economic developments following growth and boom of 1950s and 1960s. Japanese shipping related to Japanese economy, that of South-East Asia and world shipping industry as a whole.

393 Ferraz, Joao Carlos. Determinants and consequences of rapid growth in the Brazilian shipbuilding industry. *Maritime Policy and Management* 13(1), October–December 1986, 291–306.

Paper takes qualitative approach to assess influence of certain factors upon direction of economic growth and levels of technological development achieved by Brazilian shipbuilding industry post-1958.

394 Ferraz, J. C. Technological development and conditioning factors: The case of the Brazilian shipbuilding industry. Thesis submitted for D.Phil, University of Sussex, 1985.

395 Harrison, R. T. The politics of industrial decline and restructuring in the Northern Ireland shipbuilding industry. In: David Gibbs (ed.) *Government policy and industrial change.* London: Routledge, 1989, pp. 117–53.

396 Hilditch, P. J. Management strategy in the British shipbuilding industry, 1945–1986, with special reference to Scotland and the North East of England. Thesis submitted for PhD, University of Kent at Canterbury, 1986.

Identifying a series of explanations for the decline of the UK shipbuilding industry, it concludes that management strategy is of greater importance than the scale, size or location of individual shipyards.

397 Hogwood, B. W. The politics of industrial change: Government involvement in the U.K. shipbuilding industry, 1959–73. Thesis submitted for PhD, University of Keele, 1977.

398 Hwang, H.-D. Preferential credit and export competitiveness: A case study of the preferential credit scheme provided to the shipbuilding industry in South Korea, 1962–1982. Thesis submitted for D.Phil, University of Sussex, 1986.

This argues that the operation of a preferential credit scheme within the current industrial strategy has generated a bias in favour of a limited number of large firms. It argues that policy action should be taken to create a more competitive environment.

399 Lorenz, Edward H. *Economic decline in Britain: Shipbuilding industry, 1890–1970.* Oxford: Oxford University Press, 1991. 166 pp.

Presents explanation for decline in British shipbuilding. Draws on recent developments in behavioural economics and industrial sociology. Contrasts given explanation with other principal interpretations of decline. Considers implication of argument for other British industries that have coped poorly with foreign competition in the twentieth century.

400 McGoldrick, James. Industrial relations and the division of labour in the shipbuilding industry since the War. *British Journal of Industrial Relations* 21(2), July 1983, 197–220.

This argues that the industry was forced to change by the contingencies of crises, but that these changes were mediated by the interpretations attached to them by the workforce and that changes in industrial relations were foreshadowed to a significant extent by changes in the division of labour.

401 Nagatsuka, Seiji. *China's shipbuilding industry: Present and future.* Tokyo: Japan Maritime Research Institute, 1988. 54 pp.

Part of series covering maritime topics in Japan and abroad. Chinese economy. Maritime transport of China. China's shipbuilding industry. Business environment of shipbuilding. Future prospects and problems of China's shipbuilding industry.

402 Nagatsuka, Seiji. *Present and future of the U.S. shipbuilding industry: Shipbuilding capacity and possibility of its revival.* Tokyo: Japan Maritime Research Institute, 1990. 46 pp.

Part of series covering maritime topics in Japan and abroad. History of the US shipbuilding industry. US shipbuilding policy. US shipping. Volume of new buildings. American shipyards. Workforce, productivity, etc.

403 Nagatsuka, Seiji. *The present situation and future prospects in third shipbuilding countries: Upheaval in Eastern Europe and possible revival.* Tokyo: Japan Maritime Research Institute, 1990. 54 pp.

Part of series covering maritime topics in Japan and abroad. Changes in shipbuilding world-wide and in third shipbuilding countries. Definition of third shipbuilding countries. Trend of orders for new ships and completion in third shipbuilding countries. Shipbuilding facilities in third shipbuilding countries.

404 Peck, Francis and **Townsend, Alan.** Contrasting experience of recession and spatial restructuring: British shipbuilders, Plessey and Metal Box. *Regional Studies* 18(4), August 1984, 319–38.

Contrasting experiences of recession and spatial reconstruction – British shipbuilding Plessey and Metal Box trends indicators in recent recessions conceal mock variations in the experience of different industrial corporations within the meso-economic sector of the UK economy, it argues.

405 Todd, D. Strategies of growth, diversification and rationalization in the evolution of concentration in British shipbuilding. *Regional Studies* 18(1), February 1984, 55–67.

Examines strategies of growth, diversification and rationalization in the evolution of concentration in British shipbuilding.

406 Totten, George O. The reconstruction of the Japanese shipbuilding industry. In: Robert L. Friedheim and others. *Japan and the new ocean regime.* Boulder (Colorado) and London: Westview Press, 1984, pp. 130–72.

407 Tsurumaki, Yasuo. *Japan shipbuilding industry foundation: Its goal for the improvement of mankind.* Pergamon Press, 1989. 184 pp.

408 Tzamouzakis, Frangiscos G. and **Spathi, Sophia K.** *Shipbuilding and shiprepairing industry in Greece.* Athens: Centre of Planning and Economic Research, 1991. 123 pp.

409 Vambery, Robert G. Proposals for a long-range maritime transport programme for the United States. *Journal of Transport Economics and Policy* IV(2), May 1970, 121–43.

US government's approach to maritime industry. Regulation. Subsidies. Merchant Marine Act 1936 – impact, problems. Reasons for decline of US shipbuilding.

410 Whitehurst, Clinton H. *United States shipbuilding industry: Past, present and future.* Annapolis (Maryland): Naval Institute Press, 1987. 300 pp.

411 Yinghui, Min and **Jixiong, Piao.** *Prospects for the Korean shipbuilding industry and the industry efficiency improvement.* South Korea: Korea Industry Institute and Shipbuilding, 1987.

412 Yoshida, Shigeru. *General overview of Chinese shipbuilding.* Tokyo: Japan Maritime Research Institute, 1985. 63 pp.

State of China's social economy at time of writing. History of shipbuilding in China. Shipping industry of China. Future of shipbuilding in China.

Conversion; repair; drydocking

413 Bescoby, J. H. and **Potts, A.** Productivity agreements in Tyneside shiprepair, 1969–72. *Durham University Journal* 66, March 1974, 185–202.

414 Drewry Shipping Consultants. *The future of world shiprepairing.* London: Drewry Shipping Consultants, 1974. 67 pp.

Report – part of series covering world's bulk shipping. Changing world trade pattern. Development of world fleet. Demand for shiprepairing. Future demand for shiprepairing.

415 Drewry Shipping Consultants. *Repairing and drydocking large carriers.* London: H. P. Drewry, 1974. 65 pp.

Report – part of series covering world's bulk shipping. Growth and development of the large carrier fleet. Future drydocking requirements of the large carrier fleet. Supply of large carrier drydocking facilities. Supply and demand for large repair dock facilities.

416 Drewry Shipping Consultants. *Shiprepair and conversion: A market outlook to the mid-1990's.* London: Drewry Shipping Consultants.

417 Drewry Shipping Consultants. *The world ship repair facilities: Outlook for the next ten years.* London: H. P. Drewry, 1977. 64 pp.

Report – part of series covering world's bulk shipping. Development of world shiprepair facilities – past, present and future. Demands for drydocking – growth of large vessel fleet; components of repair demand. Repair dock supply and demand.

418 Drewry Shipping Consultants. *World shiprepair and conversion markets.* London: Drewry Shipping Consultants, 1991. 158 pp.

419 Moustakas, P. Productivity of Greek shiprepair yards: A study in efficiency. Thesis submitted for MSc, UWIST, Cardiff, 1985.

420 Nagi, A. A. The ship repair market. Thesis submitted for MSc, University of Strathclyde, 1984.

421 P. A. Management Consultants Limited. *U.K. shiprepair industry: Report to the Department of Industry.* London: HMSO, 1974. 64 pp.

Safety

422 Meek, M., Brown, W. R. and **Fulford, K. G.** A shipbuilder's view of safety. *Maritime Policy and Management* 12(4), October–December 1985, 251–62.

Paper looks at question of ship safety and suggests that the responsibilities for safety in the longer term need consideration. Rapid growth in safety legislation in contrast with modest actual improvement. Effective reduction in casualty rates.

Loss and Scrapping; Demolition

General

423 Burrell, D. C. E. *Scrap and build.* Kendal: World Ship Society, 1983. 79 pp.

British shipping. Scrap and build – the British Shipping (Assistance) Act, 1935. Shipping loan – the British Shipping (Assistance) Bill, 1939. Fleet lists.

424 *Decommissioning and demolition.* T. Telford, 1990.

425–34 No entries.

435 Drewry Shipping Consultants. *Influence of obsolescence, scrapping and losses on the tanker fleet.* London: H. P. Drewry, 1980. 44 pp.

Part of series covering world's bulk shipping. Trends during the 1970s. The economics of scrapping. Tanker fleet projections.

436 Drewry Shipping Consultants. *Ship obsolescence and scrapping: Problems and perspectives in the 1990's.* London: Drewry Shipping Consultants, 1992. 96 pp.

Market

437 Buxton, I. L. The market for ship demolition. *Maritime Policy and Management* 18(2), 1991, 105–12.

Explores fundamentals of ship demolition market and trends from 1960s to present. In that time market has moved from West Europe to Asia, especially Taiwan, now back to Indian subcontinent.

Scrapping activity

438 Drewry Shipping Consultants. *The scrapping of ships.* London: Drewry, 1973. 50 pp.

An economic study, part of a series covering world shipping. Examines some of the factors which influence the supply of ships for breaking. Estimates the gross tonnage which was perceived as likely to be withdrawn for scrapping 1971–80 period. Status of the shipbreaking industry.

439 Nagatsuka, Seiji. *Study about the extension of the scrapping age of tankers: Problems and measures as viewed from shipping and shipbuilding.* Tokyo: Japan Maritime Research Institute, 1989. 66 pp.

Definition of scrappage, and present status. Factors affecting demolition (1965–88). The shipbreaking capacity. Extension of the scrapping age of the ship, and matters to be considered.

6 International Seaborne Trade

General

General

440 Allocating risks in international trade. *Freight Management*, January 1976, 21–3.

441 Benabi, B. An analysis of British inter-zonal freight flows: An application of linear programming. Thesis submitted for MSc, Loughborough University of Technology, 1978.

442 British Shippers' Council. *Shipping 2000: The evolution of maritime trade in the next 10 to 25 years: An international conference held under the auspices of the British Shipper's Council, London Hilton – June 19 and 20, 1979.* London: Charterhouse Conferences, 1979. 124 pp.

Papers from an international conference in London, June 1979. Papers consider changing economic climate and impact on future developments and prospects for the shipping industry. A full report of conference proceedings.

443 Coyne, J. Cargoes in bulk. *Geographical Magazine* 53, April 1981, 450–4.

444 Drewry Shipping Consultants. *Caribbean oil trades.* London: H. P. Drewry, 1981.

445 Drewry Shipping Consultants. *Host government participation in the oil trade.* London: Drewry Shipping Consultants, 1974.

446 Drewry Shipping Consultants. *Outlook for the ore-bulk-oil carriers in oil and dry bulk trades.* London: Drewry Shipping Consultants, 1981.

447 **Fearnresearch.** *World bulk trades.* Oslo: Fearnleys, 1980.

A comprehensive review of total seaborne trade and cargo movement by tankers, combined carriers, bulk carriers. Covering all major commodities such as crude oil, iron ore, coal, grain, bauxite, phosphates, etc. An annual publication that includes comments, statistics, trade maps.

448 **Frankel, E. G.** *The effects of changes in energy balance and impact on maritime transportation.* Haifa: Israel Shipping Research Institute, 1974. 29 pp.

Analyses the effect of changes in energy balance upon maritime transportation. Reviews energy supply, demand and reserves. Cost of conventional energy and energy transport. Energy consumption. Impact of energy cost and availability on maritime transportation. Effects on demand for shipping capacity.

449 **Nagatsuka, Seiji.** *The global environmental problems and their relationships with shipping and shipbuilding: Changing industrial structures and their impacts on seaborne transport and ships.* Tokyo: Japan Maritime Research Institute, 1991. 58 pp.

Examines the connections between shipping/shipbuilding and the world's environmental problems. Likely change in future seaborne transport. Problems surrounding ship operation and shipbuilding.

450 **Nagatsuka, Seiji.** *The Pacific Rim Era and the shipping: Effects of the economic developments of Asian NIC's on seaborne trade and the shipping.* Tokyo: Japan Maritime Research Institute, 1987. 68 pp.

What is the Pacific Rim Era? Economies of major countries. Changes in industrial and trade structure. Relationship with shipping.

451 **Packard, William V.** *Sea trading. Vol. 2.: Cargoes.* London: Fairplay Publications, 1985. 153 pp.

Description of shipping world. Guide for shipbroker on origins of goods carried by sea; basic characteristics and factors which must be considered when drawing up charterparties. Stowage, carriage and handling.

452 *Patterns of trade and ship cargoes.* Glasgow: Brown, Son and Ferguson.

453 **Peebles, D. R.** A study of the transport of cargo from the United Kingdom and continental Europe to Canada. Thesis submitted for MSc, University of Birmingham, 1971.

454 Peters, H. J. *Seatrade, logistics and transport.* Washington (DC): The World Bank, Policy and Research Series, 1989.

455 *Ports, waterways, intermodal terminals, and international trade transportation issues.* Transportation Research Board, 1989. 72 pp.

456 *Ports, waterways, rail, and international trade issues, 1990.* Transportation Research Board, 1990. 110 pp.

457 Schmitthoff, Clive M. *Schmitthoff's export trade: The law and practice of international trade.* Revised edition by J. N. Adams. London: Stevens and Sons, 1990. 798 pp.

Law and practice of international trade. International sale of goods; marketing organization abroad; finance of exports; insurance, transportation of exports; international commercial dispute settlement; construction and long-term contracts; customs law.

458 Shipping and world trade. *Barclays Review* 49, November 1974, 84–6.

459 Stopford, R. M. Inter regional seaborne dry cargo trade, 1965–85: A disaggregated commodity analysis. Thesis submitted for PhD, University of London, 1980.

460 United Nations Conference on Trade and Development. *Bulking of cargoes.* New York: United Nations, 1975.

461 Vallega, A. Mediterranean maritime transport: Fields and strategies. *Maritime Policy and Management* 14(3), 1987, 227–48.

Growing cities, industrial polarization, a growing upper tertiary sector, offshore oil and gas industry mark that today's Mediterranean relationships between sea and land transport are changing.

462 Yeats, Alexander J. *Shipping and development policy: An integrated assessment.* New York: Praeger Publishers, 1981. 171 pp.

Examines the influence of transportation costs on developing country trade. Transport costs and development prospects. Ocean freight rates and economic development. Institutional factors. Transport costs and developing country exports. Influence of transportation costs on resource-based industrialization in developing countries. Policy perspectives on transport problems.

463 Yoshida, Shigeru. *Changes in Japan's foreign trade structure.* Tokyo: Japan Maritime Research Institute, 1987.

Freight and transport costs

464 Brodie, Peter R. *Freight charges.* London: Shaw & Sons, 1990. 111 pp.

Identifies and explains the elements which make up the international carriage of goods by sea – both liner and charter, conventional and containerized. Identifies the different facets of freight and ancillary charges and surcharges with which shippers are faced.

465 Bryan, Ingrid. Regression analysis of ocean liner freight rates on some Canadian export routes. *Journal of Transport Economics and Policy* VIII(2), May 1974, 161–73.

Examines statistically variations of ocean freight rates on outward bound Canadian liner cargo in 1969. Emphasizes influence of stowage, distance, quantity shipped, numbers of competitors on route, in addition to unit values.

466 Jayawickrama, S. S. Devaluation surcharges in ocean freight rates. *Journal of Transport Economics and Policy* VIII(2), May 1974, 148–50.

Considers issues involved in device of devaluation of surcharge or currency adjustment factor used by conferences, shipping lines to compensate for devaluation of tariff currencies. Actual effect of devaluation. Tendency to become general rate increase without notice.

467 Koo, W. W., Thompson, S. R. and **Larson, D. W.** Effects of ocean freight rate changes on the U.S. grain distribution system. *Logistics and Transportation Review* 24(1), March 1988, 85–100.

468 Laing, E. T. Shipping freight rates for developing countries: Who ultimately pays? *Journal of Transport Economics and Policy* XI(3), September 1977, 262–76.

Refutes common belief that developing countries bear burden of freight rates on both imports and exports. Concludes that a reduction in export rates would be detrimental as it would lead to increased freight rates charged to importers unable to pass it on.

469 Sampson, Gary P. and **Yeats, Alexander J.** Tariff and transport barriers facing Australian exports. *Journal of Transport Economics and Policy* XI(2), May 1977, 141–54.

Evaluates incidence of transport costs facing Australian exports to US and compares their magnitude with the existing level of most-favoured-nation tariff.

470 Suthiwartnarueput, K. The exploration of sea transport efficiency with a concentration on the case of Thailand. Thesis submitted for PhD, UWIST, Cardiff, 1988.

It argues that from the consumer's point of view shipping freights are more important than the quality of service rendered and that the efficiency levels of Thai and overseas lines are similar. For the port user, the Port of Bangkok and the quality of service are more important than the level of charges. It makes recommendations on the appropriate government policy towards Thai shipping lines and the more efficient operations for port authorities in Thailand.

Bulk wet

471 Drewry Shipping Consultants. *60,000–175,000 dwt crude oil carriers: Employment and prospects.* London: Drewry Shipping Consultants, 1981. 59 pp.

Report – part of series covering world's bulk shipping. Market conditions for medium-sized crude oil carriers. The fleet and its characteristics. Carrier performance and employment. Market prospects for medium-sized crude oil carriers.

472 Drewry Shipping Consultants. *Crude oil.* London: Drewry Shipping Consultants, 1989. 55 pp.

Report – part of Seaborne Trade and Transport series covering world's bulk shipping. Industry structure. Pattern of trade. Sea transport. Projects, developments and trends. Forecast seaborne trade (updates previous titles of same name).

473 Drewry Shipping Consultants. *Forecast world oil trades.* London: Drewry Shipping Consultants, 1991. 101 pp.

Report – part of Seaborne Trade and Transport series covering world's bulk shipping. World oil market trends to 1996. Forecast developments in the main oil importing and exporting regions. Bilateral oil trades to 1996.

474 Drewry Shipping Consultants. *Liquefied gas.* London: Drewry Shipping Consultants, 1988. 98 pp.

Report – part of Seaborne Trade and Transport series covering world's bulk shipping. Industry structure. The pattern of trade. Sea transport. Projects, developments and trends. Forecast LNG shipping.

475 Drewry Shipping Consultants. *Liquid cargo crude oil.* London: Drewry Shipping Consultants, 1989.

476 **Drewry Shipping Consultants.** *Liquid chemicals.* London: Drewry Shipping Consultants, 1989. 100 pp.

Report – part of Seaborne Trade and Transport series covering world's bulk shipping. Industry structure. Pattern of trade. Sea transport. Projects, developments and trends. Forecast seaborne trade.

477 **Drewry Shipping Consultants.** *Oil trades: Secondary transportation modes and tanker demand.* London: Drewry Shipping Consultants, 1992. 83 pp.

Report – part of Seaborne Trade and Transport series covering world's bulk shipping. Regional oil balances. Bilateral oil trades. Secondary transportation modes. Oil carrier employment and demand – the outlook to 1997. Future influences.

478 **Drewry Shipping Consultants.** *Outlook for oil trades to 1995.* London: Drewry Shipping Consultants, 1991. 89 pp.

Report – part of Seaborne Trade and Transport series covering world's bulk shipping. Oil demand. World oil supply. Oil trades.

479 **Drewry Shipping Consultants.** *Secondary oil transport modes.* London: H. P. Drewry, 1979. 68 pp.

Report – part of series covering world's bulk shipping markets. Outline of specific secondary oil transport modes. Secondary transport modes and tanker demand. Prospects for the future.

480 **Drewry Shipping Consultants.** *Trade and shipping in the new low oil price era 1986–1990.* London: Drewry Shipping Consultants, 1986.

481 **Drewry Shipping Consultants.** *Unconventional trading opportunities for tankers.* London: Drewry Shipping Consultants, 1983. 74 pp.

Report – part of series covering world's bulk shipping. Summary of unconventional trades. Potential for reducing the oil carrier surplus.

Bulk dry

482 **Alba-Barriga, T. E.** Chartering for international sea transport of coal. Thesis submitted for MSc, UWIST, Cardiff, 1987.

This examines historical and statistical information and different factors that have influenced coal trade patterns during the last two decades, mentioning the export and import countries and providing figures in the world coal trade, forecasts of our future coal consumption and information about emergent coal suppliers.

483 Clarkson Research Department. *Seaborne coal trade, 1980–1985.* London: H. Clarkson and Company, 1980. 37 pp.

Report on seaborne coal trade – importers; exporters; world trade; trade matrices.

484 Drewry Shipping Consultants. *Bulk shipping costs and commodity markets.* London: Drewry Shipping Consultants, 1985. 66 pp.

Report – part of series covering world's bulk shipping. Commodity prices and freight costs. Iron ore. Coal. Grain. Bulk shipping costs.

485 Drewry Shipping Consultants. *Coal trade and shipping in the 1980's.* London: Drewry Shipping Consultants, 1983. 85 pp.

Report – part of series covering world's bulk shipping. Coal trade and shipping. Coal markets and demand. Coal exports and supply. Coal trade projections. Coal shipping developments.

486 Drewry Shipping Consultants. *Combined carriers.* London: Drewry Shipping Consultants, 1989. 77 pp.

Report – part of Seaborne Trade and Transport series covering world's bulk shipping. The combined carrier and its role. Employment and the demand for combined carriers. Utilization of the present combined carrier fleet. Combined carrier fleet supply. Future developments and market trends.

487 Drewry Shipping Consultants. *Dry cargo fertilisers and fertiliser raw materials.* London: Drewry Shipping Consultants, 1990.

488 Drewry Shipping Consultants. *Dry cargo iron ore and coking coal.* London: Drewry Shipping Consultants, 1989.

489 Drewry Shipping Consultants. *Fertilisers and fertiliser raw materials.* London: Drewry Shipping Consultants, 1989. 88 pp.

Report – part of Seaborne Trade and Transport series covering world's bulk shipping. Industry structure. Pattern of trade. Sea transport. Projects, developments and trends. Forecast seaborne trade.

490 Drewry Shipping Consultants. *Global prospects for iron ore and coking coal.* London: Drewry Shipping Consultants, 1990. 90 pp.

Report – part of Seaborne Trade and Transport series covering world's bulk shipping. The steel industry and its raw materials. Steelmaking raw material inputs and future demand. Steelmaking raw material supply and trade patterns. Forecast to 1995. Sea transport and steelmaking raw material costs.

491 Drewry Shipping Consultants. *Global prospects for steam coal trade: Trade flows and ship demand to 2000.* London: Drewry Shipping Consultants, 1992. 96 pp.

Report – part of Seaborne Trade and Transport series covering world's bulk shipping. Energy demand, electricity supply and the market for steam coal in the 1990s. Global prospects for steam coal demand and imports. Supply prospects and export availabilities. Global steam coal trade and the growth of seaborne traffic. Prospects.

492 Drewry Shipping Consultants. *Grain.* London: Drewry Shipping Consultants, 1988. 82 pp.

Report – part of Seaborne Trade and Transport series covering world's bulk shipping. Industry structure. Pattern of trade. Sea transport. Projects, developments, trends. Forecast seaborne trade.

493 Drewry Shipping Consultants. *Grain trade and shipping in the 1980's.* London: H. P. Drewry, 1981. 64 pp.

Pattern of grain production and distribution; development of international grain trade; organization of international grain trade; limitations on grain transportation, handling and distribution. Shipping grain by sea. Outlook for grain trade in 1980s.

494 Drewry Shipping Consultants. *The growth of steam coal: A review and forecast of international trade in thermal coal and shipping requirements, 1980–1990.* London: Drewry Shipping Consultants, 1980.

495 Drewry Shipping Consultants. *Growth of steam coal trade.* London: H. P. Drewry, 1980.

496 Drewry Shipping Consultants. *Growth prospects for iron ore and coking coal.* London: Drewry Shipping Consultants, 1991. 101 pp.

Report – part of Seaborne Trade and Transport series covering world's bulk shipping. The steel industry and its raw materials. Steelmaking raw material inputs and future demand. Steelmaking raw material supply and trade patterns. Forecast seaborne trade to 1995. Iron ore mine-capacity developments. Market structure, pricing, costs.

497 Drewry Shipping Consultants. *Iron ore and coking coal.* London: Drewry Shipping Consultants, 1989. 73 pp.

Report – part of Seaborne Trade and Transport series covering world's bulk shipping. Steel industry structure. Pattern of iron ore and coking coal trade. Sea transport. Projects, developments and trends. Forecast seaborne trade.

498 Drewry Shipping Consultants. *Iron ore shipping and trade through to the 1980's.* London: H. P. Drewry, 1975. 70 pp.

Report – part of series covering world's bulk shipping. The growth of world seaborne iron ore trade. Structure of world seaborne iron ore trade. Ocean transport of iron ore. The ports shipping or receiving iron ore. Trends.

499 Drewry Shipping Consultants. *Minor ores and minerals (including bauxite/alumina).* London: Drewry Shipping Consultants, 1988. 79 pp.

Report – part of Seaborne Trade and Transport series covering world's bulk shipping. Bauxite/alumina. Other metal ores. Chemical industry ores and minerals. Construction industry minerals. Clays. Processed products.

500 Drewry Shipping Consultants. *The movement of minerals by sea.* London: Drewry Shipping Consultants, 1973. 50 pp.

Report – part of series covering world's bulk shipping. Growth and pattern of mineral shipping; bulk transportation; optimum ship sizes; costs; value of mineral commodities. Shipping and chartering methods employed by minerals industry. Mineral trades and their importance.

501 Drewry Shipping Consultants. *Ocean shipping of coal.* London: Drewry Shipping Consultants, 1981.

502 Drewry Shipping Consultants. *Ocean trade and transportation of dry bulk cargo: A review of trends in major and minor bulk trades, 1965–1985.* London: H. P. Drewry, 1976. 170 pp.

Part of series covering world's bulk shipping. Bulk cargo in international trade. The iron ore trade. The coal trade. The bauxite and alumina trade. The phosphate rock trade. The minor bulk trades. The growth of world dry bulk trade 1975–85.

503 Drewry Shipping Consultants. *Operation of dry bulk shipping: Present and prospective trading costs in the context of current and future market trends.* London: H. P. Drewry, 1979. 42 pp.

Report – part of series covering world's bulk shipping. Review of demand for and supply of dry bulk shipping. Definition of dry bulk cargo. Structure of dry bulk carrier costs. Dry bulk carrier revenues. The return to equilibrium in the dry bulk carrier market.

504 Drewry Shipping Consultants. *The organisation and structure of the deep-sea fertiliser trades.* London: Drewry Shipping Consultants, 1984. 87 pp.

Report – part of series covering world's bulk shipping. The active international trade in fertilizer raw material, intermediate and finished products. The products. The fertilizer industry. Major trade routes. Shipping. Market trends.

505 Drewry Shipping Consultants. *Outlook for phosphate shipping and trade.* London: Drewry Shipping Consultants, 1982.

506 Drewry Shipping Consultants. *Outlook for steam coal shipping and trade.* London: Drewry Shipping Consultants, 1990. 86 pp.

Report – part of Seaborne Trade and Transport series covering world's bulk shipping. Energy trends and the role of steam coal. Steam coal trade and the growth of seaborne trade. Supply and demand factors in the steam coal market. Future trade prospects and forecast seaborne steam coal flows.

507 Drewry Shipping Consultants. *Outlook for world grain trade.* London: Drewry Shipping Consultants, 1972. 36 pp.

Report – part of series covering world's bulk shipping. World demand for grain. World supply of grain. Seaborne grain trades.

508 Drewry Shipping Consultants. *The prospects for seaborne iron ore trade and transportation.* London: H.P. Drewry, 1979. 61 pp.

International iron ore trade; sea transport of iron ore; iron ore supply; iron ore consumption and import demand; forecast growth of world seaborne iron ore trade and shipping requirements.

509 Drewry Shipping Consultants. *Steam coal.* London: Drewry Shipping Consultants, 1988. 62 pp.

Report – part of Seaborne Trade and Transport series covering world's bulk shipping. Industry structure. Pattern of trade. Sea transport. Projects, developments and trends. Forecast seaborne trade.

510 Drewry Shipping Consultants. *Steam coal: Forecast trade and shipping developments, 1987–1992.* 2nd rev. edn. London: Drewry Shipping Consultants, 1988. 62 pp.

511 Drewry Shipping Consultants. *Steel making raw materials.* London: Drewry Shipping Consultants, 1988. 72 pp.

Report – part of Seaborne Trade and Transport series covering world's bulk shipping. Industry structure. Pattern of trade. Sea transport. Projects, developments and trends. Forecast seaborne trade.

512 Drewry Shipping Consultants. *World iron ore trade: Shipping and cost to 2000.* London: Drewry Shipping Consultants, 1992. 104 pp.

Report – part of Seaborne Trade and Transport series covering world's bulk shipping. Steel industry trends. Iron ore markets and future import demand. Iron ore exporters: Performance and capacity. Forecast seaborne iron ore trade and shipping patterns to 2000. Iron ore market organization. Iron ore chartering, freights, shipping costs.

513 Gloersen, Erik and **Johnsen, Erik.** *The seaborne trade in phosphate rock.* Bergen: Institute for Shipping Research, 1971. 134 pp.

514 Henrion, F. M. Optimum systems for the economic transport of iron ore by sea. Thesis submitted for MSc, University of Newcastle, 1973.

515 Lambert Brothers Shipping. *A statistical survey of seaborne trade in dry bulk commodities.* London: Lambert Brothers Shipping, Research and Development Department, 1979.

516 Nagatsuka, Seiji. *The demand and supply of grain in the world and its impact on shipping.* Tokyo: Japan Maritime Research Institute, 1986. 47 pp.

Part of series covering maritime topics in Japan and abroad. Demand and supply of grain. Grain trade and marine transportation.

517 Lambert Brothers Shipping Limited. Research and Development Department. *A statistical survey of seaborne trade in dry bulk commodities: Report.* London: Lambert Brothers Shipping, 1979. 42 pp.

Study which determines the volume of seaborne trade in dry bulk commodities between (i) OECD countries; (ii) developing countries; (iii) Eastern bloc countries.

518 Rogers, P., Strange, J. and **Studd, B.** *Coal: Carriage by sea.* London: Lloyd's of London Press, 1991. 250 pp.

Reference into all aspects of the carriage of coal – the coal trades, vessels, carriage, sampling, analysis and inspection. Illustrations, tables and graphs.

Major producers and importers. Specifications and specific problems of vessels. Analysis and handling. Role of inspector/surveyor. Problems.

519 **Soondrum, S.** The economics and operations of seaborne transportation of coal to Mauritius. Thesis submitted for MSc, University of Manchester, 1982.

Transportation costs by sea are influenced by factors like the high fluctuating freight rate market, fuel costs, and total port waiting time. This analyses the influences of freight rates on the delivery of steam coal to Mauritius.

520 **United Nations Conference on Trade and Development Secretariat.** *Maritime transportation of iron ore: A report.* New York: United Nations, 1974. 95 pp.

Analyses conditions of demand for and supply of tonnage for the maritime transport of iron ore in the light of developments, in particular over preceding decade. Assessment of level of freight costs and their incidence on the trade in the commodity. Solutions to problems of ocean transport facing the iron ore industry.

521 **Viscencio-Brambilla, Hector** and **Fuller, Stephen.** Effects of port user fees on export grain flow patterns. *Southern Journal of Agricultural Economics* 18(2), December 1986, 25–37.

522 **Yang, Chin Wei** and **Labys, Walter C.** Stability of Appalachian coal shipments under policy variation. *Energy Journal* 2(3), July 1981, 111–28.

Bulk 'mini' and 'neo'

523 **Drewry Shipping Consultants.** *Car bulk carriers.* London: H. P. Drewry, 1971.

524 **Drewry Shipping Consultants.** *Changes in 'neo bulk' trade, transport and handling.* London: H. P. Drewry, 1981.

525 **Drewry Shipping Consultants.** *Dry bulk carrier profitability.* London: Drewry Shipping Consultants, 1986.

526 **Drewry Shipping Consultants.** *Dry cargo forest products.* London: Drewry Shipping Consultants, 1989.

527 **Drewry Shipping Consultants.** *Forest products.* London: Drewry Shipping Consultants, 1989. 110 pp.

528 **Drewry Shipping Consultants.** *Forest products: Forecast trade and shipping developments, 1988–1993.* London: H. P. Drewry, 1989. 110 pp.

529 **Drewry Shipping Consultants.** *Sea transport of motor vehicles: The outlook for vehicle production, trade and shipping to 1990.* London: Drewry Shipping Consultants, 1984.

530 **Drewry Shipping Consultants.** *Seaborne cement trade and shipping.* London: Drewry Shipping Consultants, 1986.

531 **Drewry Shipping Consultants.** *World seaborne cement trade, transport and handling.* London: H. P. Drewry Shipping Consultants, 1983. 71 pp.

Report – part of series covering world's bulk shipping. Cement and the international cement trade. The international seaborne cement trade. Cement shipping operations. The cement shipping fleet. Cement handling equipment.

532 **Sparks, A.** *Steel: Carriage by sea.* London: Lloyd's of London Press, 1988. 193 pp.

Provides a ready reference into all aspects of the carriage of steel products by sea, required by ship master and world of marine insurance. Covers care of cargo, from loading through voyage to discharge. Corrosion, ventilation of cargo, maintenance of steel hatches, mate's receipts, bills of lading. Underlying theme is avoidance of claims.

533 **Yoshida, Shigeru.** *Status of cement exports/imports in the world and Japan as well as the shipping industry.* Tokyo: Japan Maritime Research Institute, 1987. 48 pp.

Part of series covering maritime topics in Japan and abroad. Situation surrounding cement in the world. Status of major cement importing countries. Situation surrounding cement in Japan. Status of Japan's exports and imports of cement. Seaborne transport of cement. Outlook for cement exports and imports and Japanese shipping.

Other

534 **Bohme, Hans.** The seaborne transport of automobiles: Growth, change and adjustment. *Marine Policy* 13(2), April 1989, 155–65.

Discusses the transport of automobiles by sea. Traces the development of the industry and looks at its growth and the related technological changes and methods of remaining competitive.

535 Drewry Shipping Consultants. *LNG shipping in the eighties.* London: H. P. Drewry, 1980. 78 pp.

Report – part of series covering world's bulk shipping. Changes to seaborne LNG trades: End 1975–end 1979. LNG trades by country of origin. Analysis of the LNG carrier fleet trading performance in 1978 and 1979. Prospects for LNG seaborne trades in the 1980s.

536 Drewry Shipping Consultants. *The market for refrigerated ships: An analysis of the trade and transportation of perishable commodities.* London: H. P. Drewry, 1979. 127 pp.

Report – part of series covering world's bulk shipping. Characteristics of the refrigerated fleet. The trade in perishable commodities. Future trade in perishable commodities to 1985. Operation and employment of, and general outlook for, refrigerated ships.

537 Drewry Shipping Consultants. *Minor agricultural commodities.* London: Drewry Shipping Consultants, 1987. 100 pp.

Report – part of Seaborne Trade and Transport series covering world's bulk shipping. Sugar and rice. Tapioca. Oil seeds and meals.

538 Drewry Shipping Consultants. *The outlook for bauxite/alumina trade and shipping.* London: Drewry Shipping Consultants, 1984. 64 pp.

Report – part of series covering world's bulk shipping. Aluminium – processes, raw materials – production, export. Trends in the aluminium industry. Consumption and import demand. Trade and shipping of aluminium raw materials.

539 Drewry Shipping Consultants. *Outlook for trade and ocean shipping of forest products.* London: Drewry Shipping Consultants, 1991. 110 pp.

Report – part of Seaborne Trade and Transport series covering world's bulk shipping. Forest products industries – overview of market influences. Forest products trade and traffic patterns. Sea transport. Forest product ports. Forecast seaborne trade.

540 Drewry Shipping Consultants. *Sea trade, transport and handling of steel products.* London: Drewry Shipping Consultants, 1986. 66 pp.

Report – part of series covering world's bulk shipping. Structural changes in steel industry over preceding two decades. Shifts in geographical spread. Trade in steel products. Shipping operations. Handling of steel products. Industry trends.

541 Drewry Shipping Consultants. *Seaborne transportation of liquefied natural gas, 1977–1985: Trades, costs and revenues.* London: H. P. Drewry, 1977. 116 pp.

542 Drewry Shipping Consultants. *Structure of bauxite/alumina trade and trends in ocean transportation.* London: H. P. Drewry, 1980. 69 pp.

Report – part of series covering world's bulk shipping. The aluminium industry and its raw materials. Supply of aluminium raw materials. Pattern of trade in bauxite and alumina. The demand for aluminium raw materials. Shipping aluminium raw materials.

543 Drewry Shipping Consultants. *Sugar trade and shipping trends in the 1980's.* London: Drewry Shipping Consultants, 1985. 92 pp.

Report – part of series covering world's bulk shipping. World sugar industry. International sugar trade. Sugar trade flows. Shipping patterns. Shipping trends.

544 Drewry Shipping Consultants. *World seaborne trade in LPG and chemical gases.* London: Drewry Shipping Consultants, 1974.

545 Howard, C. A. The strategic role of maritime transportation in the development of selected major West German and Japanese export car markets: A comparative analysis (shipping). Thesis submitted for PhD, Indiana University, 1992.

546 Jones, Philip N. and **North, John.** Car imports in the U.K.: The geography of an expanding traffic. *Regional Studies* 24(1), February 1990, 1–13.

Information is gathered concerning the origin, flows and port of entry of imported cars and the development in both the motor car manufacturing industry and the organizational characteristics of deep-sea and near-sea car-carrying trades.

547 Laskey, Steve. *Car carriers: A Fairplay guide to the pure car carrier fleet at March 1986.* London: Fairplay Information, 1986. 58 pp.

Reviews development of specialist ships for handling of export of cars. Lists vessels by name, by owner, by shipbuilder. Vessels on order at time of writing. Statistical tables. Owning companies with addresses, telex numbers.

548 United Nations Conference on Trade and Development. *The maritime transportation of cotton.* New York: United Nations, 1977.

549 United Nations Conference on Trade and Development. *The maritime transportation of jute.* New York: United Nations, 1971.

550 United Nations Conference on Trade and Development Secretariat. *Maritime transportation of natural rubber: Report.* New York: United Nations, 1970. 116 pp.

A study on the ocean transportation of natural rubber. The supply of rubber. The consumption of natural rubber. The natural and synthetic rubber industries. Market organization. The international trade in natural rubber. The ocean transportation of natural rubber. Conference freight rates in rubber trades. Alternative patterns of rubber transportation.

551 Westinform. *Bauxite and alumina cargoes: Forecasts of annual quantities.* London: Westinform, 1977. 59 pp.

Economic Theory and Modelling

General

552 Dutton, C. M. J. Modelling the international trade in steam coal. Thesis submitted for PhD, University of Cambridge, 1982.

553 Frankel, Ernst G. and **Marcus, Henry S.** *Ocean transportation.* Cambridge (Massachusetts) and London: MIT Press, 1973. 822 pp.

A world review of the demand and supply of shipping based on a research project of the MIT Sea Grant Program 1970–72. Ocean transportation technology; developing and using data on trade commodity flows; shipbuilding costs; merchant marine subsidies.

554 Goss, R. O. Economics and the international regime for shipping. *Maritime Policy and Management* 11(2), 1984, 135–45.

Examines economic ideas from theory of international trade and assesses their relevance to international shipping, now and in the future.

555 Kappel, Robert. Global trends in international shipping: Overcapacities, liberalization and national state interference. In: Karl Wohlmuth (ed.) *Structural adjustment in the world economy and east–west–south economic co-operation.* Bremen: University of Bremen, Institute for World Economics and International Management, 1989, pp. 319–55.

556 Pak, M. S. Participation of developing countries in world shipping. Thesis submitted for PhD, University of Liverpool, 1991.

Considers the question with particular reference to liner trades and concludes that the non-selective participation of developing countries in shipping should be restrained in relation to the comparative advantage theory of modern shipping.

557 *Trade and shipping policy issues for world business: Third International Shipping Conference. International Chamber of Commerce, Bath.* London: I.C.C., 1979.

558 Warf, B. and **Cox, J. C.** The US–Canadian free trade agreement and North American maritime trade. *Maritime Policy and Management* 19(1), 1992, 19–29.

FTA implemented in 1988 and affects maritime trade. Spatial structure of water-borne trade examined, with investigation of short-run impacts of FTA on US port system and changes since 1988.

559 White, Lawrence J. *International trade in ocean shipping services: The United States and the world.* Cambridge (Massachusetts): Harper and Row, Ballinger, 1988. 121 pp.

560 Wohlmuth, Karl (ed.) *Structural adjustment in the world economy and east–west–south economic co-operation.* Bremen: University of Bremen, Institute for World Economics and International Management, 1989. 546 pp.

Tariffs

561 Matz, James R. Generalised tariff preferences for the developing countries. *Journal of Maritime Law and Commerce* 2(3), April 1971, 645–59.

What are the generalized preferences? Why given? Who are potential preference-giving countries? What do they propose to do? Likely effects of such preferences?

Economic nationalism

562 Vambery, Robert G. Nationalism in shipping. *Maritime Studies and Management* 1(4), April 1974, 243–6.

Part of series of essays reflecting opinions on current issues: Economic nationalism; use and restriction of flags of convenience; balance of payments advantage argument.

Forecasting

563 Abdalla, A. A. F. Application of some forecasting techniques to shipping trade in Sudan. Thesis submitted for MSc, UWIST, Cardiff, 1983.

564 Beth, Hans Ludwig and **Volk, B.** *Shipping markets in 1980 and seaborne trade forecasts.* Bremen: Institute of Shipping Economics, 1981. 35 pp.

Part of series looking at world shipping. Considers the development of the demand side of the shipping market and its interrelation with the economic development in general.

565 Drewry Shipping Consultants. *Forecast tanker profitability 1989–1993.* London: Drewry Shipping Consultants, 1989. 68 pp.

Report – part of Seaborne Trade and Transport series covering world's bulk shipping. Recent developments in the oil and tanker markets. Forecast oil market developments. Forecast oil carrier demand and supply. Forecast shipping costs. Forecast freight rates and tanker profitability.

566 No entry.

567 Gilman, Sidney. Extrapolation and models in the prediction of developments in the marine transport industry. *Maritime Studies and Management* 3(2), October 1975, 69–70.

Editorial beginning to journal. Suggests that both statistical and intuitive extrapolation provide a poor basis for prediction of maritime industry development.

568 **Lloyd's Shipping Economist.** *Tankers: The challenge of the 1990's. Conference Proceedings.* London: Lloyd's of London Press, 1988.

569 **Lu, W. G.** The containerization of China's seaborne trade. Thesis submitted for PhD, University of Liverpool, 1990.

This is an investigation of the relationship between trade development and economic growth in China. It concludes that during a very long period, China's trade and its economy grew hand in hand. A sophisticated forecasting technique is then employed to predict China's seaborne container traffic up to 1995.

570 *Review and outlook for the shipping market 1980–1988.* London: Terminal Operators Ltd, 1981.

571 **Schrier, E., Nadel, E.** and **Cleary, C.** Testing the forecasters. *Journal of Maritime Law and Commerce* 8(1), October 1976, 121–5.

Summarizes an analysis (from a larger study) of the accuracy of nineteen published forecasts of a well-defined subject of economic data: Of liner cargo movements into and out of the US for the years 1969–73.

572 **Ullman, H. K.** and **Pettavino, P. J.** *Forecasts for United States maritime industries in 1989: Balancing national security and economic considerations.* University Publications of America, 1987. 52 pp.

573 **United Nations Conference on Trade and Development.** *Perspective and problems in world shipping.* New York: United Nations, 1971.

574 **Volk, B.** *Seaborne trade forecasts 1984/85: A synoptical review.* Bremen: Institute for Shipping Economics, 1985. 37 pp.

575 **Wing, John F.** Methods of forecasting ocean trade. *Journal of Maritime Law and Commerce* 4(4), July 1973, 681–90.

Presents methods used in developments of forecasts to project fleet requirements. Types and purposes of trade forecasts. Techniques for trade forecasting. Trend exploitation. Economic factor analysis. Input–output analysis. Judgements of experts. Data sources.

Balance of Payments

General

576 Abrahamsson, B. J. A note on national shipping and the balance of payments. *Malayan Economic Review* 15(1), April 1970, 74–82.

577 Gardner, Bernard. Investment grants and balance of payments tests. *Maritime Studies and Management* 2(4), April 1975, 221–30.

Identification of policy objectives which might induce UK Government to reintroduce investment grants for shipping. Assesses whether objectives might be achieved without harming balance of payments.

578 Great Britain. Department of Transport. *United Kingdom shipping industry international revenue and expenditure, 1989.* London: HMSO, 1990. 23 pp.

579 Haralambides, Hercules E. The shipping account in the balance of payments: The case of Greece. Thesis submitted for PhD, UWIST, Cardiff, 1985.

580 Haralambides, Hercules E. Investment in shipping and the balance of payments short-run equilibrium freight rate. *International Journal of Transport Economics* 13(3), October 1986, 351–7.

581 Hawkins, J. E. *Shipping subsidies and the balance of payments.* London: Lloyd's of London Press, 1989. 92 pp.

582 Nicholson, Sir John. London's foreign business. 4. United Kingdom shipping and the balance of payments. *Banker* 121, February 1971, 187–98.

Attempts to estimate the importance of UK shipping industry to the invisible balance of payments in the light of the Report by the Committee of Invisibles and the Chamber of Shipping in that year.

583 Shimojo, Tetsuji. An input–output table for evaluation of the shipping activities. *Kobe Economic and Business Review* 27, 1981, 49–68.

Discusses and proposes a slight modification of the input–output table for the sake of correct evaluation of contribution of the shipping industry to the national and international economy of a country. Investigation for the appropriate method to evaluate the *raison d'être* and contribution of ocean shipping industries on the national economy and the international balance of payments.

584 Shneerson, Dan. The profitability of investment in shipping: The balance of payments criterion. *Journal of Transport Economics and Policy* XVII(3), September 1983, 285–98.

Case-study showing investment in shipping profitable for Israel, making net contribution of circa 25 per cent to balance of payments. Applies 'Domestic Resource Costs' concept to evaluate national profitability of investment in shipping.

585 Wood, N. J. Shipping and the U.K. economy. *Maritime Studies and Management* 3(2), October 1975, 71–9.

Considers the evaluation of shipping's place in the UK economy. Examines the efficiency of world shipping in respect of benefit to the UK. Examines the part played by shipping in contributing to the UK's balance of payments.

7 Market Structures

General

Theory

586 **Abrahamsson, Bernhard J.** *International ocean shipping: Current concepts and principles.* Boulder (Colorado): Westview Press, 1980. 232 pp.

Provides a basic text in international ocean shipping – discusses how the industry functions and the problems and issues that arise because of its international character. Explains how various aspects of ocean shipping relate to each other.

587 **Beenstock, Michael** and **Vergottis, Andreas.** The interdependence between the dry cargo and tanker markets. *Logistics and Transportation Review* 29(1), March 1993, 3–38.

Previous research by these authors in 1989 is developed by incorporating econometric models for the dry cargo and tanker markets.

588 **Bohme, Hans.** *Restraints on competition in world shipping.* London: Trade Policy Research Centre, 1978. 85 pp.

A paper for the Trade Policy Research Centre, looking at problems of international shipping. Restrictive practices and policies in marine transport. Analysis of the restraints on international shipping. Disintegrating tendencies in post-war shipping policies. Future of the international shipping order.

589 **Canner, W. H. P.** Air-cargo market penetration. *Maritime Policy and Management* 8(3), 1981, 191–6.

Examines depth of air-cargo market penetration made in UK following deregulation of US domestic market in 1978. Compares sea and air costs using distribution costs analysis.

590 **Casson, Mark.** The role of vertical integration in the shipping industry. *Journal of Transport Economics and Policy* XX(1), January 1986, 7–30.

Outlines main incentives/disincentives of vertical integration by theory of the firm. Empirical evidence on extent of integration into shipping varies across industries. Contractual arrangements in shipping industry where integration absent.

591 Fearnresearch. *Monthly report.* Oslo: Fearnleys.

A monthly report with updated information on the world economy, chartering markets, sale and purchase, contracting, demolition, etc. The report covers tankers, dry bulk, general cargo and gas carriers. 'Frankly speaking' expresses personal views of key staff on current topics.

592 Lin, B. and **Chen, H.** Fuzzy probability simulation research on the world ship market. *Maritime Policy and Management* 19(3), 1992, 223–37.

On new method for studying world ship market, using fuzzy probability simulation theory. Shows that market forecasts have controllable results. FPS theory applied to market simulation.

593 *Maritime industry: High risk/low return? – Seascape Conference Proceedings, 1986.* Newcastle: North East Coast Institution of Engineers and Shipbuilders, 1987. 167 pp.

594 *Medium to long-term analysis of the shipping market i) 1984–1995 ii) 1986–2000 iii) 1990–2005.* Tokyo: Japan Maritime Research Institute, 1984, 1987, 1991.

Three separate publications – part of the 'Shipping Market' analysis series. The general situation confronting the shipping market. Basic composition of models. Estimate of shipping demands by commodity. Average shipping distances. Simulation results and analysis.

595 Muller, E. J. Ocean shipping: Battle for market share continues. *Distribution* 89(7), July 1990, 84–94.

596 Packard, William V. *Sea-trading. Vol. 3.: Trading.* London: Fairplay Publications, 1986. 153 pp.

Description of shipping world. Covers trading of raw materials, the world of liner conferences and tramp shipping. Charterparties, bills of lading.

597 Perakis, A. N. and **Papakidis, N.** Fleet deployment optimization models. Part 1. *Maritime Policy and Management* 14(2), 1987, 127–44.

Studies problem of minimum-cost operation of a fleet of ships carrying a specific amount of cargo between two ports in a given time period for a specific, fixed contract price.

598 **Perakis, A. N.** and **Papakidis, N.** Fleet deployment optimization models. Part 2. *Maritime Policy and Management* 14(2), 1987, 145–55.

Computerized solution to fleet deployment problem for the one-origin, one-destination fixed-price contract requiring transport of given amount of cargo within given period is covered.

599 **Vergottis, A. R.** An econometric model of the world shipping markets. Thesis submitted for PhD, London: City University, 1988.

This presents an aggregated econometric model of the world's shipping markets. It distinguishes between dry cargo and tankers and also between freight markets, second-hand markets for ships, shipbuilding markets and scrap markets. It argues that the model can be used for real world forecasting purposes and scenario planning.

600 **Warf, B.** Air and water cargo transport mode substitutions among US customs districts. *Maritime Policy and Management* 16(3), 1989, 247–56.

Nature of increased competition between air and water cargo transportation shown with reference to exports and imports of thirty commodity groups in six major US customs districts and whole of the USA.

Optimum ships

601 **Deere, Derek H.** (ed.) *Barge carriers: Proceedings of Bargecon '77 the world's first conference on barge carrying ships.* London: Benn Group, 1977. 148 pp.

The full text of the papers presented at the conference, Hong Kong, June 1977, plus the discussion.

602 **Drewry Shipping Consultants.** *Deep sea towage, salvage and heavy lift.* London: Drewry Shipping Consultants, 1985. 84 pp.

Report – part of series covering world's bulk shipping. The vessels and their market. The fleet. The companies. Market structure and directions.

603 **Drewry Shipping Consultants.** *Deep-sea towage, salvage and heavy lift markets.* London: Drewry Shipping Consultants, 1992. 56 pp.

Report – part of Seaborne Trade and Transport series covering world's bulk shipping. The vessels and their market. The fleet. The companies. Market structure. Market trend.

604 Drewry Shipping Consultants. *Evolution, economics and employment on multi-purpose dry cargo ships (10–25,000 dwt.)* London: H. P. Drewry, 1977. 122 pp.

Part of series covering world's bulk shipping. The origin of the multi-purpose cargo ship. Principal types of small dry cargo ship. The concept of the multi-purpose cargo ship. Choice, characteristics of, growing fleet of, costs, revenues, employment, role of multi-purpose cargo ships.

605 Garrod, Peter and **Milklius, Walter.** The optimal ship size: A comment. *Journal of Transport Economics and Policy* XIX(1), January 1985, 83–90.

Extends analysis of Jansson and Shneerson, 'The optimal ship size', *Journal of Transport Economics and Policy* XVI(3), September 1982, 217–38. Faster loading, unloading rates. Greater optimal ship size. Significant effect on users.

606 Gilman, S. The choice of ship size on deep-sea general cargo routes. *Maritime Studies and Management* 3(2), October 1975, 95–102.

Ship size is determined by a set of interactions between handling performance, route length, traffic flow, itinerary, requirements for frequent and regular service, port costs and general system organization. To identify major influences and prospects for further growth.

607 Heaver, Trevor D. and **Studer, Keith R.** Ship size and turnround time: Some empirical evidence. *Journal of Transport Economics and Policy* VI(1), January 1971, 32–50.

Quantitative trial *re* loading rate and larger vessels. Relationship between loading rate, size of vessel. Effects on port turnround time of factors exogenous to vessel itself. Implications of results for costs of loading vessels.

608 Jansson, Jan Owen and **Shneerson, Dan.** The optimal ship size. *Journal of Transport Economics and Policy* XVI(3), September 1982, 217–38.

Presents a model for determining optimal size of a ship to minimize total costs (at sea and in port) per ton of cargo. Model shows how optimal size varies as result of changes in route characteristics, factor prices.

609 Kendall, P. M. H. A theory of optimum ship size. *Journal of Transport Economics and Policy* VI(2), May 1972, 128–46.

Discusses optimum ship size theory developed to explain macroeconomics of some aspects of shipping practices for dry bulk cargo. Conceptual development of model for optimum ship size. Its applicability.

610 **Pope, James A.** and **Cross, Edward M.** The optimal load size for ocean shippers. *Logistics and Transportation Review* 24(4), December 1988, 299–315.

611 **Pope, James A.** and **Talley, Wayne K.** Inventory costs and optimal ship size. *Logistics and Transportation Review* 24(2), June 1988, 107–20.

612 **Robinson, Ross.** Size of vessels and turnround time: Further evidence from the port of Hong Kong. *Journal of Transport Economics and Policy* XII(2), May 1978, 161–78.

Using recent US import statistics, paper assesses relative importance of tariffs and transport charges on UK exports to US. Incidence of transport costs on exports from UK to US greater than that of tariffs.

613 **Shimojo, Tetsuji.** Effects of ships' size and cargo lots on the ocean freight rates. *Kobe Economic and Business Review* 25, 1979, 43–64.

Examines reasons for the expansion of ship size. World scale system and ship size. Ship size and shipping markets, and effect on individual rates. Application to the market model.

Freight rates; markets

614 **Arab–British Chamber of Commerce.** *Shipping and freight to the Arab countries.* 1985.

615 **Booz-Allen and Hamilton Inc.** *A study of ocean rate disparities: Final report.* Washington (DC): US Department of Transportation, 1978.

616 **Dabrowski, K.** Comments on the mechanisms of the world shipping market. *Maritime Policy and Management* 8(2), 1981, 85–98.

After ten years (1964–73) of global stability, world shipping experienced a sudden breakdown, followed by deep depression. Analyses weaknesses in the market mechanism.

617 **De O. Goncalves, Franklin.** Freight futurers and chartering: A

contingent claims analysis approach to the optimal operation and investment decision in bulk shipping. In: K.M. Gwilliam (ed.) *Current issues in maritime economics.* Dordrecht: Kluwer Academic Publishers, 1993, pp. 145–67.

Develops optimal policies for chartering ships in the market for transportation services of bulk commodities by applying the methodology of contingent claims analysis of financial economics and stochastic control theory, taking advantage of the market for freight futures.

618 Drewry Shipping Consultants. *Freight rates and bulk shipping costs: Market analysis and forecast to 2000.* London: Drewry Shipping Consultants.

619 Faust, Peter. *The influence of exogenous factors on freight rate development.* Bremen: Institute of Shipping Economics, 1976. 21 pp.

Lecture forming part of series. Development of freight markets since 1950. Characteristics of tanker and tramp markets. Influence of exogenous factors on freight rates. Post-war development of the tanker market. Influence of exogenous factors on dry cargo markets and liner freight rates.

620 Fearnresearch. *Fearnleys Review.* Oslo: Fearnleys.

A comprehensive annual survey of freight markets, sale and purchase, contracting, etc. The tonnage balance, estimated future fleet, trade developments highlighted.

621 Heaver, T. D. *The inbound/outbound rate controversy.* Vancouver: Columbia Transport Studies, University of British Columbia, 1972.

622 Laing, E. T. *Freight rates in developing countries: Who ultimately pays?* Liverpool: Marine Transport Centre, University of Liverpool, 1975.

623 Larner, R. Public policy in the ocean freight industry. In: A. Phillips (ed.) *Improving competition in regulated markets.* Washington (DC): The Brookings Institution, 1975, pp. 113–17.

624 Marshall, R. and **Sletmo, G.** *Analysis of competitive practices in the maritime industry.* Kings Point (New York): National Maritime Research Center, 1978.

625 Miyashita, Kunio. *Market behaviours in competitive shipping markets.* Bremen: Institute of Shipping Economics, 1977. 22 pp.

Part of series of lectures. Analyses the market behaviours in competitive shipping markets consisting of the tramp and tanker markets. Tests the hypotheses deduced from the microeconomic and traditional shipping theories.

626 Pace, Malcolm N. *Determination of ocean freight rates: Towards a new formula.* Bremen: Institute of Shipping Economics, 1979. 175 pp.

The international freight market. The relation of costs to ocean freight rates. Establishing a new formula of freight-rate determination. Miscellaneous aspects of ocean freight rates. Freight rates and the role of governments.

627 R. S. Platou Economic Research AS. *The Platou report.* Hovik, Norway: R. S. Platou Economic Research AS, 1970.

A report covering international markets in shipping and offshore. An annual publication.

628 Shimojo, Tetsuji. Expectations in the Shipping Exchange. *Kobe Economic and Business Review* 22, 1976, 57–73.

Examines the forecasting of ocean freight levels. Commences with assumption that the connection between voyage charter rates and time charter rates is informative about the future levels of the market. Examines this theory and proposes a method of extracting the information and using for the forecast.

629 Shimojo, Tetsuji. Some bargaining process models in the Shipping Exchange. *Kobe Economic and Business Review* 23, 1977, 69–84.

Proposes a way of explanation for causes of differences between freight rates of similar charter contracts.

630 Shimojo, Tetsuji. Studies for the prediction of shipping markets: A survey and their applications. *Kobe Economic and Business Review* 24, 1978, 39–59.

Examines various methods of prediction of shipping markets and the fundamental studies advocating them. Considers what efforts are needed in order to build more practicable methods of prediction. Intends to raise the level of clerical prediction which can be combined with the intuitive prediction.

631 Taylor, A. J. Chartering strategies for shipping companies. *Omega* 10(1), 1982, 25–33.

A paper which describes a model which may be used in the evaluation of the relative effectiveness of policies used in the areas of ship chartering and the switching of combined carriers between the dry-cargo market and the tanker market.

632 Taylor, A. J. A model for the evaluation of some shipping company decisions. *Maritime Policy and Management* 8(2), 1981, 73–83.

Describes a model for evaluating relative effectiveness of policies in the areas of ship chartering and switching combined carriers between dry-cargo and tanker market.

633 Taylor, A. J. The modelling of shipping freight markets. Thesis submitted for PhD, University of Bradford, 1979.

634 Thuong, Le T. and **Visscher, Sue L.** The hedging effectiveness of dry-bulk freight rate futures. *Transportation Journal* 29(4), Summer 1990, 58–65.

635 United Nations Conference on Trade and Development. *Freight rates.* Geneva: UNCTAD, 1976.

636 Zerby, J. A., Conlon, R. M. and **Kaye-Kucharzewski, S. L.** *Major determinants of ocean freight rates for Australia's exports to selected areas.* Sydney: School of Economics, University of New South Wales, 1981.

Shipping enterprise

637 Van Rijn, Hendrik. The trans-Siberian railway: Already a looming threat. In: Curtis Cate (ed.) *The challenge of Soviet shipping.* London: Aims of Industry (in association with the National Strategy Information Center, New York), 1985, pp. 30–36.

A brief introduction to development of Soviet merchant marine from 1956. Leads on to importance of trans-Siberian railway in the shipment of manufactured goods between the Far East and Western Europe. Development of the railway. Importance of it to Western shipping interests involved in trade between the Far East and Europe.

Intermodal and multimodal

638 **Ashar, A.** On-off terminal vessel-to-rail intermodal transfer and the case of Long Beach Port. *Maritime Policy and Management* 17(4), 1990, 235–47.

Discusses various aspects of intermodal, on/off terminal rail-to-vessel alternatives, using example of port of Long Beach, California.

639 **Cadwallader, F. J. J.** Uniformity in the regulation of combined transport. *Journal of Business Law* 1974, 193–210.

Written against background of 1972 Genoa International Seminar on Intermodal Transport. Outlines general form of rules as adopted by International Chamber of Commerce – Uniform Rules for a Combined Transport Document.

640 **Diplock, Kenneth.** A combined transport document. *Journal of Business Law* 1972, 269–79.

Introductory remarks to the Recommendations for a contractual Combined Transport Document at the May 1972 Genoa Seminar on Combined Transport.

641 **Eyre, J. L.** The containerships of 1999. *Maritime Policy and Management* 16(2), 1989, 133–45.

Intermodal transport is fast and efficient but ports are expensive and time consuming. Ports of 1999 will work round the clock and be massively automated. This change will influence ship design.

642 **Eyre, J. L.** The Third World and intermodalism. *Maritime Policy and Management* 14(3), 1987, 177–84.

Intermodal rail/truck systems linked with containerships give door-to-door service. Why do Third World cargo owners pay more than twice that charged to those from rich countries?

643 **Fleming, D. K.** On the beaten track: A view of US West Coast container port competition. *Maritime Policy and Management* 16(2), 1989, 93–107.

On efforts of US Pacific seaboard ports, Los Angeles, San Francisco, Portland and Seattle, to remain on well-used routes for container shipments between Asia and Eastern United States.

644 **Hayuth, Yehuda.** Freight modal-split analysis of air and sea transportation. *Logistics and Transportation Review* 21(4), December 1985, 389–402.

645 Hayuth, Yehuda. *Intermodality: Concept and practice – structural changes in the ocean freight transport industry.* London: Lloyd's of London Press, 1987. 149 pp.

Evaluates changes in direction and emphasis that have taken and are still taking place in international and domestic freight transportation. Analyses impact of these changes on maritime transportation and related industries. Characteristics of functions of principal components of transport system.

646 Kondratowicz, L. J. Methodological solutions for increased efficiency of modelling and simulation of seaports and inland freight terminals. *Maritime Policy and Management* 19(2), 1992, 157–64.

Outlines research intended to develop efficient simulation modelling capability for seaports and inland freight terminals. Produced the TRANSNODE decision-support expert system.

647 Maritime Transportation Research Board. *Legal impediments to international intermodal transportation. (The Slait Report).* Washington (DC): National Academy of Sciences, 1971. 184 pp.

648 Oda, Hisashi. *Advent of the large scale intermodalism and underlying problems.* Tokyo: Japan Maritime Research Institute, 1988. 23 pp.

Examines the diverse phenomena prevailing in the intermodal market with some assumptions. Technological development and its background. Social background of the large-scale intermodalism. Problems in the mass transport system.

649 Thuong, L. T. From piggyback to double-stack intermodalism. *Maritime Policy and Management* 16(1), 1989, 69–81.

Intermodal transportation in USA (integrated containerized rail, highway and ocean transport) is growing fast. Carriers in all modes will have to realign to offer more efficient provision.

650 Ullman, Gerald H. The ICC's decision in Ex Parte 261: Its residual value. *Journal of Maritime Law and Commerce* 4(3), April 1973, 455–65.

The interstate commerce commission's tariff filing rules for rail and motor carriers and relationship with vessel-operating common carriers by water. Draft convention on combined transport of goods.

651 United Nations Conference on Trade and Development. *Evalu-*

ation of the technical co-operation programme in the field of maritime and multimodal transport. Geneva: UNCTAD, 1990.

652 United Nations Conference on Trade and Development. *International multimodal transport in developing countries: Issues and references.* New York: United Nations, 1979.

653 United Nations Conference on Trade and Development. *Technological development on multimodal transport.* New York: United Nations, 1975.

654 Wei, J. A. U.N. multimodal transport convention. *Journal of World Trade Law* 15(4), July/August 1981, 283–304.

Review of 1980 UN convention of international multimodal transport of goods. History of the convention. General features. The modified uniform liability system. Agent and independent contractor and 'the other person'.

655 Yamada, Hideo. *Strategy of shipping companies for international physical distribution.* Tokyo: Japan Maritime Research Institute, 1990. 81 pp.

Examines the international multimodal transport system. Quality of transportation service. Multimodal transport operator (MTO). International multimodal transport as a systems industry. Ocean shipping and land transportation within a system.

Transhipment

656 Arthur D. Little Limited. *Transhipment in the seventies.* London: National Ports Council, 1970. 48 pp.

657 Browne, M., Doganis, R. and **Bergstrand, S.** *Transhipment of U.K. trade.* London: British Ports Federation, 1989. 216 pp.

Study of causes and effects of deep-sea transhipment. Analysis of major factors underlying decisions to tranship UK deep-sea trade via continent. Effects of transhipment on UK transport sector.

658 Heaver, T. D., Taplin, J. H. E. and **Aandahl, C. L.** Terms of shipment and efficiency in overseas trade. *Maritime Policy and Management* 8(4), 1981, 235–52.

Examines the interests of buyers and sellers in alternative shipping terms and sets out principles expected to influence terms of shipment, using examples from Canadian trades.

Dry Bulk Trades

General and theory

659 **Alexandrakis, A. J.** A multi-stage process for the optimization of the bulk cargo transportation system. Thesis submitted for MSc, University of Newcastle, 1979.

660 **Ariel, A.** Delphi forecast of the dry bulk shipping industry in the year 2000. *Maritime Policy and Management* 16(4), 1989, 305–36.

Delphi method used to obtain views of industry leaders on many issues relating to dry bulk shipping, so as to project a scenario of the industry in the year 2000.

661 **Ariel, A.** The effect of inventory holding costs on the optimal payload of bulk carriers. *Maritime Policy and Management* 18(3), 1991, 217–24.

Shipowners in captive (industrial) trades have different views of transportation costs from those owners involved in 'pure' shipping. Proposes 'Economic Costs of Transport' as measure, enabling owners to optimize carrying capacity.

662 **Beenstock, Michael** and **Vergottis, Andreas.** An econometric model of the world market for dry cargo freight and shipping. *Applied Economics* 21(3), March 1989, 339–56.

Beenstock's theoretical model is applied empirically to the world dry cargo market, assuming that investors in shipping have 'rational expectations'. The theoretical base and properties of the model are explored, the conclusion being that a reasonable statistical account can be taken from the dry cargo market since 1960 using the model.

663 **Bes, J.** *Bulk carriers: A Practical guide to the subject for all concerned with the shipping business.* London: Barker and Howard, 1972. 145 pp.

Gives a picture up to 1972 of the world bulk carrier fleet with particular emphasis on the shipping angle by including information about the actual operation of bulk carriers.

664 **Binkley, James K.** and **Harrer, Bruce.** Major determinants of

ocean freight rates for grains: An econometric analysis. *American Journal of Agricultural Economics* 63(1), February 1981, 47–57.

665 Damigos, E. A disequilibrium analysis of the dry bulk shipping market. Thesis submitted for PhD, University of Newcastle, 1991.

Using econometric modelling, this thesis sets out to identify and examine developments that are radically reshaping both the setting and operations of the shipping industry, with particular reference to the dry bulk cargo sector.

666 De Borger, B. and **Nonneman, W.** Statistical cost functions for dry bulk carriers. *Journal of Transport Economics and Policy* XV(2), May 1981, 155–66.

Estimates sea transportation cost functions for dry bulk carriers on basis of market data. Direct statistical estimates of main determinants of sea transport costs presented for dry bulk carriers operating in grain, coal, iron ore trades.

667 Drewry Shipping Consultants. *Deep sea trade and transportation of forest products.* London: Drewry Shipping Consultants, 1979.

668 Drewry Shipping Consultants. *Dry bulk carrier operating costs: Past, present and future.* London: Drewry Shipping Consultants, 1984. 47 pp.

Report – part of a series covering world's bulk shipping. Dry bulk operating costs. Other dry bulk costs. Development of costs 1973–83. Future development of costs.

669 Drewry Shipping Consultants. *Dry bulk charter market: Developments and trends.* London: Drewry Shipping Consultants, 1983. 72 pp.

Report – part of a series covering world's bulk shipping. Organization and structure of the dry bulk shipping industry. Chartering methods and alternatives. Dry bulk charter market activity, and market operators. Factors affecting freight rates. Outlook for the dry bulk charter market.

670 Drewry Shipping Consultants. *Forecast dry bulk carrier profitability 1990–1995.* London: Drewry Shipping Consultants, 1990. 106 pp.

Report – part of Seaborne Trade and Transport series covering world's bulk shipping. Dry bulk trade and market developments. Forecast dry bulk trade. Bulk carrier fleet. Forecast supply and demand. Development of dry bulk shipping costs. Forecast dry bulk carrier profitability, 1990–95. (Updates previous titles on same subject.)

671 **Drewry Shipping Consultants.** *The future of the container/bulk carrier.* London: Drewry Shipping Consultants, 1984. 72 pp.

Report – part of series covering world's bulk shipping. Historical background of the container/bulk carrier concept. The container/bulk carrier fleet. Types of carriers. Container/bulk carrier operations. Future prospects.

672 **Drewry Shipping Consultants.** *The market for handymax (35–50,000 dwt) dry bulk carriers.* London: Drewry Shipping Consultants, 1985. 95 pp.

Report – part of series covering world's bulk shipping. The handymax bulk carrier. Employment and utilization. Trading opportunities. Costs and revenues. Fleet structure. Trends and prospects.

673 **Drewry Shipping Consultants.** *The market for handymax dry bulk carriers in the 1990's.* London: Drewry Shipping Consultants, 1991. 90 pp.

Report – part of Seaborne Trade and Transport series covering world's bulk shipping. The market role of the handymax carrier and employment and utilization. Trading opportunities and demand for handymax bulk carriers. Costs, revenues and profitability. Handymax bulk carrier fleet supply. Market prospects.

674 **Drewry Shipping Consultants.** *Options for bulk carrier trading: Cargo opportunities, trade practices, size preferences and port constraints in the 1990's.* London: Drewry Shipping Consultants.

675 **Drewry Shipping Consultants.** *Organisation and structure of the dry bulk shipping industry.* London: H. P. Drewry, 1978.

676 **Drewry Shipping Consultants.** *Panamax bulk carriers (50–80,000 dwt).* London: Drewry Shipping Consultants, 1989. 109 pp.

Report – part of Seaborne Trade and Transport series covering world's bulk shipping. 'Panamax' bulk carriers and their changing market role. Employment and the market for panamax bulk carriers. Panamax bulk carrier fleet supply. Panamax costs, revenues, profitability. Market prospects.

677 **Drewry Shipping Consultants.** *Prospects for dry bulk carrier demand and profitability 1991–96.* London: Drewry Shipping Consultants, 1991. 108 pp.

Report – part of Seaborne Trade and Transport series covering world's bulk shipping. Dry bulk trade and market developments. Forecast dry bulk trade. Dry bulk vessel supply. Forecast dry bulk shipping supply and demand.

Development of dry bulk shipping costs. Forecast dry bulk carrier profitability 1991–96.

678 Drewry Shipping Consultants. *The rise of dry cargo shipping in the Far East.* London: H. P. Drewry, 1982. 54 pp.

Report – part of series covering world's bulk shipping. Far Eastern shipping and trade developments. Country reviews: Hong Kong, India, South Korea, China, Taiwan, Philippines, Singapore, Malaysia, Indonesia, and others.

679 Drewry Shipping Consultants. *Specialized types of bulk carrier.* London: H. P. Drewry, 1978. 74 pp.

Report – part of series covering world's bulk shipping. Lumber carriers. Car bulk carriers. Self-unloading bulk carriers. Wood chip carriers. Ore carriers.

680 Drewry Shipping Consultants. *Trading prospects for dry bulk carriers.* London: Drewry Shipping Consultants, 1979.

681 Drewry Shipping Consultants. *Trading prospects for the dry bulk carrier fleet: Demand, employment rates and profitability.* London: Drewry Shipping Consultants, 1992. 116 pp.

Report – part of Seaborne Trade and Transport series covering world's bulk shipping. Dry bulk trade and market developments. Forecast dry bulk trade. Dry bulk vessel supply. Forecast dry bulk shipping supply and demand. Development of dry bulk shipping costs. Forecast dry bulk carrier profitability, 1992–2000.

682 Fearnresearch. *World bulk fleet.* Oslo: Fearnleys.

A report with details on total world fleet of tankers, bulk carriers and combined carriers, existing and on order as well as estimated future fleet. Data refer to 1 January and 1 July and include vessels over 10,000 d.w.t.

683 Ghang, J. H. Market imperfections in Korean dry bulk trades. Thesis submitted for PhD, University of Liverpool, 1990.

This argues that Korean dry bulk shipping markets have been generally subject to various market imperfections and that as a result almost all Korean dry bulk shipping operators were worse off than they would have been if the Korean market had been an open one.

684 Goss, R. O. and **Jones, C. D.** The economies of size in dry bulk carriers. In: R. O. Goss (ed.) *Advances in maritime economics.* Cambridge: Cambridge University Press, 1977, pp. 90–137.

Quantifies effects of using ships of various sizes to carry cargo over various route lengths. Model developed for assessing capital and operating costs using discounting methods. This applied to series of round voyages with cargo one direction only.

685 Logan, J. A. Economics of controllable ballast in bulk carriers. Thesis submitted for M.Phil, University of Newcastle, 1989.

686 Marlow, Peter B. and **Gardner, Bernard.** Some thoughts on the dry bulk shipping sector. *Journal of Industrial Economics* 29(1), September 1980, 71–84.

This examines the possibility of the shipping industry providing the economy with certain indirect benefits and suggests a method of quantifying these benefits for some. It also examines the effects of an increase in the size of a nation's fleet generated by a subsidy programme.

687 Metaxas, B. N. *The economics of tramp shipping.* London: Athlone Press, 1971. 268 pp.

Tramp shipping industry. Analysis of industry. Significant economic phenomena of industry. What demand? Supply of tramp services. Factors affecting demand, incremental nature, relationship of demand to costs. Effects of technical progress? Fluctuations in freight rates.

688 Metaxas, B. N. The fixture of the tramp shipping industry. *Journal of Transport Economics and Policy* VI(3), September 1972, 271–80.

Basic economics realities and characteristics of tramp shipping. Relationship of neighbouring industries of bulk carriers, tankers, liner vessels. Future of tramp shipping.

689 Möncke, Helmuth and **Faust, Peter.** *Barge carrier supply–demand analysis.* Bremen: Institute of Shipping Economics, 1979. 34 pp.

690 Norman, Victor D. Market strategies in bulk shipping. In: Einar Hope (ed.) *Studies in shipping economics: In honour of Professor Arnljot Stromme Svendsen.* Oslo: Bedriftsokonomens Forlag A/S, 1981. 16 pp.

Portfolio rules for chartering. Gains from good timing. Sale and purchase policies.

691 Pullat. *Dry bulk shipping.* London: Lloyd's of London Press, 1988.

692 Rowlinson, M. P. A political economy of bulk shipping, with special reference to the relations between the developed and less developed economies since 1970. Thesis submitted for M.Phil, CNAA, Liverpool Polytechnic, 1986.

Using a political economy approach, it aims to give an accurate description of the industry and to provide an analytical framework for the ideas of those developing nations intent on bulk shipping expansion.

693 Thomas, A. J. A study of intra E.E.C. dry bulk shipping markets. Thesis submitted for M.Phil, CNAA, 1980.

694 *Trading prospects for container/bulk carriers.* Worcester Park, Surrey Cargo Systems Research Consultants, 1981. 135 pp.

695 Weale, John. *Tramp shipping arithmetic: The application of the programmable calculator to commercial problems.* London: Fairplay Publications, 1982. 120 pp.

A collection of well-tried calculator programmes designed to deal with everyday problems of commercial ship operation. Explanation of the principles and methods involved. Divided into sections relating to voyage estimating, speed and fuel consumption, and financial calculation. An appendix deals with iterative calculations.

696 Westinform. *Bulk carrier cargoes for the tramp trades on less than sq. ft. draft.* 1977.

697 *World liner and tramp fleets.* (2 vols.) The Hague: Maritime Economic Research Centre, 1974.

698 Wright, G. Freight rates in the tramp shipping market. *International Journal of Transport Economics* 18(1), February 1991, 47–54.

Freight rates; markets

699 Barty-King, Hugh. *The Baltic Exchange: The history of a unique market.* London: Hutchinson Benham, 1977. 431 pp.

A comprehensive study which details the story of the Baltic Exchange. From the initial dealings in the Baltic Coffee-House in Threadneedle Street to the forming of the body which, after moving to St Mary Axe in 1903, became the Baltic Exchange, the shipping and commodity centre of the world.

700 *BIFFEX and INTEX: Futures trading.* London: Fairplay Publications, 1985. 6 pp.

Market report on the Baltic International Freight Futures Exchange (BIFFEX). Indications for commencement of trading in 1985.

701 BIMCO. *BIMCO 1905–1980: 75 years of international co-operation.* Copenhagen: Baltic and International Maritime Council, 1980. 63 pp.

A publication to commemorate 75th anniversary of BIMCO. Important historical events, principal activities and achievements.

702 Croxson, R. C. Charter planning in bulk shipping companies under conditions of uncertainty. Thesis submitted for M.Phil, CNAA, Plymouth Polytechnic, 1985.

Dry cargo shipping markets were investigated to identify the main factors creating financial difficulties within shipping companies. Showing how the interrelationship between shipowners' chartering policies, purchasing policies and the world economic climate were the major influences.

703 Cullinane, K. A short-term adaptive forecasting model for BIFFEX speculation: A Box–Jenkins approach. *Maritime Policy and Management* 19(2), 1992, 91–114.

Develops a model for short-term forecasting as tool for continuous speculation on future markets, using Box–Jenkins method. Applies model to BIFFEX-type speculation.

704 Cullinane, K. Who's using BIFFEX? Results from a survey of shipowners. *Maritime Policy and Management* 18(2), 1991, 79–91.

How effective is Baltic International Freight Futures Exchange in providing shipowners with a powerful, flexible and viable mechanism for hedging their market risk? Results of survey.

705 Cullinane, K. P. B. The application of modern portfolio theory to hedging in the dry bulk shipping markets. Thesis submitted for PhD, CNAA, Polytechnic South West, 1989.

Using the methodology based on modern portfolio theory, it attempts to analyse the impact of BIFFEX on reducing risk in shipping. Participants in shipping can now hedge against risk in the physical market by taking a position on the new futures market. It argues that successful implementation of modern portfolio theory allows the deduction of a subjectively optimal portfolio of shipping market investments.

706 No entry.

707 Drewry Shipping Consultants. *Forecast dry bulk carrier profit-*

ability: Supply/demand, future costs and revenues, 1988–1991. 3rd rev. edn. London: H. P. Drewry, 1989. 98 pp.

708 Drewry Shipping Consultants. *The market for capesized (100–150,000 dwt) bulk carriers.* London: Drewry Shipping Consultants, 1990. 99 pp.

Report – part of Seaborne Trade and Transport series covering world's bulk shipping. Changing market role of capesize bulk carriers. Demand, employment and the market for capesize bulk carriers. Capesize costs, revenues and profitability. Capesize bulk and combined carrier fleet supply. Market prospects.

709 Drewry Shipping Consultants. *The market for large bulk carriers (over 50,000 dwt).* London: Drewry Shipping Consultants, 1975.

710 Ewart, W. O. *Bulk carriers.* London: Fairplay Publications, 1983. 64 pp.

Explains rationale of bulk carrier operations and examines application of the ship in relation to the world's freight markets.

711 Gray, James. *BIFFEX manual: Baltic international freight futures market.* London: Fairplay Publications, 1988. 122 pp.

712 Hale, C. and **Vanags, A.** Spot and period rates in the dry bulk market: Some tests for the period 1980–86. *Journal of Transport Economics and Policy* XXIII(3), September 1989, 281–92.

Expectations hypothesis provides explicit model of relationship between spot and period rates for dry bulk market. Some tests of the hypothesis applied to data for period 1980–86. Results provide little/no support for expectations hypothesis.

713 Miyashita, K. Structural changes in the freight level determination of tramp shipping markets. *Maritime Policy and Management* 11(1), 1984, 57–67.

Substantiates econometrically how the oil crisis caused major structural changes in setting tramp shipping freight levels. Did it alter important role of theory of lay-up points?

714 Nagatsuka, Seiji. *The demand and supply of grain in the world and its impact on shipping.* Tokyo: Japan Maritime Research Institute, 1986. 52 pp.

Examines trend of grain demand and supply and its effects on maritime transportation. Investigates the preceding ten years and considers latest factors for changes.

715 *New worldwide tanker nominal freight scale applying to the carriage of oil in bulk.* London and New York: Worldscale Association, 1988. 438 pp.

Sets out the 'New Worldscale' – Worldwide Tanker Nominal Freight Scale, introduced with effect from 1 January 1989.

Demand

716 Drewry Shipping Consultants. *Future role of the large bulk carrier (100,000 dwt+).* London: H. P. Drewry, 1981. 73 pp.

Report – part of series covering world's bulk shipping. The large bulk carrier fleet and changing market role. Structure of the fleet. Operation of the large bulk carrier in the dry bulk charter market. Shipping and chartering practices in the commodity trades. Employment of large bulk carriers. Demand.

717 Eriksen, I. Erik. The demand for bulk ship services. In: Einar Hope (ed.) *Studies in shipping economics: In honour of Professor Arnljot Stromme Svendsen.* Oslo: Bedriftsokonomens Forlag A/S, 1981. 8 pp.

Effect on trade pattern of freight rates and other prices from effects on quantities traded internationally.

718 Veldhuizen, Paul J. *Freight futures: Targeting the 90's.* London: Lloyd's of London Press, 1988. 118 pp.

An analysis of freight futures – traditional methods of hedging, viability of the futures concept, historic development of freight futures. How freight futures exchange works. Trends and developments in the dry bulk shipping markets and their impact on freight futures.

Supply

719 *Bulk carrier register.* London: Clarkson Research Studies.

Record of all bulk and combined carriers having a deadweight of 10,000 tons and above.

720 Drewry Shipping Consultants. *Changing ship type/size preferences in the dry bulk market.* London: H. P. Drewry, 1981. 69 pp.

Developments in the transport and handling of dry bulk cargo; shipping type/size preferences in major and minor dry bulk trades; future trends.

721 Drewry Shipping Consultants. *Fleet sectors Panamax bulk carriers.* London: Drewry Shipping Consultants, 1990.

722 Drewry Shipping Consultants. *Future role and profitability of the capesize bulk carrier fleet.* London: Drewry Shipping Consultants, 1992. 93 pp.

Report – part of Seaborne Trade and Transport series covering world's bulk shipping. Capesize bulk carriers – a fleet in crisis? Demand for capesize bulk carriers. Capesize bulk carrier employment and chartering options. Capesize bulk carrier costs, revenues and profit. Capesize dry bulk vessel supply. Market prospect to 2000.

723 Drewry Shipping Consultants. *Future role of the large bulk carrier.* London: H. P. Drewry, 1981.

724 Drewry Shipping Consultants. *Handy-sized (25–40,000 dwt) dry bulk carriers in the 1980's.* London: H. P. Drewry, 1982. 61 pp.

Report – part of series covering world's bulk shipping. The handy-sized (25–40,000 d.w.t.) dry bulk carrier fleet and its changing market role. The market for handy-sized dry bulk carriers. Structure of the fleet. Costs and revenues. Trends and prospects.

725 Drewry Shipping Consultants. *Impact of inflationary costs tendencies on dry cargo shipping operations.* London: Drewry Shipping Consultants, 1982. 124 pp.

Part of series covering world's bulk shipping. Vessel acquisition cost. Development of operating costs 1973–81. Voyage costs. Fully built-up costs. Impact of currency fluctuations on shipping costs. Impact of inflationary cost pressures. The dry cargo shipping market of the 1980s.

726 Drewry Shipping Consultants. *The trading prospects for 'Panamax' (50–80,000 dwt): Bulk carriers and OBO's.* London: Drewry Shipping Consultants, 1980. 55 pp.

Report – part of series covering world's bulk shipping. Size and structure of present fleet of 'Panamax' dry bulk carriers. Market role of the 'Panamax' dry bulk fleet. Employment pattern. Trading costs. Future prospects.

727 Drewry Shipping Consultants. *World bulk fleets: Growth and* *developments 1983–88.* 3rd rev. edn. London: Drewry Shipping Consultants, 1988. 78 pp.

728 **Laskey, Steve R.** *Fairplay world fleets, Vol. 1: Bulk carriers over 12,000 dwt.* London: Fairplay, 1986.

Regulation

729 **Georgandopoulos, E. A.** The proposed international regulations in the dry bulk cargo markets. *Maritime Policy and Management* 13(3), July–September 1986, 223–33.

Examines problems of terms under which bulk commodities exported. Solution suggested to be through progressive change in terms of shipment. Aims at industry operating in international markets without restrictions in order to offset disadvantages of home country's economy.

730 **Packard, William V.** *Shipping pools.* London: Lloyd's of London Press, 1989. 152 pp.

A realistic look at the workings and administration of a typical shipping pool, how liabilities are shared, income equally weighted and distributed and the group, or pool, promoted and marketed.

Wet Bulk Trades (including Combined Carriers)

General and theory

731 **Artz, John C.** The economics of tanker slowdown: An observation. *Maritime Studies and Management* 2(4), April 1975, 244–9.

Factors causing near collapse of tanker charter market following Arab–Israeli war of late 1973. The trade off between high prices and low W. S. rates. The relationship between speed reduction and spot rates at which owner breaks even.

732 **Bartlett, Paul.** Bulk shipping: Few bright spots in a generally dull scene. In: *EEC Shipping: A guide to the sea transport industries of the European Economic Community.* Colchester: Seatrade Publications, pp. 31–3.

Disarray caused by Iranian revolution and usual supply of crude oil being stopped. Impact on shipping world. One encouraging area of tanker market is steady progress towards equilibrium interrupted by hike in price of crude and consequent restrictions in imports.

733 **Beenstock, Michael** and **Vergottis, Andreas.** An econometric model of the world tanker market. *Journal of Transport Economics and Policy* XXIII(3), September 1989, 263–80.

Model which stresses interdependence of market for oil freight and market for tankers. Rational expectations ensure that when shocks anticipated some adjustment takes place beforehand.

734 **Bremer, W. M.** and **Perakis, A. N.** An operational tanker scheduling optimization system: Model implementation, results and possible extensions. *Maritime Policy and Management* 19(3), 1992, 189–99.

Describes scheduling system implementation on University of Michigan IBM 3090-600E computer, generating feasible schedules for each vessel, and attempts to provide optimal overall schedule.

735 **Champness, Michael** and **Jenkins, Gilbert.** *Oil tanker databook 1985.* New York: Elsevier Applied Science Publishers, 1985. 347 pp.

Comprehensive collection of statistics of the world oil tanker fleet and related topics since the 1950s. Tanker data presented as the position at 30 June and 31 December each year. Vessels classified according to size and ownership. Additions to fleets. Scrappings. Combined carriers. LNG, LPG carriers. Freight rates. Oil statistics.

736 **Clarkson Research Studies.** *Tanker register.* London: Clarkson Research Studies.

Record of all combined carriers and tankers having a deadweight of 5,000 tons and above.

737 **Coulson, E. C.** *A guide for tanker brokers.* 2nd rev. edn. London: Clarkson Research Studies, 1991. 69 pp.

Basic principles of voyage estimating. Concept of Worldscale explained. Charterparty terms and conditions and their legal interpretation. Activities of the tanker broker.

738 **Drewry Shipping Consultants.** *Changes in tanker chartering.* London: H. P. Drewry, 1981. 75 pp.

Report – part of series covering world's bulk shipping. The world's oil markets and tanker demand. Emergence of new charterers in the tanker market. Structure of the tanker charter market. Tanker charter market in the 1980s.

739 Drewry Shipping Consultants. *Combined carriers: Future role and profitability of ore/oil and OBO carriers.* London: H. P. Drewry, 1989. 77 pp.

740 Drewry Shipping Consultants. *Fleet sectors combined carriers.* London: Drewry Shipping Consultants, 1990.

741 Drewry Shipping Consultants. *The future of the million barrel tanker.* London: Drewry Shipping Consultants, 1990. 94 pp.

Report – part of Seaborne Trade and Transport series covering world's bulk shipping. The fleet. Employment fleet performance in 1989. Costs and revenues. New building and sale and purchase market. Market prospects.

742 Drewry Shipping Consultants. *The market for medium-sized tankers (90–175,000 dwt).* London: Drewry Shipping Publications, 1985. 133 pp.

Survey – part of series covering world's bulk shipping. The fleet. Employment in 1983. Fleet performance 1983. Costs and revenues. New building and sale and purchase market. Market prospects.

742a Drewry Shipping Consultants. *The market for Panamax (50–80,000 dwt) dry bulk carriers.* London: Drewry Shipping Consultants, 1984. 69 pp.

Report – part of series covering world's bulk shipping. The Panamax bulk carrier fleet and its changing market role. Structure of the fleet. Marketing of Panamax bulk carriers. Demand, costs and revenues for Panamax bulk carriers. Future prospects.

743 Drewry Shipping Consultants. *The market for tankers below 20,000 dwt.* London: H. P. Drewry, 1982. 54 pp.

Report – part of series covering world's bulk shipping. The fleet of tankers in the range 5,000–19,999 d.w.t. The employment of the small tanker fleet. Costs and freight rates.

744 Drewry Shipping Consultants. *The oil exporting countries and shipping.* London: Drewry Shipping Consultants, 1982.

745 Drewry Shipping Consultants. *The role of independent tanker owners.* London: Drewry Shipping Consultants, 1976.

746 Drewry Shipping Consultants. *The tanker crisis: Causes, effects and prospects to 1985.* London: H. P. Drewry, 1976.

747 Drewry Shipping Consultants. *The tanker fleets of the international oil companies.* London: H. P. Drewry, 1979. 92 pp.

748 Drewry Shipping Consultants. *Tanker industry 1981: Market trends.* London: H. P. Drewry, 1982. 84 pp.

Report – part of series covering world's bulk shipping. Oil scene in 1981. Tanker demand and supply. The charter market in 1981. Operating and trading costs.

749 Drewry Shipping Consultants. *Trends in tanker operations and economics.* London: Drewry Shipping Consultants, 1979. 44 pp.

Report – part of series covering world's bulk shipping. Transportation costs in the 1970s. Technical and operational developments. Implications of developments.

750 Everett, J. and others. Optimization of a fleet of large tankers and bulkers: A linear programming approach. *Marine Technology,* October 1972.

751 Glen, D. R. Differentiation in the oil tanker market 1970–1978. Thesis submitted for PhD, London Graduate School of Business Studies, 1987.

752 H. Clarkson and Company. *Tanker prospects to 1985.* London: H. Clarkson and Company, 1975. 117 pp.

Report which examines a two-part question: (a) what are the prospects for the tanker market as a whole in the period up to 1985? (b) within that framework, what is the outlook for large tonnage, i.e. 200,000 d.w.t. and over?

753 Hills, W. The effects of economics on ship design, with special reference to large tankers. Thesis submitted for M.Phil, CNAA, 1969.

754 Jenkins, Gilbert and others. *Clarkson oil tanker databook: Key statistics from the Clarkson oil tanker registers and research publications over the last 30 years.* London: Clarkson Research Studies, 1993. 368 pp.

755 Kalindaga, Y. C. Estimation of capacity utilization in world shipping. *Maritime Policy and Management* 17(1), 1990, 41–67.

Paper examines selected methods used to compute tanker tonnage and proposes other methods. Notes conceptual and practical problems in agreeing suitable base periods from which to start.

756 Khanti-Akom, Sophon. Essays on the economic analysis of oil tanker transportation. Thesis submitted for PhD, Texas A & M University, 1991.

757 Nagatsuka, Seiji. *Tonnage supply and demand forecast for large-sized tanker fleet and new buildings.* Tokyo: Japan Maritime Research Institute, 1988. 44 pp.

Examines the medium-term outlook for tonnage supply/demand of VLCCs in light of environmental changes.

758 Pedrick, John L. Tankship design regulation and its economic effect on oil consumers. *Journal of Maritime Law and Commerce* 9(3), April 1978, 377–95.

Considers economic factors in oil tanker design. The double bottom/hull controversy and associated cost increases; segregated ballast and its associated cost increases; indirect cost factors; balancing consumer costs involved with benefits derived.

759 Perakis, A. N. and **Bremer, W. M.** An operational tanker scheduling optimization system: Background, current practice and model formulation. *Maritime Policy and Management* 19(3), 1992, 177–87.

Reviews tanker scheduling problem as it applies to Chevron Shipping Company. Develops model for scheduling and describes an integer programming formulation for schedule optimization.

760 Ratcliffe, Mike. *Liquid gold ships: A history of the tanker 1859–1984.* London: Lloyd's of London Press, 1985. 181 pp.

A history of the development of the oil tanker and its international trade. Key trends in the complex evolution of the industry. Social and economic changes.

761 Renouf, A. Accounting for the tanker crisis. *Accountancy* 86, October 1975, 46–50.

This is a plea for a better understanding of how the shipping industry really works, particularly in considering the cyclical nature of the tanker freight rates.

762–5 No entries.

766 Shell Briefing Service. *The tanker industry in the 1990's. (No. 5 – 1991).* London: Shell International Petroleum Company. 9 pp.

Review of the growth of the world tanker fleet since the early twentieth century. Emphasizes importance of oil to international seaborne trade. Fundamentals of business – charter market; freight; flags of registration; international conventions. Challenges facing tanker industry, especially environmental challenge.

767 Shipping Economist Conference. *Tankers ... The new era.* London: Lloyd's of London Press, 1980.

767a Shipping Economist Conference. *Tankers: Where's the miracle?* London: Lloyd's of London Press, 1983.

768 Talley, W. K., Agarwal, V. B. and **Breakfield, J. W.** Economies of density of ocean tanker ships. *Journal of Transport Economics and Policy* XX(1), January 1986, 91–100.

Finds that elasticities of cost and economies of density vary for different sizes of ships. Optimum ship size for given type of ship should not be derived from single ship elasticity of operating costs.

769 *Tankers: The challenge of the 1990's.* London: Lloyd's of London Press, 1988.

770 Thuong, Le T. and **Elvey, Lisa A.** The end of the supertanker era. *Transportation Journal* 25(2), Winter 1985, 4–17.

771 Trew, Antony. *Death of a supertanker.* London: Collins, 1978. 220 pp.

772 Vielvoye, Roger. World tanker industry maintains momentum from Persian Gulf War. *Oil and Gas Journal* 89(23), 10 June 1991, 12–15.

773 Whyte, I. M. A study of a service performance analysis system for a tanker fleet. Thesis submitted for MSc, University of Newcastle, 1974.

774 Zannetos, Zenon S. Oil tanker markets: Continuity amidst change. In: Richard L. Gordon, Henry D. Jacoby and Martin B. Zimmerman (eds) *Energy: Markets and regulation. Essays in honour* *of M. A. Adelman.* Cambridge (Massachusetts) and London: MIT Press, 1987, pp. 235–57.

775 Zannetos, Zenon S. *The theory of oil tankship rates: An economic analysis of tankship operations.* Cambridge (Massachusetts) and London: MIT Press, 1966. 336 pp.

Oil tankship transportation economics. Essential characteristics of the tankship markets. Factors that influence behaviour of tanker owners and tanker charterers. Factors affecting rates in short and long run. Impact. Economics of tankship building and market operation.

Freight rates; markets

776 Beth, Hans Ludwig. *The world tanker market situation: 1974–1976.* Bremen: Institute of Shipping Economics, 1975. 23 pp.

A paper which considers the impact on oil transport trends of the development of the organization and pattern of the trade.

777 Drewry Shipping Consultants. *The Drewry tanker market review.* London: Drewry Shipping Consultants, 1983.

778 Drewry Shipping Consultants. *Forecast tanker profitability: Supply/demand, future costs and revenues, 1988–1992.* 3rd rev. edn. London: Drewry Shipping Consultants, 1988. 78 pp.

779 Drewry Shipping Consultants. *Risks and rewards in the tanker market: Implications for owners and charterers.* London: Drewry Shipping Consultants, 1992.

780 Drewry Shipping Consultants. *The role of VLCC's in the 1980's.* London: Drewry Shipping Consultants, 1984. 73 pp.

Report – part of series covering world's bulk shipping. Supply/demand 1979–83. Market review 1979–83. Tracking analysis 1982. Demand and supply forecasts to 1990. Market outlook to 1990.

781 Drewry Shipping Consultants. *Tanker charter market: Structure, participants and trends.* London: Drewry Shipping Consultants, 1992. 88 pp.

Report – part of Seaborne Trade and Transport series covering world's bulk shipping. Oil trades and tanker market. Recent chartering history 1986–91. Tanker charterers.

782 Drewry Shipping Consultants. *Tanker chartering: Trends and prospects.* London: Drewry Shipping Consultants, 1985. 82 pp.

Report – part of series covering world's bulk shipping. 1984 charter market review. Market structure – past and present. Chartering activity by company. Factors affecting tanker chartering. Future outlook.

783 Drewry Shipping Consultants. *Tanker chartering through the 1970's.* London: H. P. Drewry, 1977. 67 pp.

Report – part of series covering world's bulk shipping. Historical tanker chartering. Prospects for the tanker charter market. Types of charter. Freight scales.

784 Drewry Shipping Consultants. *Tanker freight costs and crude oil prices.* London: H. P. Drewry, 1980. 50 pp.

Changes in the freight cost element in landed crude oil prices; cost of transportation; trend of freight rates 1973–8; methods of tanker cost recovery.

785 Drewry Shipping Consultants. *The tanker market: Five year forecast of demand, supply and profitability.* London: Drewry Shipping Consultants, 1992. 98 pp.

Report – part of Seaborne Trade and Transport series covering world's bulk shipping. Tanker profitability. The nature and development of costs. Oil and tanker market trends. The trading environment.

786 Drewry Shipping Consultants. *Tanker market profitability to 1995.* London: Drewry Shipping Consultants, 1990. 94 pp.

Report – part of Seaborne Trade and Transport series covering world's bulk shipping. The oil market 1985–90. The tanker market 1985–90. Forecast market developments to 1995. Outlook for tanker demand and oil carrier supply. Forecast tanker costs. Forecast tanker market profitability.

787 Drewry Shipping Consultants. *Trends in U.S. oil and tanker supply.* London: Drewry Shipping Consultants, 1976.

788 Glen, D. The emergence of differentiation in the oil tanker market, 1970–1978. *Maritime Policy and Management* 17(4), 1990, 289–312.

Traditional assumption that oil tanker market is homogeneous is no longer valid because market differentiation by size and route emerged during 1970s.

789 Glen, D., Owen, M. and **Van Der Meer, R.** Spot and time charter rates for tankers, 1970–77. *Journal of Transport Economics and Policy* XV(1), January 1981, 45–58.

Examines changes in pattern of timechartering activity. Model relating spot and charter rates developed and tested at disaggregated level. Sheds light on probable nature of formation of 'rate expectations' in tanker market.

790 Hawdon, D., Pearson, P. and **Stevens, P.** (eds) *Prospects for the oil tanker market.* Guildford: University of Surrey, Department of Economics, 1989. 300 pp.

Proceedings of a half-day workshop at the Surrey Energy Economics Centre, 16 November 1988.

791 Hawdon, David. Tanker freight rates and profitability. In: David Hawdon, Peter Pearson and Paul Stevens (eds) *Prospects for the oil tanker market.* Guildford: University of Surrey, Department of Economics, 1989. 15 pp.

Review of tanker market developments 1980–88. Assesses prospects for rates and tanker profitabilities over following twelve months.

792 Hawdon, D. Tanker freight rates in the short and long run. *Applied Economics* 10(3), September 1978, 203–17.

The process of spot freight rate determination is viewed as a series of interactions between the market for tanker services and the market for tankers. The conclusion is that there is little the independent tanker owners can do by way of modifying their tanker investment decisions to significantly influence the stability of the industry. This will continue as long as the old companies continue to tie their investment decisions solely to anticipated volumes of oil exports.

793 INTERTANKO. *Tanker chartering: Questions and answers.* Oslo: INTERTANKO, 1990. 128 pp.

Presents a selection of questions from members and advice given by the INTERTANKO office during the few years up to publication. Safety and commercial trends. Practical guidance.

794 INTERTANKO. *Tanker chartering clauses.* Oslo: INTERTANKO, 1984. 38 pp.

Provides guidance in daily tanker chartering activities in respect of safety and commercial trends. Includes usage of tankers as floating storage where INTERTANKO has drafted chartering provisions to deal with events which may arise; lays down model provision for laytime and demurrage in respect of Crude Oil Washing operation and MARPOL 73/78.

795 *New worldscale: A new chartering tool.* Oslo: INTERTANKO, 1988. 55 pp.

796 Smith, Leigh. Prospects for the oil tanker market. In: David Hawdon, Peter Pearson and P. Stevens (eds) *Prospects for the oil tanker market.* Guildford: University of Surrey, Department of Economics, 1989. 9 pp.

Political appraisal. Oil prices. Production of oil and impact on market. Demand for biggest vessels. Prospects.

797 Waaler, Claus. Prospects for the oil tanker market. In: David Hawdon, Peter Pearson and Paul Stevens (eds) *Prospects for the oil tanker market.* Guildford: University of Surrey, Department of Economics, 1989. 6 pp.

Political appraisal. What do events of the early 1980s mean? What is the future?

Demand

798 Apostolides, Michael. VLCC/ULCCs: The dying giants. *Maritime Policy and Management* 10(2), April–June 1983, 87–91.

Commentary raising question of what happened to large crude carriers after the oil crises following the Yom Kippur War, 1973.

799 Bayley, A. Gordon. *World tanker prospects 1980–1983.* Liverpool and London: Tilney & Co., 1979. 82 pp.

Assesses the supply of and demand for ocean-going tankers in the free world over years 1980–83. Tanker supply and demand during each year. The balance between supply and demand.

800 Drewry Shipping Consultants. *Crude oil: Forecast trade and crude carrier demand.* 3rd rev. edn. London: Drewry Shipping Consultants, 1988. 61 pp.

801 Drewry Shipping Consultants. *Crude oil supply and tanker demand.* London: H. P. Drewry, 1979. 86 pp.

World oil production, consumption and seaborne trade in 1977; future prospects for oil production; trends in tanker demand up to 1985; implications for tanker employment.

802 Drewry Shipping Consultants. *Seasonal fluctuations in crude,* *products and tanker demand.* London: Drewry Shipping Consultants, 1975. 49 pp.

Report – part of series covering world's bulk shipping. Seasonal fluctuations in product demand: (i) the US; (ii) Western Europe; (iii) Japan. Seasonal factors and crude oil production. Seasonal fluctuations in oil demand and the tanker market.

803 Gould, H. R. Offshore petroleum development and potential. *Maritime Studies and Management* 2(3), January 1975, 181–9.

Summarizes world-wide energy situation, brought to a critical point by the 1973–4 cut-back in Middle East oil production, with regard to offshore oil and implications for total energy picture.

804 Jeffrey, Angus Cuthbert Trevelyan. The Caribbean oil trades: The demand for medium-size crude oil carriers. Thesis submitted for MSc, CNAA, Plymouth Polytechnic, 1984. 140 pp.

Examines why Caribbean is a significant area within world's oil trade. Examines structure of and advantages possessed by key countries in that area. Assesses demand for medium-size crude oil carriers along region's oil route. Links between economic growth and energy consumption. Tanker demand in face of developments.

805 Strandenes, Sir I. Pettersen. Demand substitution between tankers of different sizes. In: Einar Hope (ed.) *Studies in shipping economics: In honour of Professor Arnljot Stromme Svendsen.* Oslo: Bedriftsokonomens Forlag A/S, 1981. 16 pp.

Size distribution of tanker fleet. Correlation analysis. Estimation methods, problems, cross-pricing elasticities. Estimates of reduced form parameters.

806 White, Brian. World LNG trade and tanker availability. *Gas Energy Review* 18(11), November 1990, 21–4.

LNG trade grew significantly over the latter half of the 1980s. Modifications will be needed from the world's LNG tanker fleet to meet demand.

Supply

807 Drewry Shipping Consultants. *The future of the supertanker.* London: Drewry Shipping Consultants, 1991. 104 pp.

Report – part of Seaborne Trade and Transport series covering world's bulk shipping. The fleet. Employment in 1989. Fleet performance 1989–90. New building and second-hand markets. Costs and revenues. Market prospects.

808 Drewry Shipping Consultants. *Investment in tankers.* London: Drewry Shipping Consultants, 1978.

809 Drewry Shipping Consultants. *Japan, trends in crude, products and tanker supply.* London: Drewry Shipping Consultants, 1974.

810 Drewry Shipping Consultants. *Tanker operating cost trends: 1970–81.* London: Drewry Shipping Consultants, 1983. 60 pp.

Report – part of series covering world's bulk shipping. Trends in inflation and exchange rates. Operating costs. Other tanker costs. The development of fully built-up tanker costs 1970–81. Future influences on tanker costs.

811 Drewry Shipping Consultants. *Trading outlook for very large tankers.* London: H. P. Drewry, 1975. 63 pp.

Report – part of series covering world's bulk shipping markets. The tracking of VLCC movements during 1973–4. The market for VLCCs in 1973 and 1974. Port facilities. VLCC supply/demand balance in 1977 and 1980. Future VLCC trading.

812 Drewry Shipping Consultants. *ULCC trading opportunities.* London: Drewry Shipping Consultants, 1977.

812a Drewry Shipping Consultants. *VLCC/ULCC employment, costs and revenues.* London: H. P. Drewry, 1979. 74 pp.

Report – part of series covering world's bulk shipping. The economics of VLCC/ULCCs. The VLCC/ULCC fleet and characteristics. VLCC/ULCC employment and performance in 1978. Observations on the future.

812b Hawdon, David. Transport costs and capacity adjustment in the tanker market. In: David Hawdon (ed.) *Oil prices in the 1990's.* New York: St Martin's Press, 1989, pp. 80–104.

813 INTERTANKO. *Storage of coal in tankers: A feasibility study.* Oslo: INTERTANKO, 1982. 17 pp.

A summary of two studies which examine using VLCCs as a floating coal-terminal. Exposes the potentials to ease the congested US coal port situation.

814 INTERTANKO. *A survey of the combined carrier fleet and its trade.* Oslo: INTERTANKO, 1981.

815 Shimojo, T. *Size effects in tanker shipping.* Bremen: Institute of Shipping Economics, 1981. 28 pp.

Discusses the effects of ship size expansion in tanker shipping from three points of view: (i) the shipbuilding costs; (ii) the operating costs; (iii) the profits of operation.

816 Terminal Operators Ltd. *World oil demand and tanker markets 1978–81.* London: Terminal Operators Ltd, 1978. 84 pp.

Oil demand and supply and the market for carriers. The supply of crude and products carriers. Moving towards balanced supply and demand.

817 Tolofari, S. R., Button, K. J. and **Pitfield, D. E.** An econometric analysis of the cost structure of the tank sector of the shipping industry. *International Journal of Transport Economics* 14(1), 1987, 71–84.

818 *VLCC's: A market perspective.* London: Fairplay Publications.

819 *World tanker fleet review.* London: John I. Jacobs PLC.

A combination of analysis and statistical data on the world tanker market produced every six months.

Pollution

820 Clark, R. B. Tanker wrecks and the human element. *Marine Pollution Bulletin* 9(5), 1978, 113–

821 Drewry Shipping Consultants. *Impact of new tanker regulations.* London: H. P. Drewry Shipping Consultants, 1981. 61 pp.

Report – part of series covering world's bulk shipping. Outline of regulations. Safety and anti-pollution measures. Compliance. Effects on tanker transportation costs. Effects on world tanker fleet. Tanker profitability.

822 Drewry Shipping Consultants. *Marine pollution and safer ships: Implications for the tanker industry.* London: Drewry Shipping Consultants, 1992. 110 pp.

823 Embiricos, Philip. *The quest for the environmental ship.* Oslo: INTERTANKO, 1991. 143 pp.

An INTERTANKO discussion paper – part of series dealing with cleaner seas and the commercial implications. Object of paper is to derive comparative methodologies to assess the relative merits of alternative tanker designs, having regard to the various factors which can give rise to oil spills.

824 Forsyth, C. J. Factors affecting tanker safety. *Maritime Policy and Management* 18(4), 1991, 313–19.

On factors affecting oil tanker safety, using views of maritime labour officials, industry experts and US Coast Guard reports. Analyses size and age of vessels, crew qualifications, manning levels, etc.

825 International Tanker Owners' Pollution Federation. *Response to marine oil spills. The International Tanker Owners' Pollution Federation Ltd.* 2nd edn. London: Witherby & Co., 1989. 136 pp.

First published in 1986 as five booklets. Comprehensive review of the problems posed by marine oil spills and the response measures that can be implemented. The oil spill. Containment and recovery. The use of dispersants. Shoreline cleanups. Planning and operations.

826 M'Gonigle, R. Michael and **Zacher, Mark W.** *Pollution, politics and international law: Tankers at sea.* University of California Press, 1981. 416 pp.

827 Maybourn, Ralph. Large tankers: Their safety and their impact on the marine environment. *Royal Society of Arts Journal* 128, July 1980, 511–27.

The paper given and discussion following at the Thomas Gray Memorial Lecture, March 1980, to the Society. Considers the risks inherent in tanker operations.

828 *Oil cargo losses and problems with measurement.* Oslo: INTERTANKO, 1986. 15 pp.

Booklet aims to show that many of apparent cargo losses either did not, in fact, occur – and thus the carriers were victims of measurement inadequacies – or that such losses were, in fact, accountable.

829 Pritchard, S. Z. Load on top: From the sublime to the absurd. *Journal of Maritime Law and Commerce* 9(2), January 1978, 185–224.

Considers concept of 'load on top' as a major anti-pollution method for tankers. Development of LOT.

830 Van Poelgeest, F. M. *Sub standard tankers.* Rotterdam: Netherlands Maritime Institute, 1978. 68 pp.

Ascertains in which section of the tanker fleet sub-standard tankers can be found. Determines trends in size, age, flag, ownership, type of accident, etc.

Considers consequences of sub-standard tankers. Suggestions for improvement.

Product carriers

831 Abecassis, D. W. Risk sharing in LNG shipping: Some aspects of the relationships between owners and charterers of LNG ships and third parties. *Lloyd's Maritime and Commercial Law Quarterly*, February 1982, 138–51.

Considers some aspects of the relationships between owners and charterers of LNG ships and third parties, and illustrates the theme of risk sharing. General observations about LNG shipping business followed by consideration of each aspect in turn.

832 Clarkson Research Studies. *Chemical Tanker Register.* London: Clarkson Research Studies.

833 Clarkson Research Studies. *Liquid gas carrier register.* London: Clarkson Research Studies.

Record of all seagoing vessels having at least 5 per cent of their total cargo capacity designed to carry gas in liquid form.

834 Corkhill, Michael. *LNG carriers: The ships and their market.* London: Fairplay Publications, 1975. 114 pp.

Examines short and long-term prospects for LNG industry. Supply and demand situation assessed. Potential of the various projects being planned at the time. Development potential of LNG market and evaluation of LNG carriers. Factors affecting ship safety. Future requirements for LNG carriers.

835 Corkhill, Michael. *LPG and chemical gas carriers: The ships and the market.* London: Fairplay Publications, 1976. 113 pp.

Report tracing development of various types of gas carrier and examining those aspects which set them apart from other vessels. Cargo-handling procedures. Safety problems associated with carriage of hazardous cargoes. Prospects for LPG industry. LPG fleet requirements.

836 Corkhill, Michael. *Product tankers and their market role.* London: Fairplay Publications, 1978. 200 pp.

837 Drewry Shipping Consultants. *Chemical carrier market: Pros-*

pects for trade and shipping in the 1990's. London: Drewry Shipping Consultants, 1991. 186 pp.

838 Drewry Shipping Consultants. *The chemical carrier trades.* London: Drewry Shipping Consultants, 1983.

839 Drewry Shipping Consultants. *Future development of the LPG and chemical gas trades.* London: Drewry Shipping Consultants, 1991. 99 pp.

Report – part of Seaborne Trade and Transport series covering world's bulk shipping. The LPG and chemical gas industries. Seaborne trade and LPG carrier demand. The LPG fleet and the outlook for supply to 1996. The LPG and chemical gas freight markets. LPG carrier costs, revenues and profitability.

839a Drewry Shipping Consultants. *The growth of the chemical carrier fleet.* London: H. P. Drewry, 1981. 56 pp.

Report – part of series covering world's bulk shipping. Chemical tanker operation. The chemical tanker fleet and the future.

840 Drewry Shipping Consultants. *Liquefied gas: Forecast LNG/LPG trade and gas carrier demand, 1987–1992.* 2nd rev. edn. London: Drewry Shipping Consultants, 1988. 98 pp.

841 Drewry Shipping Consultants. *Liquid cargo liquid chemicals.* London: Drewry Shipping Consultants, 1989.

842 Drewry Shipping Consultants. *Liquid cargo petroleum products.* London: Drewry Shipping Consultants, 1989.

843 Drewry Shipping Consultants. *Market conditions and tanker economics.* London: H. P. Drewry, 1976. 73 pp.

Report – part of series covering world's bulk shipping. Background to the present tanker market. Long-term implications of the present tanker surplus.

843a Drewry Shipping Consultants. *The market for LPG carriers in the 1980's.* London: Drewry Shipping Consultants, 1982.

844 Drewry Shipping Consultants. *Market prospects for products tankers.* London: Drewry Shipping Consultants, 1983. 80 pp.

Report – part of series covering world's bulk shipping. Products balances

1981 and 1982. World refining capacity. Trade and employment in 1982. Future trade and employment. Products tanker supply 1983–90. Market prospects.

845 Drewry Shipping Consultants. *The parcel tanker trades.* London: Drewry Shipping Consultants, 1977.

846 Drewry Shipping Consultants. *Product tanker prospects: Outlook for trade and profitability.* London: Drewry Shipping Consultants, 1991. 176 pp.

847 Drewry Shipping Consultants. *The products tanker fleet.* London: H. P. Drewry, 1981. 44 pp.

Report – part of series covering world's bulk shipping. The small tanker fleet by d.w.t. size tranche. The total fleet – present and future structure.

848 Drewry Shipping Consultants. *Products tanker trade and employment prospects.* London: Drewry Shipping Consultants, 1985. 90 pp.

Report – part of series covering world's bulk shipping. Petroleum products demand. Refining capacity throughput. Trade in petroleum products. Products tanker employment. Products carrier supply. Market prospects.

849 Drewry Shipping Consultants. *Prospects for LNG and LPG shipping in the 1990's.* London: Drewry Shipping Consultants, 1990. 110 pp.

Report – part of Seaborne Trade and Transport series covering world's bulk shipping. Natural gas and the world energy economy. LNG – the producers, consumers and international trade. LNG shipping. The outlook for LNG trade and shipping in the 1990s.

850 Drewry Shipping Consultants. *Prospects for U.S. tanker requirements.* London: Drewry Shipping Consultants, 1984. 68 pp.

Report – part of series covering world's bulk shipping. US energy policies. US oil consumption, production and trade. US tanker demand 1982. US tanker demand 1983–90. US flag tanker fleet.

851 Drewry Shipping Consultants. *Refined petroleum carriers.* London: Drewry Shipping Consultants, 1989.

852 Drewry Shipping Consultants. *Refined petroleum products.* London: Drewry Shipping Consultants, 1989. 67 pp.

Report – part of Seaborne Trade and Transport series covering world's bulk shipping. Industry structure. Pattern of trade. Sea transport. Projects, developments and trends. Forecast products carrier employment.

853 Drewry Shipping Consultants. *Trading in LNG and natural gas: Global patterns and prospects.* London: Drewry Shipping Consultants, 1992. 90 pp.

Report – part of Seaborne Trade and Transport series covering world's bulk shipping. Terminology and conversion factors. A growing role for natural gas in the world energy economy. Regional natural gas markets. Pipeline trade. LNG trades. The LNG fleet and transportation costs.

854 Drewry Shipping Consultants. *World LNG trade and shipping.* London: H. P. Drewry, 1976. 70 pp.

Report – part of series covering world's bulk shipping. Mid-1973 operating and planning seaborne LNG trades reviewed and updated to end 1975. Individual LNG exporting countries (geographic region) at end of 1975. Analysis of the LNG trading performance 1975–6. Prospects.

855 Fariday, Edward K. *LNG Review 1977: A study of marine operations, ship technology, project finance, LNG prices and market prospects for LNG.* Wokingham: Energy and Economics Research, 1977. 136 pp.

856 Ffooks, Roger C. (ed.) *Gas carriers.* London: Fairplay Publications, 1984. 108 pp.

Examines the design and operation of gases fleet and the trades in which such ships are engaged. Developments in the future.

857 Ffooks, Roger. *Natural gas by sea: The development of a new technology.* 2nd rev. edn. London: Witherby & Co., 1993. 245 pp.

An introduction to LNG transportation technology. Development of marine technology, advanced marine technology. Gas conferences. Potential dangers, safety, reliability. Increase in LNG fleet.

857a Grey, Michael (ed.) *Chemical/parcel tankers: Their cargoes, markets and design.* 3rd edn. London: Fairplay Publications, 1984. 152 pp.

Traces continuing development of the chemical and parcel tanker sector. Cargoes and their markets. Cargoes and their constraints. Ship design. Ship operations. Chemical parcel tanker development.

858 *Leading developments in the products tanker market.* London: Lloyd's of London Press, 1992. 124 pp.

859 *Oil, chemical and combined carriers: 10,000 dwt and above.* London: E. A. Gibson Shipbrokers, 1992. 92 pp.

Contains details of all oil, chemical and combined carriers known to be in existence or on order at shipyards with a deadweight of 10,000 tons or above.

860 Partridge, Christopher. *Product tankers and their market.* London: Fairplay Publications, 1993.

861 Scott, Robert and **Grey, R. J. M.** *Product tankers and their market role.* 2nd rev. edn. London: Fairplay Publications, 1982. 140 pp.

862 Wooler, R. G. *Marine transportation of LNG (liquefied natural gas) and related products.* Centreville (Maryland): Cornell Maritime Press, 1975. 359 pp.

863 Yoshida, Shigeru. *World's LNG situation and LNG carriers.* Tokyo: Japan Maritime Research Institute, 1988. 73 pp.

Booklet, part of series covering shipping activities. LNG import projects in the world. Signs of changes in LNG export and import systems. Summary of LNG import projects for Europe, USA, South Korea and Taiwan. Summary of LNG import projects and prices for Japan. Outlook for LNG demand in Japan. New building requirements.

Regulations

864 Drewry Shipping Consultants. *Tanker regulations: Enforcement and effect.* London: Drewry Shipping Consultants, 1985. 92 pp.

Report – part of series covering world's bulk shipping. Defines I.M.O. regulations and state and port regulations on crude, products, chemical, gas carriers. Level of compliance. Level of enforcement. Effects of regulations on vessel design, fleet development, trading economics. Future developments.

865 Drewry Shipping Consultants. *Tanker regulations: Implications for the market.* London: Drewry Shipping Consultants, 1991. 96 pp.

Report – part of Seaborne Trade and Transport series covering world's bulk shipping. I.M.O. regulation. Additional state and port authority regulations. Conformity. Tanker regulations and market costs. Effect of tanker regulations on operating practices and fleet development, and development of chemical and gas fleets.

866 International Tanker Nominal Freight Scale Association and **Association of Shipbrokers and Agents.** *The development of tanker rate schedules.* London: International Tanker Nominal Freight Scale Association and Association of Shipbrokers and Agents, 1973. 23 pp.

Pre-Second World War freight rate schedules. MOT Schedule. USMC rates. Scales numbers 1, 2 and 3. The American Tanker Rate Schedule, 1956. Worldscale – September 1969 and onwards.

Liner Trades

General and theory

867 Alfred, I. Technical and economic analysis of ship choice in the general cargo/container trades in Nigeria. Thesis submitted for MSc, UWIST, Cardiff, 1987.

Uses discount flow analysis techniques to examine the choice of ships in the general cargo/container trade of Nigeria.

868 Anand, J. C. Liner shipping under developing countries' flag. In: Hans Ludwig Beth (ed.) *Liner shipping in the eighties: Report on the International Symposium, Bremen, October 24th–26th 1979.* Bremen: Institute of Shipping Economics, 1980, pp. 319–35.

Problems faced in establishing merchant fleets by developing countries. The general view of the developing nations in relation to liner services for their trades.

869 Australian Shippers Council. *Future of Australia's overseas liner shipping. International Symposium.* Sydney: Australian Shippers Council, 1984.

870 Bank, R. K. Panel on U.S. shipping policy. In: Hans Ludwig Beth (ed.) *Liner shipping in the eighties: Report on the International Symposium, Bremen, October 24th–26th 1979.* Bremen: Institute of Shipping Economics, 1980, pp. 237–44.

International aspects of liner shipping in the 1980s, viewed from the vantage point of diplomatic and commercial considerations. The UNCTAD Code; antitrust matters; the Omnibus Bill and regulation itself. The need for mid-Atlantic, mid-Pacific solutions to international shipping problems.

871 **Bennathan, Esra** and **Walters, A. A.** Shipping conferences: An economic analysis. *Journal of Maritime Law and Commerce* 4(1), October 1972, 93–115.

Examination of rationalization for conference in shipping industry – to support high freight rates, and to keep competitors out.

872 **Beth, H. L.** Fluctuations of exchange rates: Their impact on liner shipping. In: Hans Ludwig Beth (ed.) *Liner shipping in the eighties: Report on the International Symposium, Bremen, October 24th–26th 1979.* Bremen: Institute of Shipping Economics, 1980, pp. 38–60.

Impact of the fixation to a given freight rate, printed in US dollars, being also subject to market fluctuations. The commensurate value in national or home currency which is exposed to a medium-term fluctuation trend.

873 **Beth, Hans Ludwig** (ed.) *Liner shipping in the eighties: Report on the International Symposium held at Bremen, October 24th–26th, 1979.* Bremen: Institute of Shipping Economics, 1980. 386 pp.

Papers presented at the International Symposium held at Bremen, 24–26 October 1979. World-wide shipping policies and problems, liner shipping. Opening statements set the scene – developments of liner shipping and impact of economic developments. Discussion led by chairmen following each paper.

874 **Brooks, M. R.** An alternative theoretical approach to the evaluation of liner shipping. Part 1. Situational factors. *Maritime Policy and Management* 11(1), 1984, 35–43.

Price/competition/regulation cannot be divorced from the service perspective if liner shipping policy is to reflect shipper interest. Do different buying situations exist for shippers?

875 **Brooks, M. R.** An alternative theoretical approach to the evaluation of liner shipping. Part 2. Choice criteria. *Maritime Policy and Management* 12(2), 1985, 145–55.

Evaluates shipper choice criteria for the rebuy situation of liner services.

876 **Byington, Russell** and **Olin, Gary.** An econometric analysis of freight rate disparities in U.S. liner trades. *Applied Economics* 15(3), June 1983, 403–7.

877 **Canadian Transport Commission.** *Supply of liner shipping to*

Canada. Ottawa: Canadian Transport Commission Research, 1983. 94 pp.

878 **Casson, Mark.** *The quality of shipping services: An economic appraisal of strategies for the European liner shipping industry.* Reading: University of Reading, Department of Economics, 1986. 56 pp.

Impact of threat to European shipping firms from new sources of foreign competition. The valuation of quality in the market for shipping services. Impact on shipping industry of 'quality revolution'. Production quality distinguished from transaction quality. Efficient supply of production quality and dealing with 'system effects'.

879 **Collison, F. M.** Market segments for marine liner services. *Transportation Journal* 24, 1984, 40–54.

880 **Collison, F. M.** North to Alaska: Marketing in the Pacific Northwest–Central Alaska liner trade. *Maritime Policy and Management* 11(2), 1984, 99–112.

Case analysis and shipper survey used to describe trade in this area and to analyse the role of marketing. Article assesses the extent to which marketing has developed in each of the lines and its impact on shipping line successes.

881 **Conlon, R. M.** and **Zerby, J. A.** *Public policy in the liner shipping industry.* University of New South Wales, 1984.

882 **Davies, J. E.** An analysis of cost and supply conditions in the liner shipping industry. *Journal of Industrial Economics* 31(4), June 1983, 417–35.

This examines the nature of costs and supply conditions incurred by companies engaged in the provision of scheduled cargo minor services. It concludes that the liner conference system, despite being over a century old, still has a valuable role to play in the servicing of world trade and a role that is likely to remain for many years to come.

883 **Davies, J. E.** Competition, contestability and the liner shipping industry. *Journal of Transport Economics and Policy* XX(3), September 1986, 299–312.

Claims that international liner shipping industry can be appropriately described as highly contestable. Canadian data to show that liner shipping industry is almost perfectly contestable and fulfils all requirements of the theory.

884 Davies, J. E. *Contestability and the Australian liner trades.* Canberra: Bureau of Transport Economics, 1986.

885 Davies, J. E. Contestability of liner shipping and markets. *Maritime Studies* 22, 1985.

886 Davies, J. E. Contestability theory and liner shipping: Some doubts on Pearson's doubts. *Maritime Policy and Management* 15(1), January–March 1988, 15–18.

Response to paper 'Some doubts on the contestability of liner markets' by R. Pearson in the same journal 14(1) 1987, pp. 71–8. Defence against Pearson's critique. Justifies contestability theory.

887 Davies, J. E. Destructive competition and market unsustainability in the liner shipping industry. *International Journal of Transport Economics* 17(3), October 1990, 227–45.

888 Davies, J. E. Impediments to contestability in liner markets. *Logistics and Transportation Review* 25(4), December 1989, 325–42.

889 Davies, J. E. On the nature and sources of controversy in the economic analysis of the liner shipping industry. *Maritime Policy and Management* 14(3), 1987, 249–61.

How to evaluate performance of liner shipping industry in order to establish current levels of efficiency. What are the social effects of the industry's overt cartelization?

890 Davies, J. E. *Pricing in the liner shipping industry: A survey of conceptual models.* Ottawa: Canadian Transport Commission, 1984.

891 Davies, J. E. Promoting competition and market performance in liner shipping. *Marine Policy* 14(6), November 1990, 477–83.

Explores the ways in which competition and economic performance may be enhanced in the liner shipping industry. Uses the theory of contestable markets to isolate those conditions in the market environment having most impact on effective competition. Traces the implications of these conditions for policy.

892 Davies. J. E. A rejoinder. *Journal of Transport Economics and Policy* XXIII(2), May 1989, 203–8.

Rejoinder to Jankowski's comments in same journal, pp. 199–203, on Davies's article 'Competition, contestability and the liner shipping industry', XX(3), September 1986, pp. 299–312. Suggests Jankowski's comments only relevant from a singularly academic viewpoint. Supports original contention as suggested in title.

893 Davies, J. E. *The theory of contestable markets and its application to the liner shipping industry.* Ottawa: Canadian Transport Commission, 1986.

894 Davies, John. Peak load pricing in liner shipping: A reply. *International Journal of Transport Economics* 13(2), June 1986, 235–6.

895 Doyer, S. Panel on U.S. shipping policy. In: Hans Ludwig Beth (ed.) *Liner Shipping in the eighties: Report on the International Symposium, Bremen, October 24th–26th 1979.* Bremen: Institute of Shipping Economics, 1980, pp. 229–36.

Changing circumstances in liner shipping – containerization; cargo flows; infrastructure; fast progress in technology leading to multipurpose unit loaders, ro-ro vessels, lash carriers, etc. Resulting need for co-operation between shipping lines in order to meet risks connected with high replacement and investment costs.

896 Drewry Shipping Consultants. *The future of liner shipping.* London: Drewry Shipping Consultants, 1984.

897 Driehuis, W. An econometric analysis of liner freight rates. *Weltwirtschaftliches Archiv* 104(1), 1970, 96–119.

898 Ellsworth, Robert A. Competition or rationalization in the liner industry? *Journal of Maritime Law and Commerce* 10(4), July 1979, 497–517.

Object is to demonstrate that unfettered competition is unworkable and that the rationalization of the liner industry, through the decision-making process of the carriers, offers the only economically efficient method of keeping ocean freight rates at a minimum.

899 Evans, J. J. Concerning the level of liner freight rates. *Maritime Policy and Management* 9(2), 1982, 103–14.

Looks at general level of freight rates in liner shipping. Concludes that liner conferences' policy is towards attainment of 'normal' profits commensurate with risks of the business.

900 Evans, J. J. Liner freight rates: Theory and practice. Thesis submitted for MSc, UWIST, Cardiff, 1982.

901 Evans, J. J. and **Behnam, A.** A fork tariff system for liner freight rates. *Journal of Transport Economics and Policy* IX(1), January 1975, 62–6.

How conference rate-making could be made more flexible and how this would lead to economic gain. Suggests commodity tariff/classification wherein rates fluctuate between defined upper and lower limits. Shipowners respond to demand charges by set basic rates variation.

902 Fox, N. R. An empirical analysis of ocean liner shipping. *International Journal of Transport Economics* 19(2), June 1992, 205–25.

903 Franck, Bernard and **Bunel, Jean Claude.** Contestability, competition and regulation: The case of liner shipping. *International Journal of Industrial Organization* 9(1), 1991, 141–59.

904 Frankel, E. G. Strategic decision making in shipping. *Marine Policy* 6, October 1982, 259–64.

905 *The future of liner shipping: Report on the International Symposium held at Bremen, September 23rd–25th, 1975.* Bremen: Institute of Shipping Economics, 1976. 352 pp.

906 Gallagher, F. D. and **Meyrick, S. J.** *ASEAN–Australian liner shipping: A cost-based simulation analysis.* Kuala Lumpur and Canberra: ASEAN–Australia Joint Research Project, 1984.

907 Gardner, Bernard. Some thoughts on normal-cost price theory and its application to liner shipping. *Maritime Policy and Management* 13(3), July–September 1986, 235–44.

Paper examines view that normal cost-price theory can be reconciled with static marginalist micro-equilibrium theory of the firm. How theory differs from other pricing models. Modifications to accommodate discriminatory rate structure in liner shipping.

908 Gardner, B. M. and others. *Liner shipping in the U.S. trades.* Cardiff: UWIST study for CENSA, 1978.

909 Goss, R. O. Peak load pricing in liner shipping: A comment.

International Journal of Transport Economics 13(2), June 1986, 221–34.

910 Great Britain. Department of Transport. *Liner shipping and freight rates: A United Kingdom–Northern Europe comparison.* London: Department of Transport, 1986.

911 Heaver, T. D. Shippers and the rationalization of liner shipping. *International Journal of Physical Distribution* 3(1), Autumn 1972, 91–6.

This is a short article looking at the controversy over liner shipping of the early 1970s and arguing that shippers should expect and support change in the conference system.

912 Heaver, T. D. Workable competition, politics, and competition policy in liner shipping. In: K. M. Gwilliam (ed.) *Current issues in maritime economics.* Dordrecht: Kluwer Academic Publishers, 1992, pp. 68–84.

The arguments of failure in the working of competition in liner shipping are examined. It is argued that public policies should reflect liner shipping becoming more like other businesses.

913 Institute of Shipping Economics and Logistics, Bremen, and **Institute for Shipping Research, Bergen.** *Liner shipping: 3rd International Symposium.* Bremen: Institute of Shipping Economics, 1983.

914 Institute of Shipping Economics and Logistics, Bremen, and **Institute for Shipping Research, Bergen.** *Liner shipping: 4th International Symposium.* Hamburg, 1988.

915 Jankowski, W. B. Competition, contestability and the liner shipping industry: A comment. *Journal of Transport Economics and Policy* XXIII(2), May 1989, 199–203.

Questions Davies's article 'Competition, contestability and the liner shipping industry', *Journal of Transport Economics and Policy* XX(3), September 1986, pp. 299–312. Suggests that market contestability inadequate as industry model and policy framework for liner shipping.

916 Jansson, Jan Owen. Intra-tariff cross-subsidisation in liner shipping. *Journal of Transport Economics and Policy* VIII(3), September 1974, 294–311.

Argues that no recognition of any proper 'charging floor' exists in rate structures of liner conferences. Analyses liner shipping marginal cost relevant for rate-making. Comparison of freight rates and costs. Demonstrates pattern of cross-subsidization.

917 **Jansson, J. O.** and **Shneerson, D.** The design of liner shipping services: The problem of feeder services versus multi-port-calls. *Maritime Policy and Management* 9(3), 1982, 175–88.

Outlines structure of liner shipping systems and questions whether feeder services are superior in economic terms to more traditional and commonly used multiport-calling system.

918 **Jansson, J. O.** and **Shneerson, D.** *Liner shipping economics.* London and New York: Chapman and Hall, 1987. 299 pp.

Liner shipping industry. Economics of liner shipping. Economic evaluation of the conference system. Analysis of liner service optimization. Case studies, practical examples. Technological and organizational changes in liner shipping.

919 **Jansson, J. O.** and **Shneerson, Dan.** Economies of trade density in liner shipping and optimal pricing. *Journal of Transport Economics and Policy* XIX(1), January 1985, 7–22.

Economies of trade density in liner shipping. Uses a system cost model including producers and users. Model shows that economies of trade density not of large magnitude. Economies do not impede competition/optimal pricing.

920 **Jansson, Jan Owen** and **Shneerson, Dan.** The effective protection implicit in liner shipping freight rates. *Review of Economics and Statistics* 60(4), November 1978, 569–73.

This extends the theory of tariff structures to include an analysis of the structure of protection afforded by transportation charges.

921 **Jansson, Jan Owen** and **Shneerson, Dan.** A model of scheduled liner freight services: Balancing inventory cost against shipowners' costs. *Logistics and Transportation Review* 21(3), September 1985, 195–215.

922 **Jaramillo, D. I.** and **Perakis, A. N.** Fleet deployment optimization for liner shipping. Part 2. Implementation and results. *Maritime Policy and Management* 18(4), 1991, 235–62.

Optimization model described in *Maritime Policy and Management* 18(3),

1991, pp. 183–200, is applied to find best fleet composition and deployment on given set of trade routes.

923 **Krzyzanowski, M. T.** Liner shipping as an element of the international multimodal system. In: Hans Ludwig Beth (ed.) *Liner shipping in the eighties: Report on the International Symposium, Bremen, October 24th–26th 1979.* Bremen: Institute of Shipping Economics, 1980, pp. 122–33.

General criteria of the multimodal system. Shipping in the multimodal system. Basic scope of research on multimodal transport integrating inland with sea – Polish case.

924 **Lane, D. E., Heaver, T. D.** and **Uyeno, D.** Planning and scheduling for efficiency in liner shipping. *Maritime Policy and Management* 14(2), 1987, 109–25.

Presents a dynamic cost-based model for providing liner services which incorporates practical aspects of the problem by decomposing it into manageable components.

925 No entry.

926 **Lloyd's of London Press.** *Modern liner contracts.* London: Lloyd's of London Press.

927 **Mackay, J. J. D.** Some aspects of current U.S. shipping policy and the effects upon the economics of cargo liner operations in the United States/Far East trades. Thesis submitted for MSc, University of Salford, 1975.

928 **Lord Mancroft.** Can ocean travel survive? *Royal Society of Arts Journal* 119, May 1971, 384–93.

Paper given and discussion following at the lecture given to the Society, March 1971. Examines some of the reasons for the decline in passenger ocean travel and considers the implications of the trend. Forecasts in which direction and at what speed that trend might develop.

929 **Marlow, Peter B.** and **Gardner, Bernard.** The future of liner shipping: International symposium, Bremen, 23–25 September, 1975. *Maritime Studies and Management* 3(3), January 1976, 190.

A short commentary on the symposium organized by the Institute of Shipping Economics of Bremen.

930 Muller, E. J. Ocean shipping gets rational. *Distribution* 90(8), July 1991, 44–51.

Two key trends in ocean shipping: Differentiation between carriers, and ship sharing between shipping lines. Can the two be compatible?

931 Ohsuga, Hiroshi. *Transpacific liner trade: Its carrying capacity and cargo volumes.* Tokyo: Japan Maritime Research Institute, 1984. 18 pp.

Prospects regarding the issue of overtonnage, and cargo movement in the transpacific liner trade, with particular reference to eastbound shipping.

932 Pace, M. H. Liner freight rates: Towards a new formula. In: Hans Ludwig Beth (ed.) *Liner shipping in the eighties: Report on the International Symposium, Bremen, October 24th–26th 1979.* Bremen: Institute of Shipping Economics, 1980, pp. 151–78.

Examines the determination of freight rates. Sets out the various factors which can be identified as being influential in price formation. Need for efficient resource allocation. Elasticities and competition. Suggests formula for freight rate calculation.

933 Pearson, R. Some doubts on the contestability of liner shipping markets. *Maritime Policy and Management* 14(1), 1987, 71–8.

Examines theories of 'contestable markets' and 'perfect competition' as they apply to liner shipping markets and open and closed conferences.

934 Perakis, A. N. and **Jaramillo, D. I.** Fleet deployment optimization for liner shipping. Part 1. Background, problem formulation and solution approaches. *Maritime Policy and Management* 18(3), 1991, 183–200.

Develops detailed model for estimating operating costs of liner ships on various routes and offers a linear programming formulation for liner fleet deployment problem.

935 Ryder, S. C. and **Chappell, D.** *Optimal speed and ship size for the liner traders.* Liverpool: Marine Transport Centre, University of Liverpool, 1979. 104 pp.

General study concerned with distinguishing features of liner shipping and their effect on optimal ship speed and size. Emphasis on analysis of deep-sea general cargo liner operations. Reference also made to other types of cargo shipping.

936 Sayers, C. R. *Contestability theory: An assessment of its relevance* *to Australian liner shipping.* Canberra: Bureau of Transport Economics, 1986.

937 Schmeltzer, Edward and **Weiner, George J.** Liner shipping in the 1980's: Competitive patterns and legislative initiatives in the 96th Congress. *Journal of Maritime Law and Commerce* 12(1), October 1980, 25–45.

Considers that 1980s the time to analyse the prospects for the competitive structure of the US liner trades. Theoretical basis of three approaches – open competition, industry self-regulation, governmental regulation. Impact of each.

938 Shipping Economist Conference. *Liner shipping: Survival of the fittest.* London: Lloyd's of London Press, 1981.

939 Shneerson, D. Price discrimination in liner sea transport. Thesis submitted for PhD, University of London, 1973.

940 Shneerson, D. The structure of freight rates: A comparative route study. *Journal of Transport Economics and Policy* X(1), January 1976, 52–67.

Explains liner freight rates in respect of important variables – stowage (ratio volume to weight), unit value of commodities – on conference lines serving Thailand, Singapore, Israel. Hypotheses regarding effect of these variables on freight rates.

941 Singels, A. Analysing intra-regional liner transport. In: Hans Ludwig Beth (ed.) *Liner shipping in the eighties: Report on the International Symposium, Bremen, October 24th–26th 1979.* Bremen: Institute of Shipping Economics, 1980, pp. 275–304.

Examines a medium-scale case study of intra-regional shipping, carried out in 1977 and 1978 by a team of the Netherlands Maritime Institute of Rotterdam.

942 Sjostrom, William. Price discrimination by shipping conferences. *Logistics and Transportation Review* 28(2), June 1992, 207–16.

943 Stott, K. L. and **Douglas, B. W.** A model-based decision support system for planning and scheduling ocean-borne transportation. *Interfaces* 11(4), 1981, 1–10.

944 Sturmey, S. G. Pragmatic competition and liner shipping. *Marine Policy* 14, January 1990.

945 Svendsen, A. Stromme. The impact of the present crisis in shipping on liner activities. In: Hans Ludwig Beth (ed.) *Liner shipping in the eighties: Report on the International Symposium, Bremen, 24th–26th October, 1979.* Bremen: Institute of Shipping Economics, 1980, pp. 19–33.

The shipping crisis since 1974 – freight rates; ship prices; contracting of tonnage; laid-up tonnage; income and profit.

946 Tatham, Nigel and **Poynder, Anthony.** Bridging the gap. In: National Ports Council, *Containers: Their handling and transport. A survey of current practice.* London: National Ports Council, 1978, pp. 360–76.

Deals with the two main elements in bridging the gap between traditional conventional breakbulk cargo operations and the ultimate in sophisticated container ro-ro operations – the physical requirements of ship and hardware; finance.

947 Tye, William B. The applicability of the theory of contestable markets to rail/water carrier mergers. *Logistics and Transportation Review* 21(1), March 1985, 57–76.

948 Von Schirach-Szmigiel, Christopher. *Liner shipping and general cargo transport.* Stockholm: Economic Research Institute at the Stockholm School of Economics, 1979. 339 pp.

Examines the era from the 1960s of economic and political changes which has led to far-reaching restructuring of the liner industry on short-sea as well as deep-sea markets, as a result of competition from road transports (short-sea markets) and technical innovations, political blocks (deep-sea markets).

949 Waldron, A. J. The theory and practice of a carrier's liability under the Athens Convention: The English experience. *Maritime Policy and Management* 18(1), 1991, 29–41.

On the effect of regulation of carrier's liability by mandatory application of rules imposed by international convention in context of UK passenger carriers *vis-à-vis* Athens Convention.

950 Walters, A. A. *Gains from deregulation of ocean shipping. Proceeding: In search of a rational liner policy.* New York: A Northwestern University Transport Center Forum, 1978, pp. 110–15.

951 *World directory of liner shipping agents.* London: National Magazine Co.

952 Zerby, J. A. Clarifying some issues relating to contestability in liner shipping (and perhaps also eliminating some doubts). *Maritime Policy and Management* 15(1), January–March 1988, 5–14.

Paper argues that port-to-port liner markets highly contestable. Suggests information in isolation insufficient to generate reasonable conclusion. Discussion of theories in general.

953 Zerby, J. A. and **Conlon, R. M.** An analysis of capacity utilisation in liner shipping. *Journal of Transport Economics and Policy* XII(1), January 1978, 27–46.

Analyses institutional arrangements in liner shipping with respect to the resulting utilization of capacity. Evaluation of the analysis; policy recommendations.

954 Zerby, J. A. and **Conlon, R. M.** Joint costs and intra-tariff cross-subsidies: The case of liner shipping. *Journal of Industrial Economics* 31(4), June 1983, 383–96.

This looks at the problem of allocating joint costs arising in the attempt to analyse the profit maximizing behaviour of multi-product firms.

955 Zerby, J. A. and **Conlon, R. M.** Liner costs and pricing policies: A reconsideration of regulatory issues. *Maritime Policy and Management* 9(3), 1982, 207–18.

Questions assumption that liner costs are homogeneous per unit of cargo. Re-evaluates regulatory arguments that are based on this assumption.

Liner code

956 Ademuni-Odeke. The Brussels Package: The United Kingdom and the E.E.C. respond to the Code. *Maritime Policy and Management* 11(4), 1984, 233–50.

Charts progress towards a world-wide regime for liner shipping, in particular 1974 UN Convention on Code of Conduct for Liner Conferences. Brussels Package is described and implications set out.

957 Ademuni-Odeke. Implementing the UNCTAD Code of Conduct – the U.K. Merchant Shipping (Liner Conferences) Act 1982. *Marine Policy* 6(1), January 1982, 56–64.

958 Bar-Lev, Joshua. UNCTAD Code of Practice for the regulation of liner conferences. *Journal of Maritime Law and Commerce* 3(4), July 1972, 783–91.

Summarizes the criticisms of the conference system as voiced in UNCTAD. UNCTAD's proposals for transnational solution of perceived problems.

959 Brooks, Mary R. Canada and the liner Code of conduct: Consumerism and public policy. *Maritime Policy and Management* 12(2), April–June 1983, 77–85.

Paper focuses on the potential effects of the UNCTAD liner Code of Conduct's provisions for cargo reservation on consumers of shipping services, with special reference to Canada.

960 Daltung, S. Potential fleet redistributions of the Code of Conduct for Liner Conferences. *Maritime Policy and Management* 14(1), 1987, 63–70.

Developing countries sure that price fixing by liner cartels works against them. Since 1964 they pursue policies in UNCTAD, which adopted 1974 Code of Conduct for Liner Conferences.

961 Farthing, R. B. C. UNCTAD Code of Practice for the regulation of liner conferences: Another view. *Journal of Maritime Law and Commerce* 4(3), April 1973, 467–73.

Examines UNCTAD code of practice for regulation of liner conferences from standpoint of Committee of European National Shipowners' Associations. Challenges points made in article by Bar-Lev, 'UNCTAD Code of Practice for the regulation of liner conferences' in the *Journal of Maritime Law and Commerce* 3(4), July 1972, pp. 783–91.

962 Georgandopoulos, El. A. The proposed 1988 Review Conference on the Code of Conduct for Liner Conferences. *Marine Policy* 12(1), January 1988, 22–30.

Examines how effective the review conference was expected to be and looks at areas of implementation of the code. Puts forward suggested amendments to the code itself.

963 Hallock, D. S. The regulation and promotion of liner shipping for development: The code or CFB – Sri Lanka. Thesis submitted for MSc, UWIST, Cardiff, 1984.

This concludes that the Central Freight Bureau of Sri Lanka is preferable to that obtaining under the UNCTAD Code of Conduct for Liner Conferences.

964 Juda, L. World shipping, UNCTAD, and the new international economic order. *International Organization* 35(3), 1981, 493–516.

965 Juda, Lawrence. Whither the UNCTAD Liner Code: The Liner Code Review Conference. *Journal of Maritime Law and Commerce* 23(1), January 1992, 101–22.

Controversy surrounding the Liner Code. Perceptual differences in world shipping. The Review Conference: Procedural differences. The Review Conference: Substantive differences. Conference outcome.

966 Larsen, B. and **Vetterrick, V.** The UNCTAD Code of Conduct for Liner Conferences: Reservations, reactions and U.S. alternatives. *Law and Policy in International Business* 13, 1981, 223–80.

967 Levikov, G. The problem of universal acceptability of the Convention on a Code of Conduct for Liner Conferences. *Maritime Policy and Management* 15(3), July–September 1988, 247–54.

Highlights some of the major problems impinging upon universal acceptability of UN Code of Conduct for Liner Conferences.

968 Moses, Eli. *To join or not to join the UNCTAD Convention on a Code of Conduct for Liner Conferences: An examination of the Israeli position.* Haifa: Israel Shipping and Aviation Research Institute, 1983.

969 Neff, Stephen C. The U.N. Code of Conduct for liner conferences. *Journal of World Trade Law* 14(5), September/October 1980, 398–423.

Examines nature and background of liner conference system as it existed at time of writing. Overview of provisions of the 1974 UNCTAD Convention on Liner Conferences. Examines controversial provisions relating to cargo reservation. Technological revolution in liner shipping industry and perceived impact.

970 Shah, M. J. The dispute settlement machinery in the Convention on a Code of Conduct for Liner Conferences. *Journal of Maritime Law and Commerce* 7(1), October 1975, 127–68.

Examines 1974 UNCTAD Convention on Code of Conduct for Liner Conferences.

971 Shah, M. J. The implementation of the U.N. Convention on a

Code of Conduct for Liner Conferences 1974. *Journal of Maritime Law and Commerce* 9(1), October 1977, 79–110.

Examines how and where the Code's provisions could be set into motion; to whom they would be applicable; who would interpret and apply its provisions; what its principal provisions mean or imply.

972 **Shah, M. J.** The U.N. Liner Code of Conduct (1974): Some key issues regarding its implementation. *European Transport Law* 16, 1981, 491–556.

973 **Shaw, B. P.** Results and consequences from UNCTAD V. In: Hans Ludwig Beth (ed.) *Liner shipping in the eighties: Report on the International Symposium, Bremen, October 24th–26th 1979.* Bremen: Institute of Shipping Economics, 1980, pp. 105–16.

Discusses the two major developments: (i) The UN Liner Code Convention, and (ii) bulk shipping.

974 **Skinner, George** and **Noakes, Robert.** *Cargo reservation and liner conference shipping serving Canada: An analysis related to the UNCTAD Code of Conduct.* Ottawa: Canadian Transport Commission, 1979. 137 pp.

Examines probable impact of the UNCTAD Code of Conduct, Art. 2, which outlines principles concerning lines' rights to share in the conference trades. Delineates the patterns of conference and total ocean liner carriage of international winterborne shipments to and from Canada.

975 *Some economic implications of the Convention on a Code of Conduct for Liner Conferences.* Ottawa: Canadian Transport Commission, Research Branch, 1976. 285 pp.

976 **Sturmey, S. G.** Administering the Code of Conduct. *Marine Policy* 8(3), July 1984, 203–16.

Examines problems which liner conferences are likely to encounter in administering and policing the UN Code of Conduct for Liner Conferences, 1983.

977 **Sturmey, S. G.** The code of conduct for liner conferences: A 1985 view. *Maritime Policy and Management* 13(3), July–September 1986, 185–221.

Paper examines options to Code of Conduct as annexed to April 1974 'Convention on a Code of Conduct for Liner Conferences', in Geneva. Questions whether system of liner conferences should be protected.

978 Sturmey, S. G. The development of the Code for Liner Conferences. *Marine Policy* 3(2), April 1979, 133–48.

979 United Nations Conference on Trade and Development. *Code of Conduct for Liner Conferences.* New York: UNCTAD, 1975.

980 United Nations Conference on Trade and Development. *Consideration of current and long-term trends in shipping and ports in the light of relevant resolutions and other decisions of UNCTAD and the status of the Convention on a Code of Conduct for Liner Conferences: Review of maritime transport.* Geneva: UNCTAD, 1976.

981 United Nations Conference on Trade and Development. *Convention on a Code of Conduct for Liner Conferences.* Geneva: UNCTAD, 1974.

982 United Nations Conference on Trade and Development. UNCTAD: Code of Conduct for Liner Conferences. *Journal of World Trade Law* 7(5), September/October 1974, 536–74.

Code of Conduct for Liner Conferences adopted at conclusion of United Nations Conference of Plenipotentiaries on 6 April 1974. Important features of Convention.

983 Wijkman, P. M. Effects of cargo reservation: A review of UNCTAD's Code of Conduct for Liner Conferences. *Marine Policy* 4(4), 1980, 271–89.

984 Zamora, Stephen. UNCTAD III: The question of shipping. *Journal of World Trade Law* 7(1), January/February 1973, 91–115.

The third UNCTAD Conference (1972). Confrontation over reform of maritime transport system. Code of Conduct for Liner Conferences. Other shipping measures. Analysis of effective policies to alter liner conference system.

Liner conference and non-conference

985 Be Vier, Michael J. and **Lowenfield, Andreas F.** Ratemaking and shipping conferences. *Journal of Maritime Law and Commerce* 1(3), April 1970, 457–60.

Calls attention to difficulty facing ratemaking body.

986 Bromley, Raymond Curtis. Cartels as alternatives to contracts: The case of ocean shipping conferences. Thesis submitted for PhD, University of California, Los Angeles, 1987.

987 Bryan, Ingrid A. Conference freight rates and Eastern Canadian ports. *Canadian Public Policy* 10(2), June 1984, 193–200.

988 Bryan, I. A. and **Kotowitz, Y.** *Shipping conferences in Canada.* Canada: Consumer and Corporate Affairs, 1978. 106 pp.

989 Canadian Transport Commission. *Some economic implications of the Convention on a Code of Conduct for Liner Conferences.* Ottawa: Canadian Transport Commission, Research Branch, Directorate of Transport Industry Analysis, 1976. 285 pp.

Examination of the importance of liner conferences to Canada's external trade. Describes important present practices of conferences serving external trade. Assesses potential impact of the Code of Conduct for Liner Conferences on these practices and on Canadian trade.

990 Chudleigh, Peter. The economic potential of specialised wool shipping services. *Journal of Transport Economics and Policy* IX(3), September 1975, 230–54.

Study of the economics of specialized shipping services designed to carry wool as their only cargo from major wool exporting countries to major importing regions. Theoretical freight rates for wool shipped in non-conference vessels for range of wool trade routes and shipping frequencies.

991 Container Operations and Shipping Conference. *Container operations and shipping: Proceedings of a conference organised by Cargo Systems International in co-operation with the European Shippers' Councils, Amsterdam, November 27–29, 1979.* Worcester Park, Surrey: Cargo Systems International, 1979.

992 Cotton, A. R. Conference–shipper loyalty contracts: Time for a re-think? *Maritime Policy and Management* 11(2), 1984, 79–97.

Statutory limit proposed on proportion of a shipper's custom that may be demanded in a loyalty agreement with a liner conference. A 70 per cent loyalty suggested for conferences in order to enhance competitive environment of liner shipping.

993 Davies, J. E. Freight rate stability in liner shipping: Implicit peak load pricing. *International Journal of Transport Economics* 11(2–3), August–December 1984, 163–76.

994 **Davies, J. E.** The regulation of liner shipping: A study of motives and consequences. Thesis submitted for PhD, UWIST, Cardiff, 1980.

995 **Davies, J. E.** and **Evans, J. J.** Market power and competitive environment of conferences. In: Hans Ludwig Beth (ed.) *Liner shipping in the eighties: Report on the International Symposium, Bremen, October 24th–26th 1979.* Bremen: Institute of Shipping Economics, 1980, pp. 305–18.

Examines issue of why the view that conferences are monopolies has become so firmly entrenched in traditional thinking. Authors' suggestions as to why this view is inherently false.

996 **Deakin, B. M.** Shipping conferences: Some economic aspects of international regulation. *Maritime Studies and Management* 2(1), July 1974, 5–31.

Examines economic aspects and implications of various methods proposed for regulating the activities and practices of shipping conferences.

997 **Deakin, B. M.** (in collaboration with T. Seward) *Shipping conferences: A study of their origins, development and economic practices.* London: Cambridge University Press, 1973. 261 pp.

How and why shipping conferences came into existence. How they have changed and developed. How they are organized and how they operate today.

998 The case of Rochdale versus Rochdale. *Maritime Policy and Management* 12(3), July–September 1985, 177–9 (Editorial).

Discusses the Rochdale Report (Rochdale Committee, 1970) which devotes a chapter to the subject of shipping conferences.

999 **Devanny, J. W., Livanos, V. M.** and **Stewart, R. J.** Conference ratemaking and the west coast of South America. *Journal of Transport Economics and Policy* IX(2), May 1975, 154–77.

Price of transport – specifically price of ocean liner shipping – as barrier to trade. To what extent price equals social cost on liner trade route between east coast US and west coast South America. Arguments for removal or decreasing of differential.

1000 **Ellsworth, Robert A.** Liner conferences: Evolution of U.S. policy. *Marine policy* 7(4), October 1983, 254–66.

Outlines the history of US policy towards liner conferences from the Shipping Act 1916 to the Shipping Act 1983. Analyses the method of regulating the conference system contained in the 1916 Act and its impact on the next forty years of US maritime policy. The 1983 UNCTAD Code of Conduct.

1001 **Fasbender, Karl** and **Wagner, Wolfgang.** *Shipping conferences, rate policy and developing countries: The argument of rate discrimination.* Hamburg: Hamburg Institute for International Economics, 1973. 357 pp.

Study of rate discrimination by liner conferences. Lays emphasis on empirical analysis of subject. Differences of opinion between discriminating states and those discriminated against. Considers this issue. Outlook, recommendations.

1002 **Fawcett, F. C.** and **Nolah, D. C.** United States ocean shipping: The history, development and decline of the conference antitrust exemption. *North Western Journal of International Law and Business* 1(2), Autumn 1979, 537–68.

1003 **Fleming, Douglas K.** Reflections on conventional cargo liner service. *Maritime Studies and Management* 1(3), January 1974, 178–90.

Focus on three characteristics: Regularity, impartiality, and stability of freight rates. Examines carrier's behaviour in a shipping conference frame.

1004 **Fox, Nancy Ruth.** Cartel behavior: An empirical analysis of oceanliner shipping. Thesis submitted for PhD, University of Pennsylvania, 1989.

1005 **Goss, Richard.** Some financial aspects of shipping conferences. *Journal of Transport Economics and Policy* V(2), May 1971, 173–83.

Costing and accounting systems which might be employed to assist members of shipping conferences. Suggests standardization on cargo liners' voyage accounts as step towards more rational negotiation of scales of charges to encourage efficiency.

1006 **Gunaratne, M. H.** The impact of the Central Freight Bureau of Sri Lanka on liner conferences and trade patterns. Thesis submitted for MSc, UWIST, Cardiff, 1976.

1007 **Heaver, T. D.** Trans-Pacific trade, liner shipping and conference rates. *Transportation and Logistics Review,* Spring 1972.

1008 Heaver, Trevor D. The structure of liner conference rates. *Journal of Industrial Economics* 21(3), July 1973, 257–65.

This examines the structure of shipping conference tariffs and factors which account for the difference between rates on commodities in a tariff. It argues that these differences are to be expected because of the imperfect nature of the liner market and the influence which may be exerted by purchasers' associations.

1009 Heaver, Trevor D. Workable competition, politics and competition policy in liner shipping. In: K. M. Gwilliam (ed.) *Current issues in maritime economics.* Dordrecht: Kluwer Academic Publishers, 1993, pp. 68–84.

A paper dealing with questions raised during the research in respect of policy formation for the effective development of liner conferences. Reviews established positions and considers them in the context of current conditions.

1010 Herman, Amos. *Shipping conferences.* London: Lloyd's of London Press, 1983. 255 pp.

1011 Jankowski, W. B. The development of liner shipping conferences: A game theoretical explanation. *International Journal of Transport Economics* 16(3), October 1989, 313–28.

1012 Kryvoruka, Kenneth J. American ocean shipping and the antitrust laws revisited. *Journal of Maritime Law and Commerce* 11(1), October 1979, 67–108.

Shipping conferences and US antitrust laws – immunity under US Shipping Act 1916. Narrow construction by courts. Status of the exemption, and accommodation of interests. Industry practice, public national interests.

1013 Laing, E. T. The rationality of conference pricing and output policies. Part 1. *Maritime Studies and Management* 3(2), October 1975, 103–11.

Paper, to be presented in two parts, takes view that the discriminatory pricing systems of conferences are defensible for both conferences and their shippers, especially those in developing countries.

1014 Laing, E. T. The rationality of conference pricing and output policies. Part 2. *Maritime Studies and Management* 3(3), January 1976, 141–51.

Follows from Part 1 in previous issue of the same journal – pp. 103–11. This paper analyses the economic objectives of conferences and considers the implications of ending the conference pricing structures. Some recommendations offered.

1015 Llorca, Manuel R. Anti-trust exemption of shipping conferences. *Journal of Maritime Law and Commerce* 6(2), January 1975, 287–98.

History of the exemption of steamship shipping conferences from US antitrust law by Shipping Act 1916. Case law developments. Legislative developments.

1016 Marx, Daniel (Jr) *International shipping cartels: A study of industrial self-regulation by shipping conferences.* New York: Greenwood Press, 1969. 323 pp.

Describes and analyses functioning of shipping conferences. Are they necessary? Do they possess inherent vice of unregulated monopolies? Should they exist, be abolished? Regulation? Official investigations. Arrangements. Alternatives.

1017 McDonnell, John J. Unilateral regulation of international trade: U.S. Brazil coffee trade example. *Journal of Maritime Law and Commerce* 3(4), July 1972, 793–9.

Outlines history of the conference system in shipping industry and American regulation thereof. Depicts major example of problems caused by conflicting national policies.

1018 Moskowitz, Daniel B. Breaking up the old ocean rate cartel. *International Business* 5(3), March 1992, 25–6.

1019 Pearson, Roy. Freight rate disparities and loyalty contracts: An alternative explanation and solution. *Maritime Policy and Management* 12(4), October–December 1985, 305–13.

Paper deals with issue of introduction of a 70 per cent statutory limit on the proportion of a shipper's custom that may be tied to a loyalty agreement with a shipping conference. Follows on from argument 'Conference–shipper loyalty contracts: Time for a re-think?' by A. R. Cotton, Vol. 11, No.2, pp. 79–97, in the same journal.

1020 Pengilley, Warren. Shipping conferences and the Australian Trade Practices Act. *Lloyd's Maritime and Commercial Law Quarterly*, August 1991, Part 3, 342–56.

Discusses the substantive provisions of the Australian Trade Practices Act 1974 which are applicable to shipping conferences. The 1989 amendments –

the Cargo Shipping Act. Includes brief review of mechanics of registration and related matters.

1021 Robins, P. Roll-on, roll-off ships. *Freight Management*, June 1980, 20.

1022 Rose, F. D. New lines for conference lines. *Lloyd's Maritime and Commercial Law Quarterly*, November 1982, Part 4, 636–49.

Concept of conference lines. The United Nations Convention on a Code of Conduct for Liner Conferences 1974. The Merchant Shipping (Liner Conferences) Act 1982. The EEC Council Regulation May 15th, 1979.

1023 Shashikumar, N. Mandatory independent action: A legislative paradox. *Maritime Policy and Management* 15(4), 1988, 283–90.

Intraconference competition forced onto US ocean liner shipping by US government. Author suggests modifications to quell war of attrition between shippers and carriers.

1024 Shashikumar, N. Service contracts: A case of unfulfilled promises. *Maritime Policy and Management* 16(1), 1989, 13–26.

Examines major issues that arose after introduction of service contracts in ocean liner shipping in wake of US Shipping Act of 1984.

1025 Shneerson, Dan. The rationality of conference pricing and output policies. *Maritime Studies and Management* 3(4), April 1976, 245–8.

Commentary upon E. T. Laing's paper 'The rationality of conference pricing and output policies', *Maritime Studies and Management* 3(2 & 3), 1975–6. Commentary suggests that the paper breaks down before basic concepts and upon analysis of its detail.

1026 Sjostrom, William. Collusion in ocean shipping: A test of monopoly and empty core models. *Journal of Political Economy* 97(5), October 1989, 1160–79.

1027 Sjostrom, William. Monopoly exclusion of lower cost entry: Contracts in ocean shipping conferences. *Journal of Transport Economics and Policy* XXII(3), September 1988, 339–44.

Presents model showing circumstances in which exclusionary contracts can exclude lower-cost entrant from conference. If lower-cost entrant cannot meet shipper's demand for frequency, shipper will make loyalty contracts with conference.

1028 Sjostrom, William Benno. Price fixing agreements and the existence of equilibrium: The case of liner freight conferences. Thesis submitted for PhD, University of Washington, 1986.

1029 Skinner, George. *A review of foreign control over shipping conferences.* Ottawa: Canadian Transport Commission, 1978. 32 pp.

1030 Sletmo, G. K. The impact of US regulatory practices on the productivity of ocean liner services. *Maritime Policy and Management* 9(3), 1982, 189–205.

A critical review of US regulatory practices in maritime transport. They cost US trades about one billion dollars p.a. and seriously impair the liner industry's efficient functioning. Includes a definition of 'ocean/liner conferences'.

1031 Sletmo, Gunnar K. and **Williams, Ernst W.** *Liner conferences in the container age: U.S. policy at sea.* New York: Macmillan, 1981. 343 pp.

Analysis of liner conference issue from perspective of the economics of the liner industry. US Government policies. Economic effects of different forms of conference arrangements.

1032 Stonham, P. E. Conference and competition in Australian overseas liner shipping. *Economic Record* 46(114), June 1970, 197–207.

1033 Thuong, Le T. U.S. antitrust policy and liner conferences in the age of intermodalism. *Maritime Policy and Management* 10(3), July–September 1983, 181–98.

Paper examines the uncertainties concerning the antitrust immunity for collective participation by liner conferences in intermodal transportation. Includes FMC's jurisdiction under Shipping Act to authorize collective actions. Legality of intermodal dual-rate contracts. Stringency of 'public interest' justification.

1034 United Nations Conference on Trade and Development Secretariat. *The liner conference system.* New York: United Nations, 1970. 100 pp.

A study of conference practices. The conference system. Conference organization. Membership of conferences. Loyalty arrangements. Freight rates. Stability of freight rates. Pooling arrangements. Adequacy of shipping services. Way-port conferences.

1035 University of Wales, Institute of Science and Technology,

Department of Maritime Studies. *Liner shipping in the U.S. trades.* Cardiff: Department of Maritime Studies, UWIST, 1978.

1036 Wanhill, S. R. C. Freight rates, conferences, and developing countries. *Maritime Studies and Management* 2(4), April 1975, 231–43.

Looks at issue of whether developing countries bear overall burden of freight rates. What is the solution to the problem? Investigates possible theories, respective elasticities of supply and demand. Shipping conferences, freight rates.

1037 Westinform. *A review of the world conference liner fleets: An analysis of capacity, number, flag, age and change in vessel design.* London: Westinform Service, 1977.

Second report in the Westinform Fleet Survey, a constructive look at specific sectors of the merchant fleets. A comprehensive summary of material derived from examination of conventional liner fleet from a design viewpoint. Vessels selected are owned by liner conference operators.

1038 No entry.

1039 Wilson, V. E. The shippers' point of view. In: Hans Ludwig Beth (ed.) *Liner shipping in the eighties: Report on the International Symposium, Bremen, October 24th–26th 1979.* Bremen: Institute of Shipping Economics, 1980, pp. 182–92.

Considers issues expected to impact on liner shipping in US trades: Shipping Act Amendments 1979; the Ocean Shipping Act 1978; Omnibus Maritime Bill; antitrust immunity; closed conferences; shippers' councils.

1040 *World directory of freight conferences.* New Malden: Croner. Annual.

1041 Yamada, Hideo. *The significance of the formation of Trans-Pacific Discussion Agreement (TPDA) and Trans-Pacific Stabilization Agreement (TSA) and its economic evaluation.* Tokyo: Japan Maritime Research Institute, 1989. 59 pp.

Booklet, part of series covering shipping activities. Background of formation of TPDA and TSA. Is TPDA a powerful cartel? Feature of FDWS. Restructuring of conference. Modest decision. Judgement on surplus space of containerships. Inconsistency of annual load factor. Demand gap ratio.

1042 Yamada, Hideo. *Where conference and non-conference carriers* *meet: What fixed day of the week service aims at.* Tokyo: Japan Maritime Research Institute, 1987. 65 pp.

Booklet, part of series covering shipping activities. Economic growth of Asian nations. Changes in containerized liner trade operation. Load factor. Where conference and non-conference carriers meet: The Weekly Club or FDS Club.

1043 **Zerby, J. A.** Liner conferences: Villains or victims? In: *Issues in Australian shipping: Occasional papers in maritime affairs: 2.* Canberra: Australian Centre for Maritime Studies, 1983.

1044 **Zerby, J. A.** and **Conlon, R. M.** *Public policy in the liner shipping industry.* Kensington, New South Wales: Centre for Applied Economic Research, University of New South Wales, 1983.

Containers

1045 **Aila, M. T. M.** The economic effects of containerization on developing countries, with particular reference to the Sudan. Thesis submitted for MScEcon, UWIST, Cardiff, 1983.

1046 **Akiyama, Tomomasa.** *Comparative study of transportation costs by air and containership.* Tokyo: Japan Maritime Research Institute, 1984. 47 pp.

Part of series covering maritime topics in Japan and abroad. Comparison of transportation costs by air and containership. Suitable cargoes for aircraft and containership. Present situation of air cargo.

1047 **Al-Kazily, J.** Modeling containerized shipping for developing countries. *Transportation Research* Series A 16(4), 1982, 271–83.

1048 **Armstrong, Timothy J.** Packaging trends and implications in the container revolution. *Journal of Maritime Law and Commerce* 12(4), July 1981, 427–65.

Focuses on various aspects and implications of containerization with respect to package limitation, deviation and losses ashore.

1049 **Blumenhagen, D.** Containerization and hinterland traffic. *Maritime Policy and Management* 8(3), 1981, 197–206.

Rapid spread of containerization requires that ports invest in facilities suited to handling ships of increasing size and ever-changing technology. This has a decisive influence on hinterland structure of seaports.

1050 **Boffey, T. B.** and others. Two approaches to scheduling container ships with an application to the North Atlantic route. *Journal of the Operational Research Society* 30(5), 1979, 413–25.

1051 **Booker, Mark D.** *Containers: Conditions, law and practice of carriage and use.* (2 vols) Middlesex: Derek Beattie, 1987. 209 pp. and 186 pp.

Examines container safety; carriage of goods within containers; containers and insurance; container leasing; container and customs. Case law application and liability arising in respect of inland hauliers, international transporters, freight forwarders, rail transport, carriage of goods by sea.

1052 **Brookfield, H. C.** Boxes, ports and places without ports. In: B. S. Hoyle and D. Hilling (eds) *Seaport studies and spatial change: Technology, industry and development strategies.* Chichester and New York (*et al.*): John Wiley & Sons, 1984, pp. 61–79.

Impact of containerization on seaborne linkages to and within Third World countries. Considers the issue of technological innovation treated as proximate cause of spatial change. Competing innovations. Initial containerization. Diffusion of decision to containerize. Containers and places without ports.

1053 **Brooks, M. R.** Determinants of shippers' choice of container carrier: A study of eastern Canadian exporters. Thesis submitted for PhD, UWIST, Cardiff, 1983.

1054 **Brooks, Mary R.** Ocean carrier selection criteria in a new environment. *Logistics and Transportation Review* 26(4), December 1990, 339–55.

An examination of the methods used to evaluate ocean container carriers by eastern Canadian shippers. This study repeats a much larger one made in 1982, and considers whether the intervening years have seen a change in carrier selection criteria to reflect changing market conditions.

1055 **Cargo Systems International.** *Containerisation into the 1980's: An appraisal of world container demand in the medium term.* Surrey: C.S. Publications, 1978. 205 pp.

The existing container fleet. World container production 1970–78. World container production capacity at end 1978 and future development prospects. World container demand prospects in the medium term. Outlook for world container manufacturing industry.

1056 Cargo Systems Research Consultants Ltd. *Containerisation into the 1980's: Container port/terminal facilities and trade growth.* Worcester Park, Surrey: Cargo Systems Research Consultants, 1982. 152 pp.

Systematically examines, compares and contrasts the extent and scale of container port and terminal availability world-wide in the context of the growth of box traffic volumes over the previous decade.

1057 Cargo Systems International. *Containerisation into the 1980's: The market for specials.* Worcester Park, Surrey: Cargo Systems, Research/Consultancy Division, 1980. 91 pp.

Charts the historical development of the world container fleet and accompanying world container production volumes with particular reference to the evolving requirement for special containers – refrigerated/insulated units, tank containers, open tops and other special containers, as well as dry freight units in the medium-term period to 1985.

1058 Chadwin, M. L., Pope, J. A. and **Talley, W. K.** *Ocean container transportation: An operational perspective.* New York and London: Taylor & Francis, 1990. 141 pp.

The advent of containerization. The marine container terminal and its operation. Costing terminal operations and measuring capacity. Evaluating terminal operations. Ship technology and costing containership service. Networks, intermodalism and containership service. Ocean container transport in the future.

1059 Cooper, M. J. M. Container shipping and regional economic growth. *New Zealand Geographer* 30, 1974, 54–65.

1060 De Orchis, M. E. The container and the package limitation: The search for predictability. *Journal of Maritime Law and Commerce* 5(2), January 1974, 251–79.

Considers question of meaning of word 'package' as in S.4(5) COGSA in relation to containers. US case law developments.

1061 Dick, H. W. Containerization and liner conferences: A polemic. *Maritime Policy and Management* 10(3), July–September 1983, 165–80.

Paper offers a retrospective view of containerization which raises issues relevant to contemporary policy: Competition; conferences; regulation. Focuses on Europe/Australia trade.

1062 Drewry Shipping Consultants. *The advance of deep-sea, fully* *cellular container shipping.* London: H. P. Drewry Shipping Consultants, 1978. 90 pp.

Report – part of series covering world's bulk shipping. The development of deep-sea, fully cellular container shipping. Characteristics and deployment of the deep-sea, fully cellular fleet. Future prospects.

1063 Drewry Shipping Consultants. *Strategy and profitability in global container shipping.* London: Drewry Shipping Consultants, 1991. 169 pp.

1064 Drewry Shipping Consultants. *Traffic and competition on round-the-world container routes.* London: Drewry Shipping Consultants, 1986. 92 pp.

Report – part of a series covering world's bulk shipping. Containerization and conferences. Container fleet development. Deep-sea liner trades. Impact of round-the-world services.

1065 Eyre, J. L. Container shipping in 1990. *Maritime Policy and Management* 14(4), 1987, 271–8.

By 1990 express shuttle container ships will dominate USA/N. Europe and USA/Pacific trades. Twelve factors will affect progress, especially labour resistance and ship and terminal obsolescence.

1066 Gardner, Bernard. The container revolution and its effects on the structure of traditional U.K. liner shipping companies. *Maritime Policy and Management* 12(3), July–September 1985, 195–208.

Paper considers period of technological change in 1970s. Examines the causes of this technological revolution and effect on structure of established UK liner shipping companies and on organization and profitability.

1067 Gibney, R. F. *Container lines: The strategy game.* Colchester: Lloyd's of London Press, 1984. 87 pp.

A Lloyd's Shipping Economist Study. Examines the decisions facing the container industry over the following ten years. In-depth analysis of thirteen major international shipping companies.

1068 Gilman, S. *Ship choice in the container age.* Liverpool: Marine Transport Centre, University of Liverpool, 1980. 206 pp.

Report examines containership costs. Includes new costs and various components of fuel consumption. Flexible ships – basic concept, access, design of cargo spaces, operational performance. Transport geography. Size, speed and deployment of containerships. The port sector, handling performance.

1069 Gilman, S. and **Williams, G. F.** The economics of multi-port itineraries for large container ships. *Journal of Transport Economics and Policy* X(2), May 1976, 137–49.

Examines developments on major European container routes and effect on new ports and established port hinterlands. Measures relationships between marine and inland sector costs which have exerted particularly important influence.

1070 Gilman, Sidney. *Competitive dynamics of container shipping.* Aldershot, Hants: Gower Pub. Co., 1983. 135 pp.

Proceeds from basic analysis of operating characteristics of container systems, through a consideration of market and conference function, to a study of a number of container routes. Developments and issues in regulatory policy.

1071 Great Britain. Department of Trade – Shipping Policy Division. *The joint operation of container fleets.* London: HMSO, 1979.

1072 ICHCA. *Handling containers in small ports and underdeveloped situations.* London: International Cargo Handling Co-ordination Association, 1985. 20 pp.

A report of the proceedings of a seminar in London, 1985.

1073 Japan Container Association. *Prospects of demand for containerships in the second half of the 1980's.* Tokyo: Japan Maritime Research Institute, 1986. 58 pp.

Containerization – past and present. Various factors surrounding containerization. Prospects of demand for the world containership fleet.

1074 Johnson, K. M. and **Garnett, H. C.** *Economics of containerisation.* George Allen and Unwin, 1971.

1075 Kesterman, Frank. Recent findings on containership economics. *Journal of Maritime Law and Commerce* 4(2), January 1973, 360–65.

Profitability in containerships. Factors directly relating to such profitability – utilization of ships, containers, cranes, terminals. Value of speed. Changing cost structure. Cargo versus capacity. Cargo forecasts.

1076 Laing, Edward Temperley. *Containers and their competitors: The economics of deep sea general cargo shipping in the 1970's.* Liverpool: Marine Transport Centre, University of Liverpool, 1975. 89 pp.

Study originating as an input to a National Ports Council model designed to forecast United Kingdom deep-sea berth requirements in period up to 1980. Objective is to obtain general perspective on type and size of ship likely to call at UK ports. Concentrates on cost calculations.

1077 **Lambert Brothers Shipping.** *Container ships: Displacement of conventional cargo liner tonnage on major world routes.* London: Lambert Brothers Ltd, 1970. 30 pp.

1078 **Liu, S.** The Pacific container freight market: An analysis of demand, supply and ship costs. Thesis submitted for M.Phil, 1985 (university unknown).

The Pacific is the largest liner route in the world, owning a 33 per cent share in world cellular capacity. Since 1981 there has been a huge increase in containership size and the trend is still strong. This thesis is structured to attempt to develop an understanding of the dynamics of change on this route.

1079 **Maltenfort, George G.** *Performance and evaluation of shipping containers.* Plainview (New York): Jelmar Publishing Co., 1990. 488 pp.

1080 **Marti, Bruce E.** Cargo sharing through a policy of bilateralism: Effects on U.S. containerized foreign trade. *Marine Policy* 11(2), April 1987, 97–104.

Examines historical data to study the latest defects of protectionism, especially in the form of cargo-sharing under a scenario of bilateralism. Considers movements of containerized foreign trade.

1081 **Marti, Bruce E.** U.S./Pacific Basin containerized foreign trade: Flag of convenience participation. *Marine Policy* 15(3), May 1991, 193–8.

Demonstrates that economic forces rather than political forces have been the controlling factors operating in the US/Pacific Basin containerized foreign trades.

1082 **Mayo, A. H. B.** The economic effects of containerisation on the organisation of world shipping: The effects of the introduction of standard containers on liner shipping conferences. Thesis submitted for MA, University of Newcastle, 1977.

1083 **Miyajima, M.** and **Kwak, K. S.** Economic analysis of interport competition in container cargo: Peripheral ports versus Tokyo Bay ports. *Maritime Policy and Management* 16(1), 1989, 47–55.

Considers problem of interport competition for container cargo in Japan. Tokyo port has been adversely affected by competition from peripheral ports and this may become more severe.

1084 Mobereola, A. E. The impact of containerisation on ports and shipping organisations. Thesis submitted for MSc, UWIST, Cardiff, 1984.

This examines the impact of containerization on ports and shipping organizations, concentrating on changes in planning and operation. It proves that the advantages outweigh the disadvantages. In comparison with other unit transport systems, containerization has the edge in terms of efficiency.

1085 N. Y. K. Research Chamber. *World's containership fleet and its operations.* Tokyo: Tokyo News Service, 1987. 225 pp.

A report on the world containership fleet since 1968. Full – containership fleet in the world. Containership routes. Containerships in service by route.

1086 Oda, H. *The effects of containerization: The winners and the losers.* Tokyo: Japan Maritime Research Institute, 1986. 49 pp.

Report – part of series covering maritime topics in Japan and abroad. The advent of the container era. Loss of identities among shipping circles. Emergence of outsiders. Changes in the Pacific Basin countries. Winners and losers. Advice to the Third World.

1087 Pearson, R. *Containerline performance and service quality.* Liverpool: Marine Transport Centre, University of Liverpool, 1980.

1088 Pearson, Roy. *Container ships and shipping.* London: Fairplay Publications, 1988. 260 pp.

Covers a mixture of the commercial and technical aspects of shipping industry. Adopts a traditional structure and comprises a broad tripartite discussion of the basic elements of demand, supply and disequilibrium. Illustrates methodological techniques that can be applied in containerization. Case studies to show symbiotic relationship between commercial and technical aspects.

1089 Pearson, Roy and **Fossey, John.** *World deep-sea container shipping: A geographical, economic and statistical analysis.* Aldershot, Hants: Gower, 1983. 258 pp.

Changes in deep-sea container shipping. Container transport geography; containerline operating economics; structure of container carrying fleets – a statistical commentary.

1090 Roberts, W. E. The marketing of container services. Thesis submitted for MA, University of Liverpool, 1971.

1091 Robinson, Ross. Containerization in ports of third-world Asia: An overview of present patterns and the direction of future growth. *Maritime Policy and Management* 12(4), October–December 1985, 263–77.

Paper reviews status of containerization in Third World Asia and suggests that major changes reflecting a variety of factors are likely by 1990.

1092 Rutter, J. I. Distribution systems affecting the sea transport industry, with particular reference to containerisation. Thesis submitted for MA, University of Liverpool, 1970.

1093 Schmeltzer, Edward and **Peavy, Robert.** Prospects and problems of the container revolution. *Journal of Maritime Law and Commerce* 1(2), January 1970, 203–39.

Identifies benefits of containerization which have been achieved; potential benefits. Problems presented by container revolution. The impact which solutions to problems may have upon transportation industry.

1094 Schmeltzer, Edward and **Sheppard, Edward J.** Container feeder systems. *Journal of Maritime Law and Commerce* 4(2), January 1973, 215–29.

Examines legal and operational elements that encourage or constrain development of water feeder systems. Likely developments of feedership systems to ports in Canada, US, Caribbean Sea.

1095 Shachar, Yoram. The container bill of lading as a receipt. *Journal of Maritime Law and Commerce* 10(1), October 1978, 39–78.

The probative function of bills of lading and container transport. Container bills of lading – details and reservations.

1096 Simon, Seymour. Container law: A recent reappraisal. *Journal of Maritime Law and Commerce* 8(4), July 1977, 489–95.

Case law development of concept of 'package' in respect of containers in relation to COGSA 1936.

1097 Simon, Seymour. Latest developments in the law of shipping containers. *Journal of Maritime Law and Commerce* 4(3), April 1973, 441–54.

(i) Containers – are they a 'package'? Difficulties with this issue. Case law development. US COGSA 1936. (ii) Containerships – is it an 'unreasonable deviation' to stow containers on deck? Case law development.

1098 Simon, Seymour. More on the law of shipping containers. *Journal of Maritime Law and Commerce* 6(4), July 1975, 603–19.

Examines steamship companies' conversion of cargo operations from traditional break-bulk methods to the containerization concept. Consequent litigation concerning status of carriers' freight containers in connection with package limitation section of COGSA. Landmark US cases – *Shinko Boeki* v. *United States Lines* 1974; *Cameco Inc.* v. *United States Lines* 1974.

1099 Swinbank, C. *Intermediate bulk containers.* London: Institute of Packaging, 1983. 41 pp.

1100 Tabak, Herman D. *Cargo containers: Their stowage, handling and movement.* Cambridge (Maryland): Cornell Maritime Press, 1970. 386 pp.

A comprehensive guide as to container stowage, handling and movement techniques. Directed to all areas of the transportation industry involved in the movement of cargo containers by land, sea and their eventual integration to air movement.

1101 Talley, W. K. Optimal containership size. *Maritime Policy and Management* 17(3), 1990, 165–75.

Cost per container movement incurred by a containership per voyage leg on given route is reduced under three scenarios. Rationale for conclusions presented.

1102 Talley, Wayne K. A short-run cost analysis of ocean containerships. *Logistics and Transportation Review* 22(2), June 1986, 131–9.

1103 Talley, Wayne K. and **Pope, James A.** Determinants of liner conference rates under containerization. *International Journal of Transport Economics* 12(2), June 1985, 145–55.

1104 Tan, Jin Oon Francis. *Containers: The lease–buy decision.* London: ICHCA, 1983. 78 pp.

A consideration of the leasing industry in general and an examination of the advantages and disadvantages of leasing equipment. Additionally, an investigation into the lease–buy decision with reference to the container leasing industry. An attempt to identify the determinants of an optimal container leasing policy.

1105 Tombari, Henry A. Trends in oceanborne containerization and its implications for the U.S. liner industry. *Journal of Maritime Law and Commerce* 10(2), January 1979, 311–24.

Focuses on trends in containerization showing its impact on US liner trade and implications for the 1980s.

1106 Tukur, B. M. *Container explosion: The Nigerian experience.* Lagos: Nigerian Ports Authority, 1982.

1107 United Nations Conference on Trade and Development. *Unitization of cargo: Report.* Geneva: United Nations, 1970. 147 pp.

1108 Veldman, Simme. The optimum size of ship and the impact of user costs: An application to container shipping. In: K. M. Gwilliam (ed.) *Current issues in maritime economics.* Dordrecht: Kluwer Academic Publishers, 1993, pp. 112–44.

A paper which shows how user costs can be incorporated in the assessment of optimum size of ships of a liner shipping service. Develops a model to show shipping costs of a liner shipping service as a function of ship size. Uses model to consider questions on the optimal size of ships for a route.

1109 Wallace, I. Containerization at Canadian ports. *Annals of the Association of American Geographers* 65, 1975, 433–48.

1110 Whittaker, J. R. *Containerization.* 2nd edn. Washington (DC): Hemisphere Pub. Corp., 1975. 342 pp.

Records the advances in container shipping and discusses practical aspects of containerization. Rapid domination of major sea routes. Feeder services and inland shipments. New progress in Eastern Europe and the Third World. Impact on dock facilities and organizational patterns. Methods of financing new building of shipping facilities.

1111 Woods, James R. The container revolution. *Journal of World Trade Law* 6(6), November/December 1972, 661–92.

Critical legal and practical problems. Also, a background against which achievements, effects and prospects of container revolution may be measured.

1112 Yamada, Hideo. *Container shipping: Unmasking of overtonnaging.* Tokyo: Japan Maritime Research Institute, 1988. 54 pp.

So-called overcapacity. Average slot capacity utilization ratio. Seasonal fluctuations in space demand. Imbalance of round voyage, as well as TWAC. Unused space. Large volume of new building orders temporarily placed for large-capacity vessels. Recurrence of theory of industrial organization.

1113 No entry.

Ro-ro

1114 Drewry Shipping Consultants. *Market role of large container ro-ro vessels.* London: Drewry Shipping Consultants, 1985. 64 pp.

Report – part of a series covering world's bulk shipping. Unitization and the advance of ro-ro. Deep-sea fleet developments. Status of 1,000+ TEU ro-ro services. The future role of large ro-ros.

1115 Drewry Shipping Consultants. *Ro-ro shipping.* London: Drewry Shipping Consultants, 1977.

1116 Drewry Shipping Consultants. *The status of deep-sea ro-ro services.* London: H. P. Drewry, 1980. 68 pp.

General introduction; characteristics of deep-sea ro-ro fleet; status of deep-sea ro-ro services; economics of ro-ro operation and competing systems. Future prospects for deep-sea ro-ro shipping.

1117 Gilman, S., Williamson, G. A. and **Hughes, C.** *Roll-on/roll-off ships for the deep sea trades.* Liverpool: Marine Transport Centre, University of Liverpool, 1978. 107 pp.

Examines the world fleet of deep sea ro-ro ships and its deployment. Comparative space utilization. Comparative operating performance of full ro-ro ships. Cost comparisons and competitive efficiency.

1118 Grey, Michael (ed.) *Ro-ro ships and shipping.* London: Fairplay Publications, 1985. 174 pp.

Publication is a rearrangement, revision and successor to Fairplay's 1980 volume *Ro-ro ships and their market role.* Subsequent changes in scale rather than in concept – alterations have been refinements rather than new developments. Ro-ro ship types. Access equipment and its design. Cargo handling. Ship design. Services. The future.

1119 *Ro-ro survivability: Commercial retrofitting options.* London: Fairplay Publications.

Conventional cargo liners

1120 Drewry Shipping Consultants. *Modern, multi-purpose cargo ships and the market role.* London: Drewry Shipping Consultants, 1975.

1121 Jansson, J. O. *Costs and prices for the carriage and port handling of general cargo by sea.* London: ICHCA, 1971.

1122 Jansson, Jan Owen and **Shneerson, Dan.** Economies of scale of general cargo ships. *Review of Economics and Statistics* 60(2), May 1978, 287–93.

This paper presents a model of production and costs of shipping and empirical estimates of both economy and diseconomies of ship size. Its main conclusion is that while there are economies of ship size in the haulage operation there are diseconomies to size in the handling operation. This latter fact is the main check on ship sizes.

1123 Sagger, R. K. Turnround and costs of conventional cargo liners: U.K.–India route. *Journal of Transport Economics and Policy* IV(1), January 1970, 53–65.

Considers factors most important in increasing productivity of ships engaged in UK–India cargo liner trade. Large economies effected by reduction of port calls, thus increasing number of voyages.

1124 Validakis, J. E. An economic study of general cargo ships. Thesis submitted for MSc, University of Glasgow, 1978.

Cruise trade

1125 Hawkes, Kenneth Gale. Are American seaports and vessels secure? *Security Management* 36(12), December 1992, 100–02.

The US cruise ship industry, despite its growth, continues its non-compliance with the United Nations International Maritime Organisation's security recommendations. Cost is the given reason.

1126 Jones, Colin. *Cruiser management.* Helmsman Books, 1992. 128 pp.

1127 Marti, B. E. Cruise ship market segmentation: A 'non-traditional' port case study. *Maritime Policy and Management* 18(2), 1991, 93–103.

Case study of a mid-sized vessel operating from a 'non-traditional' port provides insights of the demographic and socioeconomic characteristics of a cruise population.

1128 Marti, B. E. Geography and the cruise ship port selection process. *Maritime Policy and Management* 17(3), 1990, 157–64.

Explores where new cruise ship tonnage could be deployed and its eventual impact on cruise industry. Tests premiss that geography contributes positively to cruise port selection process.

1129 McKee, David L. Some reflections on cruise ships and the economic developments of small island nations. *Canadian Journal of Development Studies* 9(2), 1988, 249–59.

1130 Mentzer, Marc S. Factors affecting cruise ship fares. *Transportation Journal* 29(1), Fall 1989, 38–43.

1131 Miller, William H. *Modern cruise ships, 1965–90.* Mineola: Dover Publications, 1992. 144 pp.

1132 Wanhill, S. R. C. Some aspects of cruise ships. *Maritime Policy and Management* 9(4), 1982, 251–7.

Reviews product trends, examines the clientele and looks at economic impact of cruise ships on tourist destinations or ports of call, using examples from the Bahamas and Bermuda.

1133 Wood, Thomas W. The economics of mixed cargo and cruise ship traffic in a port. *Journal of Transport Economics and Policy* XVI(1), January 1982, 43–54.

Provides theoretical basis for comparing economic gains for efficiencies in cruise, cargo traffic. Models competition between cruise, cargo traffic under common priority rule. Priority policies change with traffic mix change. Case study. Antigua.

Reefers

1134 Bowman, Robert J. Reefer madness: A new technology in ocean shipping is helping U.S. exporters put tough foreign markets on ice. *World Trade* 5(1), January/February 1992, 68–71.

1135 Drewry Shipping Consultants. *Reefer container market: Progress in the containerised transport of reefer commodities.* London: Drewry Shipping Consultants, 1990. 217 pp.

1136 Drewry Shipping Consultants. *The reefer market: Developments in perishable cargo. Shipping and trades.* London: Drewry Shipping Consultants, 1983.

1137 Laskey, Steve and others. *Reefer ships: A Fairplay guide to pure reefer ships as at January 1985.* London: Fairplay International Research Services, 1985. 70 pp.

Ship information from January 1967 to January 1985. Presented in sections by alphabetical order and vessel name, by reefer capacity and by shipbuilder within country of build, to identify standard designs produced by the yards.

1138 *Leading developments in reefers: Based on selected papers from the 2nd International Reefer 2000 Conference 1992.* London: Lloyd's of London Press, 1992. 116 pp.

Selection of papers from the 2nd International Reefer 2000 Conference 1992. Papers consider what the reefer industry should expect to see in the year 2000 and what the industry should prepare themselves for by 2000 and beyond.

1139 *Reefer 2000: Boom, bust or break even?* London: Lloyd's of London Press, 1990.

1140 *Reefer shipping: Global prospects for refrigerated cargo trade and shipping.* London: Drewry Shipping Consultants, 1992. 225 pp.

1141 Wild, Peter. *The reefer market: An industry report.* London: Lloyd's of London Press, 1991. 119 pp.

Examines the principal characteristics of the reefer market. An overview of the reefer market. The citrus trades. The deciduous, minor fruit and vegetable trades. The banana trades. The meat, dairy and other market sectors. Demand for reefer products. Supply-side and technical developments. The future outlook.

Coastal; Short-Sea Trades

General and theory

1142 **Bjornland, D.** The importance of short-sea shipping in European transport. In: European Conference of Ministers of Transport. *Round Table 89: Short sea shipping.* Paris: European Conference of Ministers of Transport, 1993, pp. 59–99.

1143 **Burroughs, R. H.** Offshore supply vessels: An emerging maritime industry. *Maritime Policy and Management* 11(4), 1984, 269–76.

Examines world-wide impact of offshore oil and gas activities on maritime trades, especially size of fleet, locations of fleet, managing firms, numbers of personnel involved and long-term prospects for the industry.

1144 **Canna, Elizabeth.** Short sea jitters. *American Shipper,* July 1993, 40–.

1145 *The Channel Tunnel: Prospects for users, investors and competitors.* London: Drewry Shipping Consultants. 117 pp.

1146 **Czapski, Waldemar.** Some legal aspects. In: European Conference of Ministers of Transport. *Round Table 89: Short sea shipping.* Paris: European Conference of Ministers of Transport, 1993, pp. 113–20.

Short-sea shipping – legal aspects. Different legal systems for short-sea shipping in the Baltic, the Danish straits, the North Sea, the Channel, the Mediterranean.

1147 **Dick, H. W.** Inter-island shipping: Progress, problems and prospects. *Bulletin of Indonesian Economic Studies* 21(2), August 1985, 95–114.

1148 **Dick, Howard W.** Prahu shipping in Eastern Indonesia: Part 1. *Bulletin of Indonesian Economic Studies* 11(2), July 1975, 69–107.

1149 **Dick, Howard W.** Prahu shipping in Eastern Indonesia: Part 11. *Bulletin of Indonesian Economic Studies* 11(3), November 1975, 81–103.

1150 **ECMT Economic Research Centre.** *Short-sea shipping in the economy of inland transport in Europe: Report of the Sixtieth Round Table on transport economics held in Gothenburg, Sweden, on 1st–2nd April, 1982.* Paris: European Conference of Ministers of Transport, 1983. 189pp.

A collection of case studies and summary of the discussions of the Round Table in Gothenburg, April 1982. Case studies involve Greece, Italy, Norway, Sweden and the United Kingdom.

1151 **European Conference of Ministers of Transport.** *Round Table 89: Short sea shipping.* Paris: European Conference of Ministers of Transport, 1993. 126 pp.

Conference to assess potential of short-sea shipping. Papers presented. Summary of discussions. Annex article. List of participants.

1152 **Flatman, A. J.** and **Pike, J.** *After the Channel Tunnel is built: Independent appraisal of tunnel and ferry viability.* Allied Research International, 1986. 65 pp.

1153 **Gallagher, M.** and **Crowley, J. A.** Strategic flexibility in shipping: An examination of a sector of the short-sea ferry market. *Transport Reviews* 8(4), October–December 1988, 317–30.

Principles involved in maintenance of flexibility in resource deployment. Capacity of segment of sea ferry industry to exploit strategic flexibility as means of coping with changing environment.

1154 **Garratt, Mike** and **Wingfield, Melanie.** *Traffic and competition in the short-sea market.* Liverpool: Marine Transport Centre, University of Liverpool, 1981. 112 pp.

Third in a series of monographs on subject of short-sea shipping. Determines the nature of port hinterlands, the variable penetration of unit services into the general cargo market, the trade-offs that shipping lines make between inland and sea transport. The value of market rules.

1155 **Garrod, P. V.** Application of incremental cost pricing to inter-island ocean transport services in Hawaii. *International Journal of Transport Economics* 5(3), December 1978, 243–54.

1156 **Great Britain. Department of Transport.** *Domestic waterborne freight in the U.K.* London: HMSO.

1157 **Head, G. I.** Charges and choice of mode of transport for British exports to Europe. Thesis submitted for M.Phil, London School of Economics, 1971.

1157a La Saponara, F. Short sea shipping in Europe and maritime market changes in the nineties. In: European Conference of Ministers of Transport. *Round Table 89: Short sea shipping.* Paris: European Conference of Ministers of Transport, 1993, pp. 9–58.

Short-sea shipping Euro-Arab area. Growth of inland transport systems. Tonnage supply and new EEC Regulation. Some elements of Mediterranean cabotage. The need for a consistent policy.

1158 Larkin, J. E. An integrated short-sea service. In: National Ports Council. *Containers: Their handling and transport. A survey of current practice.* London: National Ports Council, 1978, pp. 283–93.

Examines the integrated door-to-door container service and the differences from other systems. Uses the service in the short-sea trade provided by author's company as example.

1159 Loose, V. W. and **Roueche, L. R.** The impact of congestion on traffic levels: The case of ferry transportation. *Transportation Research* Series A 13, 1979, 79–82.

1160 National Economic Development Office. *Short sea shipping.* London: HMSO, 1970.

1161 Pemberton, Barry. *Australian coastal shipping.* Melbourne: Melbourne University Press, 1979. 346 pp.

1162 Pike, Jon. *The Channel Tunnel: Prospects for users, investors and competitors.* London: Drewry Business Publications. 117 pp.

1163 Rich, C. A. *Short sea shipping in the economy of inland transport in Europe: Report of the 60th Round Table on transport economics.* Gothenburg: European Conference of Ministers of Transport, 1982.

1164 Stuchtey, R. W. and **Zachcial, M.** Perspectives and trends of the European short sea shipping industry. In: K. M. Gwilliam (ed.) *Current issues in maritime economics.* Dordrecht: Kluwer Academic Publishers, 1993, pp. 36–44.

A paper which assesses the future role of European coastal shipping. Considers the structure of market supply. Evaluates ocean shipping's coming market share in the relevant European transport markets. Suggests further research activities.

1165 Tinsley, David. *Short-sea shipping.* London: Lloyd's of London Press, 1991. 225 pp.

Report, concerned with the movement of bulk cargoes and semi-finished goods between ports in Northern Europe. Reviews structure of key elements of the traffic, in terms of the supply of both cargo and ships, and to assess developments involving industrial producers and the operators and builders of carrying capacity.

1166 Uemura, Masahiro. The development of the shipping business in the Shiwaku Islands under the Ninmyo System. *Osaka Economic Papers* 33(1–2), September 1983, 50–70.

1167 *Waterborne freight in the U.K.: A report prepared for the Department of Transport by the Marine Transport Centre, University of Liverpool.* London: Department of Transport, 1985.

1168 Wijnolst, Ir. N., Peeters, C. and **Liebman, P.** (eds) *European shortsea shipping: Proceedings from the First European Research Round Table Conference on Shortsea Shipping, 26–27 November 1992. Technical University Delft, The Netherlands.* London: Lloyd's of London Press, 1993. 440 pp.

Bulk Trades

1169 Great Britain. Department of Transport. *Short sea bulk shipping: An analysis of U.K. performance.* London: Department of Transport, 1988. 158 pp.

A report which analyses the short-sea sector in terms of the performance of the UK and other major continental fleets in the light of the conditions under which they have operated. These include market opportunities, related economic parameters. Period mainly 1980–85.

1170 Drewry Shipping Consultants. *Bulk cargo in European coastal and short-sea trades.* London: H. P. Drewry, 1976. 77 pp.

Report – part of series covering world's bulk shipping. Bulk cargo carried by coastal and short-sea shipping. Operating and acquiring coastal and short-sea shipping. Structure of and competition in coastal and short-sea shipping industry.

1171 Garratt, M. G. *Economics of short-sea freight ferries.* Liverpool: Marine Transport Centre, University of Liverpool, 1980. 184 pp.

Determines the cost structures of the various components of through transport short-sea services. Considers the relationship between traffic volumes and appropriate technology, the way in which shipping lines compete and how future changes might redirect development.

1172 LeFevre, C. N. Freight transport and the cross-Channel movement of steel. Thesis submitted for M.Phil, Aston University, Birmingham, 1982.

1173 Tinsley, David. *Short sea bulk trades: Dry cargo shipping within European waters.* London: Fairplay Publications, 1984. 155 pp.

Assesses small ship fleets of countries fringing the Baltic, North Sea, and Atlantic coast of Europe. Reviews ownership structure and current design developments and changing composition of those fleets. Commercial trends. Industry's requirements. Register of small ships.

Ferries and ro-ro

1174 Alty, S. Roll-on/roll-off: The Australian scene. In: National Ports Council. *Containers: Their handling and transport. A survey of current practice.* London: National Ports Council, 1978, pp. 349–59.

Australian coastal shipping and ro-ro vessels. The Australian National Line. Background history. ANL vessels. The terminals. Overseas ramp facilities. Terminal equipment.

1175 Borrie, G. Ferry agreements taken to the Restrictive Practices Court. *Maritime Policy and Management* 16(1), 1989, 45–6.

Price fixing agreements of six ferry companies operating between Great Britain and Northern Ireland and covering passengers, freight and vehicles, referred to Restrictive Practices Court in March 1988.

1176 Bradley, M. F. Analysis of factors affecting costs in roll-on roll-off transport. *Irish Journal of Agricultural Economics and Rural Sociology* 5(2), 1974–5, 183–202.

1177 Charlier, J. J. Structural change in Anglo-continental ro-ro traffic, 1977–1987. *Maritime Policy and Management* 16(3), 1989, 233–46.

Powered vehicles dominate British exports through Channel and Straits ports, while unaccompanied trailers lead in traffic through North Sea ports. New balance at Belgian ports.

1178 Chowdhury, Khabirul Haque. Alternative operational policies in short sea ro-ro transportation systems. *Maritime Policy and Management* 15(3), July–September 1988, 193–212.

Highlights effects of several operating parameters that influence short-sea ro-ro transportation system. Output of simulation model used for economic model.

1179 *Cross channel ferries.* 3rd rev. edn. London: Key Note Publications, 1992. 67 pp.

Report provides market overview of cross-channel ferry market. Industry structure; customer profile; industry supply forum; market size, trends, future prospects; company profiles.

1180 D'Este, G. M. and **Meyrick, S.** Carrier selection in a ro-ro ferry trade. Part 1. Decision factors and attitudes. *Maritime Policy and Management* 19(2), 1992, 115–26.

Survey conducted of companies buying shipping services in overnight ro-ro ferry trade to discover priorities and understand decision-making processes. Shippers found to be conservative decision-makers with strong emphasis on service quality.

1181 D'Este, G. M. Carrier selection in a ro-ro ferry trade. Part 2. Conceptual framework for the decision process. *Maritime Policy and Management* 19(2), 1992, 127–38.

From survey of shippers, a multi-stage behavioural framework was constructed to represent stages in decision process and interaction of factors influencing shipping managers in ferry choice.

1182 Garratt, M. *Economics of short-sea freight ferries.* Liverpool: Marine Transport Centre, University of Liverpool, 1980.

1183 Great Britain. Department of Transport. *Cross-channel passenger and freight traffic.* London: HMSO, 1991. 151 pp.

1184 Lapper, R. *Ferries: Risk and liability survey.* FIBI, 1988.

1185 Mansell, Chris. European Ferries finds its haven. *Management Today*, June 1971, 58–67.

Puts the success of European Ferries down to the exploitation of the prime virtues which business conservatism can employ.

1186 **Monopolies and Merger Commission.** *Cross-channel car ferries.* London: HMSO, 1989. 121 pp.

Summarizes reports of the Commission on the question of whether a proposal to enter into agreements for the joint or co-ordinated supply of cross-channel car ferry services would be expected to operate against the public interest.

1187 **Morgan, M.** Moving ferries up-market. *International Journal of Physical Distribution and Logistics Management* 20(5), 1990, 25–32.

This argues that reinvestment in luxury ferries on the short-sea trade to retain business, in face of competition from the Channel Tunnel, will need to be financed by staff cost reductions rather than price increases.

1188 **Posner, J.** Cross-channel freight services. *Freight Management,* March 1972, 21.

1189 **Rich, C.** and **Matear, S.** Some current aspects of marketing in the near-sea ferry and liner trades, with particular reference to the Irish Sea. *Maritime Policy and Management* 16(1), 1989, 57–67.

Study considers aspects of marketing of ferry services in Irish Sea, especially three service types: Container, freight only, ro-ro. Each sector uses marketing but scope exists for greater use.

1190 **Roueche, L. R.** Ferry operations in Norway, Denmark, Britain and Canada: Some comparisons. *Maritime Policy and Management* 8(1), 1981, 1–15.

A study of some of the world's major short-sea ferry operations, and on their fuel costs, traffic forecasting, terminal operations and reservation systems.

1191 **Rowley, C., Mulley, C.** and **Whittaker, J.** *The use of roll-on/roll-off vessels for moving domestic and international freight between the Tees and the Thames areas: An economic analysis.* Hartlepool: Tees and Hartlepool Port Authority, 1983. 134 pp.

Examines the viability of coastal roll-on/roll-off services for freight haulage within the United Kingdom.

1192 **Rowley, Charles** and others. *The use of roll-on/roll-off vessels for moving freight via coastal waters within United Kingdom: An economic analysis.* Middlesbrough: Tees and Hartlepool Port Authority, 1982.

Focuses attention upon the economic viability of utilizing roll-on/roll-off sea vessels as a means of moving freight along the southern and eastern coasts of the UK – essentially between the outer-bounds of Poole and Aberdeen.

Hovercraft and other types

1193 Charlier, Jazques J. Jetfoils on the Ostend–Dover route: A technical and commercial appraisal. *Maritime Policy and Management* 17(2), April–June 1990, 123–32.

Reviews jet-foil service introduced on the 1981 Ostend–Dover route. Reviews 1982–8 operations. Based on technical and commercial data provided by operator.

1194 Kovats, L. J. *The law of hovercraft.* London: Lloyd's of London Press, 1975. 69 pp.

Hovercraft Act 1968 – brings hovercraft within Admiralty jurisdiction. Shipping legislation applicable. Definitions. Registration. Certification and maintenance. Operating permit. Application of Acts of Parliament and statutory instruments. Jurisdiction. Maritime liens. Carriage of goods. Limitation of liability.

1195 Willingham, R. Higher and higher but still a ro-ro bargain. *Freight Management*, March 1977.

Rescue systems

1196 Canner, W. H. P. Rescue at sea: Recognising the U.K. problem. *Maritime Studies and Management* 3(3), January 1976, 187–9.

A commentary in respect of the structure and *modus operandi* of the joint RAF/RN search system and the problems affecting rescue at sea around the coasts of Britain.

8 Ports

General

General

1197 Beresford, A. K. C. and **Dobson, H. W.** *Lloyd's maritime atlas of world ports and shipping places.* 17th edn. London: Lloyds of London Press, 1993. 145 pp.

Contains a 64-page full-colour map section illustrating all the world's commercial ports, region by region. Three sections of maps: World maps, ocean maps, port maps. Distances between ports, major commercial canal and river systems, inland container depots, weather conditions at sea, load-line zones. Precise latitude/longitude co-ordinates. Port facilities.

1198 Bird, James. *Seaports and seaport terminals.* London: Hutchinson & Co., 1971. 240 pp.

Reviews development and modern implications of unit transport, including impact of containers. Links with hinterlands and overseas areas. Developments in ports. Ports and planning – timewise, day to day to long term; space, regional and national and international settings.

1199 Cheka, C. N. The law relating to ports, harbours and docks in England. Thesis submitted for PhD, University of London, 1986.

1200 Cuzan, A. G. The political economy of ports in the United States and Great Britain. *Coastal Zone Management Journal* 11, 1983, 149–67.

1200a Dolman, Antony J. and **Ettinger, Jan Van** (eds) *Ports as nodal points in a global transport system: Proceedings of Pacem in Maribus XVIII, August 1990.* Oxford: Pergamon Press, 1992. 474 pp.

A report on the Pacem in Maribus Conference held in Rotterdam, August

1990. (Part of series of international conferences.) Five papers addressing different aspects of port planning and development presented.

1201 Frazer, R. J. The development of Vladivostok as a commercial port: U.S. versus Japanese development. *Maritime Policy and Management* 20(2), April–June 1993, 101–8.

Examines: (1) problems facing Vladivostock and Russian Far Eastern ports; (2) cargo throughputs and potential for growth; (3) investments by Japanese and US corporations; (4) policy of Japanese and US governments regarding economic aid and co-operation with Russia.

1202 Gifford, J. Morris. Differences in port policy between British and continental ports. *Economisch en Sociaal Tijdschrift* 28(5), October 1974, 625–34.

1203 Hoyle, B. S. and **Hilling, D.** Seaports and development strategies. In: B. S. Hoyle and D. Hilling (eds) *Seaport systems and spatial change: Technology, industry and development strategies.* Chichester and New York (*et al.*): John Wiley & Sons, 1984, pp. 461–9.

Concluding paper to those already submitted to book. Draws threads of all papers together by underlining some of the predominant themes which contributors have elaborated in respect of seaports and technologies, political factors, economic changes, problems and prospects.

1204 Knight, H. Gary. International legal aspects of deep draft harbor facilities. *Journal of Maritime Law and Commerce* 4(3), April 1970, 367–95.

Sets out the need and nature of deep-draft harbours and terminal facilities. The Louisiana deep-draft harbour facility experience. International legal implications of deep-draft harbour facilities. Alternatives. Related issues. A proposal for international action.

1205 Moon, J. R. Port costs and pricing. Thesis submitted for PhD, CNAA, 1982.

1206 Oil Companies International Marine Forum. *Marine and terminal operations: Survey guidelines.* London: Witherby & Co, 1983. 20 pp.

Guide and checklists for use by terminal operators/owners. Guidelines for surveys. General guidance on terminal facilities and procedures for survey personnel.

1207 Touche Ross and Company. *Comparison of the costs of conti-*

nental and British ports, 1974: A report. London: National Ports Council, 1974. 168 pp.

1208 **United Nations Conference on Trade and Development.** *Development and improvement of ports: Co-operation among ports in developing countries.* Geneva: UNCTAD, 1990.

1209 **Walters, A. A.** Marginal cost pricing in ports. *Logistics and Transportation Review* 12, 1976, 99–105.

A paper in which it is argued that considerable benefits may accrue from port pricing on marginal cost principles, and that this principle of charging would assist the disciplining of shipping conferences.

1210 **Yochum, G. R.** and **Agarwall, V. B.** Static and changing port economic impacts. *Maritime Policy and Management* 15(2), 1988, 157–71.

Offers general guidelines for conducting a port's economic impact analysis, and methodology for deciding degree of dependence of industry on a port, e.g. Hampton Roads.

United Kingdom

1211 **Adams, George.** *Organisation of the British port transport industry.* London: National Ports Council, 1973. 262 pp.

Provides comprehensive picture of British port industry. Thorough discussion of entire port industry. Employment; organizations; customers; authorities; administration; Customs; health; industrial relations; proposals for reorganization; related transport systems.

1212 **Arnott, D.** The last hope: Liverpool's port. *Management Today,* January 1983, 66–73.

This argues that the Port of Liverpool has had considerable difficulty keeping up with the pace of technological and other changes.

1213 **Associated British Ports.** *Strength through diversity.* London: Associated British Ports, 1983. 12 pp.

Brochure to give an insight into Associated British Ports, Britain's largest ports business.

1214 **Asteris, Michael.** Britain's seaports: Competition and transshipment. *National Westminster Bank Quarterly Review*, February 1988, 30–48.

This argues that despite dramatic increase in productivity, British ports are still not competitive when compared with foreign rivals such as Antwerp and Rotterdam. It also takes into account the potential of the fixed link.

1215 British Transport Docks Board. *Your choice of British ports.* London: British Transport Docks Board, 1981.

1216 Buck, E. F. G. An examination of the objectives, economics and technology of port operation, with special reference to current and proposed development in South East England. Thesis submitted for MSc, University of Salford, 1975.

1217 Cranfield Institute of Technology, Centre for Transport Studies. *British ports industry: A study of changing patterns of usage and capacity 1965–1980. Volume I.* London: British Ports Association. 24 pp.

Volume I of two: Provides summary of overall trade patterns by main commodity groups, by trading areas, by mode and by British region. Develops estimates of port capacity on regional basis and trends in employment.

1218 Cranfield Institute of Technology, Centre for Transport Studies. *British ports industry: A study of changing patterns of usage and capacity 1965–1980. Volume II.* London: British Ports Association. 96 pp.

Detailed analysis of data on which main findings of study of changes in British ports industry are based. Main emphasis on commodity flows and their handling. Also examines data relating to employment in the industry.

1219 Cook, T. C. C. A study of pilotage in a U.K. oil port, including work load monitoring. Thesis submitted for M.Phil, University of London, 1977.

1220 Crawshaw, B. W. Towards a rationalisation of the offshore supply ports: A study based on the experience of the ports of Grampian and Tayside regions. Thesis submitted for M.Phil, University of Edinburgh, 1976.

1221 Great Britain. Central Office of Information. *British industry today: Ports. (Reference Pamphlet No. 119.)* London: HMSO, 1974. 52 pp.

One of a series produced by the Central Office of Information for British Information Services. Describes ports industry in Britain including organization and ownership, role of Government, port administration, facilities, employment, volume and types of traffic; ports themselves.

1222 **Great Britain. Department of the Environment.** *Financial policy for ports. (Cmnd. 4794).* London: HMSO, 1971. 8 pp.

1223 **Great Britain. Department of Transport.** *Port choice and the routeing of U.K. trade. (A study of the 1978 survey of inland origin and destination of trade.)* London: Department of Transport, 1982. 107 pp.

A study which investigates the interactions between the inland and foreign distribution of trade, the routeing of traffic and the choice of UK port.

1224 **Hilling. D.** The restructuring of the Severn estuary ports. In: B. S. Hoyle and D. Hilling (eds) *Seaport systems and spatial change: Technology, industry and development strategies.* Chichester and New York (*et al.*): John Wiley & Sons, 1984, pp. 257–76.

Early history. Inter-war period. The post-war period – changing directions of national trade; improved materials and cargo-handling technology; changing patterns of industrial activity; increased significance of road transport; changing energy sources; the organization of ports.

1225 **Livesey, Robert C.** The Bathside Bay story. *Maritime Policy and Management* 17(2), April–June 1990, 133–40.

Considers whether opportunity to develop harbour facilities at strategically located site in Harwich Harbour on east coast of England missed in 1960s and 1970s.

1226 **Lowe, R.** The impact and implications of technological change at Southampton docks. Thesis submitted for M.Phil, University of Southampton, 1987.

1227 **Lundy, C. C.** Small ports in Northern Ireland: A method of study. *Maritime Policy and Management* 9(4), 1982, 273–7.

Northern Irish 'small ports' defined by tonnage of cargo handled annually, rather than by value of cargo, number of ships or berths occupied.

1228 **Moriarty, James** (ed.) *19 ports.* 5th edn. Charter in association with Associated British Ports, London: 1986. 136 pp.

1229 National Economic Development Office. *The nation's infrastructure: Sea and estuarial ports.* London: National Economic Development Office, 1986. 27 pp.

A report to consider the economic performance and prospects for Britain's ports.

1230 Newman, N. Private problem for public ports. *Management Today*, December 1980, 66–74.

Subtitled 'Denationalization'. Puts forward a plan for the privatization of British transport. Dock Board ports.

1231 *Port financial information.* London: National Ports Council, 1972. 36 pp.

1232 The Port of Southampton. *Barclays Bank Review* 65, August 1970, 47–50.

This discusses the perishable position of Southampton in British trade and points out changes which are taking place, particularly in handling methods with the adoption of container systems.

1233 Takel, R. E. *Industrial port development: With case studies from South Wales and elsewhere.* Bristol: Scientechnica Ltd, 1974. 246 pp.

Factors leading to changes in ports. Concept of industrial ports. Characteristics determining suitability of place to be a port. Technical aspects of industrial ports. Long-term prospects. Industrial port development.

1234 United Nations Conference on Trade and Development. *Measuring and evaluating port performance and productivity.* Geneva: UNCTAD, 1987.

1235 Walker, Jim. Factors affecting international trade through Scottish ports, 1967–79. *Maritime Policy and Management* 12(3), July–September 1985, 227–39.

Paper looks at level of port activity in Scotland during period of substantial structural change in both pattern of trade and type of shipping commonly associated with UK trade. Uses published data from official sources to examine factors contributing to change.

1236 Webster, P. J. K. A charges structure for the ports of the United Kingdom. Thesis submitted for PhD, London School of Economics, 1975.

Western Europe and Scandinavia

1237 **Biagini, E.** Interport competition in an integrated spatial system: Genoa–Voltr and Savona–Vado in Northern Italy. In: B. S. Hoyle and D. Hilling (eds) *Seaport systems and spatial change: Technology, industry and development strategies.* Chichester and New York (*et al.*): John Wiley & Sons, 1984, pp. 327–41.

The framework of interport competition. Stages in the competition between Genoa and Savona. A critical appraisal of the pilot project for the Ligurian ports. The return to coal and the roles of satellite ports. Developments in the early 1980s.

1238 **Centre d'Études de Transport pour la Méditerranée Occidentale (CETMO).** Data processing and telematic work and projects in western Mediterranean ports. *Maritime Policy and Management* 17(2), April–June 1990, 87–97.

Report of interviews of over 100 people working in port areas in seven countries. Aims to provide data on realities, projects, intentions to develop data processing and telematic systems in western Mediterranean ports.

1239 **Charlier, J. J.** Structural change in the Belgian port system, 1980–1986. *Maritime Policy and Management* 15(4), 1988, 315–26.

Specialization and concentration indexes are applied to the geography of the Belgian port system (Antwerp, Ghent, Zeebrugge and Ostend) to analyse structural change.

1240 **Deakin, F. R.** Anglo-Dutch trade flows, 1955–75: Their effects on and consequences for Dutch port development and planning. Thesis submitted for PhD, University of Hull, 1985.

1241 **Hoyle, B. S.** and **Pinder, D. A.** (eds) *European port cities in transition.* London: Belhaven Press, 1992. 256 pp.

Collection of papers which review the role of European ports, particularly in relation to their changing functions and structure. Particular problem of the old-established 'heavy industry' ports outlined, as well as the obsolescent ports of the emergent East European democracies. Europe-wide perspective, and addresses the problems of integration within the present and potentially wider European Community.

1242 **Panunzio, Vito.** The Italian ports and their problems. *Review of the Economic Conditions in Italy* 32(1), January 1978, 27–40.

1243 **Pinder, D. A.** Planned port industrialisation and the quest for upward economic transition. In: B. S. Hoyle and D. Hilling (eds) *Seaport systems and spatial change: Technology, industry and development strategies.* Chichester and New York (*et al.*): John Wiley & Sons, 1984, pp. 277–301.

Examines planned port development in one specific problem region (the Dutch Delta), that has been the focus of well-defined development strategies. Considers port-based industrialization and its contribution to the regional economy, and questions both its resilience to recessional pressures and its ability to reactivate industrial growth in port-recession era.

1244 **Regul, Rudolf.** *The future of the European ports. Vol. I.* Bruges: De Tempel, 1971. 517 pp.

Papers from Bruges Week 1970 conference looking at the future of European ports. The papers in Vol. I look at the function of maritime ports; future development of maritime transport; investment policy in the ports – factors determining the economic evaluations of investment.

1245 **Regul, Rudolf.** *The future of the European ports. Vol. II.* Bruges: De Tempel, 1971. 821 pp.

Papers from Bruges Week 1970 conference looking at the future of European ports. The papers in Vol. II look at investment policy in the ports: Factors determining the economic evaluations of investment.

1246 **St. Seidenfus, Hellmuth.** European ports in the context of the world economy and the European economy: Changes in sea transport. *International Journal of Transport Economics* 14(2), June 1987, 133–8.

1247 **Verbeke, A.** and **Winkelmans, W.** The illusion of Belgian sea port policy evaluation of structural effectiveness and development of a policy alternative. *International Journal of Transport Economics* 13(2), June 1986, 209–20.

1248 **Verbeke, Alain.** Assessing the effectiveness of Belgian seaport policy: A comparative institutional approach. *Annals of Public and Cooperative Economy* 59(2), April–June 1988, 157–74.

1249 **Vleugels, R. L. M.** Port economy in the European Community. *Chartered Institute of Transport Journal* 1973.

Central and Eastern Europe and Middle East

1250 **Assaad, A. F.** The provision of trained pilots for selected Gulf Arab countries and the re-organisation of the pilotage service in connection with other port marine activities. Thesis submitted for M.Phil, CNAA, 1980.

1251 **Canfield, J. L.** Independent Baltic State ports: The outlook for commercial viability. *Maritime Policy and Management* 20(2), April–June 1993, 133–52.

Purpose is to lay groundwork for in-depth analysis of port management planning under adverse political and economic conditions. Employs the 'shift-share' technique to assay status of seaports in the independent Baltic States in terms of prospects for commercial viability.

1252 **Hayut, Y.** *Shipping, ports and foreign trade of Egypt.* Haifa: Israel Institute for the Study of Shipping, 1979.

1253 **Soffer, A.** Illusion as a factor in determining policy in development of a port: The case of Haifa. *Maritime Policy and Management* 9(2), 1982, 145–52.

Changes in the Middle East over the years have deprived Haifa of most of its supposed geographical advantages over other ports in the region.

1254 **Taylor, Zbigniew.** Seaport development and the role of the state: The case of Poland. In: B. S. Hoyle and D. Hilling (eds) *Seaport systems and spatial change: Technology, industry and development strategies.* Chichester and New York (*et al.*): John Wiley & Sons, 1984, pp. 217–38.

Examines the role of the state and its impact upon seaport development in centrally planned economies. The way in which transport policies and investment in transport are part of a wider pattern of state activity.

1255 **Walker, A. R.** Oil-dependent economies and port development: The Gulf States of the Middle East. In: B. S. Hoyle and D. Hilling (eds) *Seaport systems and spatial change: Technology, industry and development strategies.* Chichester and New York (*et al.*): John Wiley & Sons, 1984, pp. 179–97.

Examines relationships between port problems and development objectives in the region. Special reference to the problems of the commercial and industrial ports of the Gulf 1971–81.

North America

1256 **Crew, J. G.** and **Horn, K. K.** The evolving functions of U.S. Gulf Coast ports. *Annual Proceedings of the Transportation Research Forum* 26, 1985, 152–60.

1257 **Fleming, D. K.** The port community: An American view. *Maritime Policy and Management* 14(4), 1987, 321–36.

Examines some internal characteristics of American port communities not usually visible. Examines interplay and levels of co-operation between private and public sectors.

1258 **Forward, C. N.** The overwhelming dominance of the port of Vancouver on Canada's West coast. In: B. S. Hoyle and D. Hilling (eds) *Seaport systems and spatial change: Technology, industry and development strategies.* Chichester and New York (*et al.*): John Wiley & Sons, 1984, pp. 343–60.

Case study which examines the effects of a unique combination of physical, political and economic factors that influenced the development of the British Columbia port system.

1259 **Goss, Richard.** *Policies for Canadian seaports.* Cardiff: UWIST, for Canadian Transport Commission, 1984. 83 pp.

1260 **Hoyle, Brian S.** Waterfront redevelopment in Canadian port cities: Some viewpoints on issues involved. *Maritime Policy and Management* 19(4), October–December 1992, 279–95.

Reports on component of research project on Canadian Dimensions of Waterfront Revitalization wherein 45 interviews held in Canadian port cities with port authority representatives, urban planners and developers.

1261 *Improving productivity in U.S. marine container terminals.* Washington (DC): National Academy Press, 1986. 205 pp.

1262 **Kuo, Tu-Cheng.** Development of a container terminal simulator model and its application in an analysis of terminal 18, Port of Seattle. Thesis submitted for PhD, University of Washington, 1992.

1263 **Marcus, Henry S.** and others. *Federal port policy in the United States.* Cambridge (Massachusetts): MIT Press, 1976. 371 pp.

1264 Marti, Bruce E. The evolution of Pacific Basin load centres. *Maritime Policy and Management* 15(1), January–March 1988, 57–66.

Paper formulates and augments general principles pertinent to load centre development. Utilizes US maritime containerized foreign trade data to identify and measure geographic concentrations of cargo movements through Pacific Basin ports.

1265 McCalla, R. J. Canadian port administration: Its future structure. *Maritime Policy and Management* 9(4), 1982, 279–93.

Considers administration, management and development of ports in Canada and their central control by the Federal government, and the Canada Ports Corporation Act abolishing the National Harbours Board.

1266 Murphy, P., Dalenberg, D. and **Daley, J.** A contemporary perspective of international port operations. *Transportation Journal* 28(2), Winter 1988, 23–32.

Results of a mail survey of 236 US and foreign ports involved in international trade activities.

1267 Randall, James E. Economic development and non-marine initiatives at American seaports. *Maritime Policy and Management* 15(3), July–September 1988, 225–40.

Role of Port Authority in fostering economic development. Economic role of public port in local community. Need for non-marine economic functions. Alternative strategies used by ports experiencing growth and decline in cargo throughput.

1268 Starr, J. T. The Port of Baltimore: Off the beaten track? *Maritime Policy and Management* 18(3), 1991, 171–81.

Explains problems and offers preliminary suggestions to improve downturn in traffic trends (falling tonnage, fewer ship calls, etc.) recently experienced by Port of Baltimore.

1269 United States Maritime Administration. *What U.S. ports mean to the economy.* Washington (DC): United States Department of Commerce, 1978.

1270 Warf, B. US imports of petroleum and related products in the 1980's. *Maritime Policy and Management* 16(4), 1989, 265–75.

Shows geographic structure of imports of three major oil-related commodities to US in 1980s. Shows crude slowing down, petroleum increasing and volatile geography of natural gas imports.

1271 **Winters, Tobey L.** *Deep-water ports in the United States: An economic and environmental impact study.* New York: Praeger, 1977. 216 pp.

Japan

1272 **Rimmer, P. J.** Japanese seaports: Economic development and state intervention. In: B. S. Hoyle and D. Hilling (eds) *Seaport systems and spatial change: Technology, industry and development strategies.* Chichester and New York (*et al.*): John Wiley & Sons, 1984, pp. 99–133.

Distinguishes throughout between (a) Japanese approach to economic development and (b) the American approach. Changing ideas on economic development. The Tokugawa Era (1603–1867). Pre-war Japan (1868–1940). Post-war Japan.

1273 **Rowe, R. P.** Port development in Japan. *Dock and Harbour Authority* 59(700), March 1979, 347.

Socialist Asia

1274 **Chiu, T. N.** and **Chu, David K. Y.** Port development in the People's Republic of China: Readjustment under programmes of accelerated economic growth. In: B. S. Hoyle and D. Hilling (eds) *Seaport systems and spatial change: Technology, industry and development strategies.* Chichester and New York (*et al.*): John Wiley & Sons, 1984, pp. 199–215.

Ideology and foreign trade in China. Port development under leftist ideology. Failure of ports to support accelerated economic growth. Port development under open-door policy.

Africa

1275 **Ainah, E. O.** The evolution and operational development of the port of Lagos (1880–1980). Thesis submitted for MSc, UWIST, Cardiff, 1984.

This supports the idea, as in other studies relating to port geography such as that of Bird and Rimmer, that a variety of external factors such as politics, physical, economic and technical changes predominate in the course of a port's growth and development.

1276 **Alabere, A. D.** Port planning and development in Nigeria. Thesis submitted for PhD, UWIST, Cardiff, 1985.

1277 **Danish, A. M. A. F. R.** The future expansion of the Port of Alexandria. Thesis submitted for MSc, UWIST, Cardiff, 1975.

1278 **Dickinson, B.** The development of the Nigerian ports system: Crisis management in response to rapid economic change. In: B. S. Hoyle and D. Hilling (eds) *Seaport systems and spatial change: Technology, industry and development strategies.* Chichester and New York (*et al.*): John Wiley & Sons, 1984, pp. 161–77.

Assesses policies and practical responses of the Nigerian Government and the Nigerian Port Authority to crisis in mid-1970s of inability of Nigerian ports to handle increase in imports and impact on ports. Development of Nigerian ports system 1970–80.

1279 **Ekpenyong, S.** The small scale labour force in Port Harcourt and Calabar. Thesis submitted for PhD, University of Birmingham, 1981.

1280 **Elawad, B. A. M.** Operational and planning factors in the development of Port Sudan and their impact on the nation's maritime transport. Thesis submitted for MSc, UWIST, Cardiff, 1981.

1281 **Falowo, I. O.** The development of the Nigerian ports, 1970–1982. Thesis submitted for PhD, University of Glasgow, 1986.

1282 **Filani, M. O.** and **Tkporukpo, C. O.** Containerization in Nigeria: Trends and patterns. *Maritime Policy and Management* 14(3), 1987, 185–95.

Started in 1960s, Nigerian containerization only reached 8.6 per cent of external trade cargo by 1981. More important for import than export trade, it is concentrated in Lagos and Port Harcourt.

1283 **Hilling, D.** The development of the Ghanaian port system. Thesis submitted for PhD, University of London, 1975.

1284 **Hilling, D.** The evolution of a port system: The case of Ghana. *Geography* 62, April 1977, 97–105.

1285 **Hilling, David.** Port specialization and efficiency: The case of Ghana. *Maritime Studies and Management* 3(1), July 1975, 13–20.

Considers development of Ghana's ports Takoradi and Tema. Analysis of development shows why both ports provide for the efficiency and low cost of external trade.

1286 **Hoyle, B. S.** and **Hilling, D.** (eds) *Seaports and development in tropical Africa.* London: Macmillan, 1970. 272 pp.

A collection of essays by professional geographers, brought together to emphasize the important role of seaports in the economic growth of underdeveloped areas and to demonstrate the increasing interest focused upon this issue by geographers in tropical Africa.

1287 **Hunter, C. M.** Port training in Africa. *Maritime Policy and Management* 8(4), 1981, 267–79.

Looks at training port labour in Africa, examines immediate training, apprenticeship training and career training, using pedagogic methods fully adapted to needs of a port.

1288 **Luke, D. F.** The political economy of labour in an African public enterprise: Dockworkers and the development of the Sierra Leone port organisation. Thesis submitted for PhD, University of London, 1982.

1289 **Ogundana, Babafemi.** Seaport development: Multi-national cooperation in West Africa. *Journal of Modern African Studies* 12, September 1974, 395–407.

1290 **Ogundana, T. T. B.** Seaport development in Nigeria. Thesis submitted for PhD, University of London, 1971.

1291 **Ogundana, T. T. B.** Some elements of changing port concentration in Nigeria. Thesis submitted for MSc, University of London, 1966.

1292 **Wanhill, S. R. C.** *Port economics in developing countries with reference to Port Louis Harbour, Mauritius.* London: International Cargo Handling Co-ordination Association, 1974.

1293 **Wiese, B.** The role of seaports in the industrial decentralisation process: The case of South Africa. In: B. S. Hoyle and D. Hilling (eds) *Seaport systems and spatial change: Technology, industry and development strategies.* Chichester and New York (*et al.*): John Wiley & Sons, 1984, pp. 415–34.

Seaports in the South African space economy: Polarization versus decentralization. National Physical Development Plan (1975), New Regional Development Incentives (1982). Seaports as decentralizating-growth points in South Africa. Deep-water bulk seaports as nuclei for metropolitan growth. Buffalo Harbour, East London: A re-evaluation. Seaports and industrial decentralization: A synthesis.

Central and South America

1294 **Chapman, J. B.** The structural evolution of seaports and transport systems in Peru. In: B. S. Hoyle and D. Hilling (eds) *Seaport systems and spatial change: Technology, industry and development strategies.* Chichester and New York (*et al.*): John Wiley & Sons, 1984, pp. 391–414.

Case study of seaport development in Peru to illustrate the technological innovations in the field of maritime transport over the last two decades and the growing awareness of a need for efficient seaports serving a vigorous foreign trade sector.

1295 **Marti, B. E.** and **Leon, P. P.** The Combined Naval Commercial Port (C.N.C.P.) in a Latin American setting. *Maritime Policy and Management* 11(4), 1984, 227–32.

Describes the Latin American version of CNCP to test the hypothesis that commercial port interests have been adversely affected by coexistence.

Asia, Australasia and Oceania

1296 **Bang, H.-S.** Factors affecting container terminal development: A critique of the approach applied in Korea Port. Thesis submitted for PhD, UWIST, Cardiff, 1984.

The present operational problems of Korean ports are the result of failures to plan developments adequately. This is manifest by the fact that existing container terminal capacity in Busan is well below demand, resulting in a high proportion of throughput still being handled by conventional facilities.

1297 Boschken, Herman L. (ed.) *Strategic design and organizational change: Pacific Rim seaports in transition.* University of Alabama Press, 1988. 250 pp.

1298 Chu, David K. Y. and **Chiu, T. N.** Laissez-faireism in port development: The case of Hong Kong. In: B. S. Hoyle and D. Hilling (eds) *Seaport systems and spatial change: Technology, industry and development strategies.* Chichester and New York (*et al.*): John Wiley & Sons, 1984, pp. 135–60.

Assesses different roles of the factors contributing to the development of the port of Hong Kong. History of port development in UK and laissez-faireism. Merits and demerits of laissez-faireism.

1299 Chu, David Kim-Yee. Planning and development of the container port of Hong Kong. [Success in exploiting container technology, privately-run shipping consortia, and *laissez-faire* tradition.] *Third World Planning Review* 11, August 1989, 275–88.

1300 Couper, A. D. Changes in port and shipping technology in the Pacific and their social impact. In: A. D. Couper (ed.) *Development and social change in the Pacific Islands.* London: Routledge, 1989, pp. 145–52.

1301 Goss, R. O. Competition in cargo-handling: Some experience from Australia. *Maritime Policy and Management* 9(1), 1982, 45–58.

Summarizes reports of the Australian Prices Justification Tribunal in 1977 on two stevedoring companies, James Patrick and Co. Proprietary Ltd and Seatainer Terminals Ltd.

1302 Gunatillake, D. S. Problems of port development in Sri Lanka, with special reference to Colombo. Thesis submitted for M.Phil, University of London, 1975.

1303 Hoyle, B. S. Seaport systems and agricultural exports in a developing regional economy: The sugarports of Queensland, Australia. In: B. S. Hoyle and D. Hilling (eds) *Seaport systems and spatial change: Technology, industry and development strategies.* Chichester and New York (*et al.*): John Wiley & Sons, 1984, pp. 361–89.

Case study of seaport development in Peru to illustrate the technological innovations in the field of maritime transport over the last two decades and growing awareness of need for efficient seaports serving a vigorous foreign trade sector.

1304 Kesterman, Frank. Shipping and port operations in the Persian Gulf. *Journal of Maritime Law and Commerce* 7(1), October 1975, 315–26.

Overview of status of maritime developments in the Gulf region. Commentary on trends, problems and constraints facing industry and governments.

1305 McDonnell, Paul. Trade and shipping needs for island economies. In: A. D. Couper (ed.) *Development and social change in the Pacific Islands.* London: Routledge, 1989, pp. 153–9.

1306 Riley, N. T. Aid for maritime developments in the Pacific. In: A. D. Couper (ed.) *Development and social change in the Pacific Islands.* London: Routledge, 1989, pp. 160–67.

1307 Robinson, Ross. Productivity of first-generation container terminals: Sydney, Australia. *Maritime Policy and Management* 12(4), October–December 1985, 279–92.

Paper explores differences and their implications for terminal productivity between Sydney's first purpose-built container facilities in Port Jackson and the second terminal at Glebe Island. Sample years of 1977, 1979 and 1981.

1308 Zerby, J. A. Competition in cargo-handling: Some comments. *Maritime Policy and Management* 9(4), 1982, 295–301.

Comments on paper by R. O. Goss in *Maritime Policy and Management* 9(1), 1982, pp. 45–58, 'Competition in cargo-handling: Some experience from Australia', suggesting other factors than those mentioned by Goss may work against proper functioning of a competitive system in cargo-handling.

Economic Theory and Price

General

1309 Alderton, Patrick M. *The quantification of port time.* London: City of London Polytechnic, 1987. 10 pp.

Discussion paper forming one of series produced through the Transport Studies Workshop of the (then) City of London Polytechnic. Examines problems of producing general model that will generate reasonable estimates for all ships going to all ports.

1310 Audo, S. Development of a dynamic model for strategic port planning and investment. Thesis submitted for PhD, Cranfield Institute of Technology, 1985.

This study attempts to relate future port demand to optimal berth capacity in an interactive dynamic fashion.

1311 Bauchet, P. Financing and charging for seaport and airport infrastructures in France. *Maritime Policy and Management* 10(2), April–June 1983, 93–101.

The paper questions the responsibility for financing of infrastructure for sea and air transport in France. Concentrating on the states' differing treatment of different modes and their infrastructure.

1312 Bennathan, E. and **Wishart, J.** *Private and public enterprise in the ports of developing countries.* Washington (DC): The World Bank, 1983.

1313 Bird, J. and **Pollock, E. E.** The future of seaports in the European Communities. *Geographical Journal* 144, 1978, 23–48.

1314 Bird, J. H. Seaport development: Some questions of scale. In: B. S. Hoyle and D. Hilling (eds) *Seaport systems and spatial change: Technology, industry and development strategies.* Chichester and New York (*et al.*): John Wiley & Sons, 1984, pp. 21–41.

Developmental complications of seaports. Development of seaport studies. Problems of seaports – on spatial and temporal scales. Problems faced by organizations in connection with the future developments of seaports. Seaports in the future.

1315 Bromich, M. Port costs: An alternative approach. *International Journal of Transport Economics* 5(3), December 1978, 227–40.

1316 Butler, Eamonn and **Davies, Warwick.** *Free port experiment.* London: Adam Smith Institute, 1986. 42 pp.

1317 Casavant, Kenneth L. and **Weeks, Eldon E.** The concept of terminal port parity as applied to the Seattle and Portland competitive port situation. *University of Washington Business Review* 31(1), Autumn 1971, 56–62.

1318 Chang, S. In defense of port impact studies. *Transportation Journal* 17, 1978, 79–85.

1319 Claessens, E. M. Port and jetty location for small-scale inter-island short-sea shipping: The warehouse approach. *Maritime Policy and Management* 16(2), 1989, 109–21.

Looks at feasibility of comprehensive jetty programmes in inter-island shipping. Alternatives are evaluated using so-called 'warehouse approach' of business logistics.

1320 Couper, Alastair. *Social consequences of maritime technological change.* Seattle: University of Washington, 1985. 17 pp.

Lecture in marine affairs, 26 November 1985. One of series. The port technological revolution. The technological revolution in shipping. Technological and social changes in fisheries.

1321 Davis, H. C. Regional port impact studies: A critique and suggested methodology. *Transportation Journal* 23, 1983, 61–71.

1322 De Lombaerde, P. and Verbeke, Alain. Assessing international seaport competition: A tool for strategic decision making. *International Journal of Transport Economics* 16(2), June 1989, 175–92.

1323 Drewry Shipping Consultants. *Bulk shipping ports and consortia.* London: H. P. Drewry, 1974.

1324 Drewry Shipping Consultants. *Crude oil pipelines and VLCC ports.* London: H. P. Drewry (Shipping Consultants), 1974. 81 pp.

Report – part of series covering world's bulk shipping. Crude oil pipelines in the major consuming areas. Centralized economies. Pipelines in the oil exporting areas. Transit pipelines. World-wide port facilities for tankers 150,000 d.w.t. and over.

1325 Drewry Shipping Consultants. *Perspectives on third world port development.* London: H. P. Drewry, 1977. 90 pp.

Report – part of series covering world's bulk shipping. Problems and constraints. Alternative lift-on lift-off systems. An overall perspective.

1326 Drewry Shipping Consultants. *Ports and terminals for large (100,000 dwt+) dry bulk carriers.* London: H. P. Drewry, 1982. 64 pp.

Report – part of series covering world's bulk shipping. Participation of large carriers in dry bulk trade. Capacity and composition of the large dry bulk carrier fleet. Commodity handling procedures and port constraints. The principal large dry bulk carrier ports. Future port developments.

1327 Drewry Shipping Consultants. *Ports and terminals for large bulk carriers.* London: H. P. Drewry, 1976. 114 pp.

Report – part of series covering world bulk shipping. Development of large bulk carrier ports and terminal facilities. Port and terminal facilities in the coal trades, in the grain trades, in the bauxite and alumina trades, in the phosphate rock trades, in the minor bulk trades.

1328 Fleming, Douglas. Safe harbours. *Maritime Policy and Management* 13(1), January–March 1986, 17–25.

Paper examines geographical patterns and rationale of shipping havens, free from heavy taxes; high costs; restrictive legislation. Purpose to find connections between safe harbours and major international shipping centres.

1329 Fleming, Douglas K. Port rivalry, co-operation, merger. *Maritime Policy and Management* 10(3), July–September 1983, 207–10.

Paper deals with transport competition and with possibilities of interport collaboration. Adopts perspective of port user – notably transport carrier and commodity shipper. Applies this perspective at various geographical scales.

1330 Foggin, J. H. and **Dicer, G. N.** Disappearing hinterlands: The impact of the logistics concept on port competition. *Annual Proceedings of the Transportation Research Forum* 15, 1984, 385–90.

1331 Ford, C. R. *Ports into the next century.* London: Thomas Telford, 1991. 276 pp.

Collection of papers given at the conference 'U.K. Ports 2000', organized by the Institute of Civil Engineers and held in Hull on 17–18 October 1990. Practical visions of future of ports industry legislation. Trading patterns. Redevelopment patterns. Design and operation. The future.

1332 Frankel, E. G. Debt–equity conversion and port privatization. *Maritime Policy and Management* 19(3), 1992, 201–9.

Reviews reasons for port privatization, different methods for port ownership financing and role debt–equity conversion could play in it.

1333 Gilman, S. and **Williams, G. F.** *Cost influences on the geography of multi-port systems.* Liverpool: Marine Transport Centre, University of Liverpool, 1974.

1334 Goss, R. O. Economic policies and seaports: 1. The economic functions of seaports. *Maritime Policy and Management* 17(3), 1990, 207–19.

Argues that economic functions of seaports are to benefit those whose trade passes through them, not to increase employment in locality or maximize profits.

1335 **Goss, R. O.** Economic policies and seaports: 2. The diversity of port policies. *Maritime Policy and Management* 17(3), 1990, 221–34.

Contrasts economic policies adopted towards seaports by different countries, especially approaches to national, regional or local control. All parties should be consulted in interests of port efficiency.

1336 **Goss, R. O.** Economic policies and seaports: 3. Are port authorities necessary? *Maritime Policy and Management* 17(4), 1990, 257–71.

Arguments made for and against public sector port authorities. Pragmatic approach advocated, tailored to needs and resources of country in question.

1337 **Goss, R. O.** Economic policies and seaports: 4. Strategies for port authorities. *Maritime Policy and Management* 17(4), 1990, 273–87.

Discusses strategies for port authorities, given that modern technology has severely limited scope for competition, e.g. minimalist, pragmatic, public sector, competitive strategies.

1338 **Goss, R. O.** The public and private sectors in ports. (Editorial.) *Maritime Policy and Management* 8(2), 1981, 69–71.

There are a multitude of possible combinations of private and public sectors in ports and most of them can be seen working today, more or less efficiently.

1339 **Hanappe, Paul** and **Savy, Michael.** *Industrial ports and economic transformations.* Tokyo: The I.A.P.H. Head Office Maintenance Foundation, 1980. 201 pp.

Impact of economies of capitalist industrialized countries on port economics. Economic trends. Future of the port industrial zones in face of these trends. Geographical field covered is of world amplitude.

1340 **Heaver, T. D.** Shipping and the market for port services. In: G. Blauwens, G. De Brabander and E. Van le Voorde (eds) *De dynamiek van een haven.* Uitgeverij Pelckmans Kapellen, 1993, pp. 227–48.

The nature of competition among port terminals is described and the implications for national port policies and for port management are considered.

1341 **Heggie, Ian.** Charging for port facilities. *Journal of Transport Economics and Policy* VIII (1), January 1974, 3–25.

Examines some existing tariffs imposed by ports. Suggests principles on which charges should be fixed. What are financial objectives of public port authorities? Suggested objectives of pricing. An example of a cost-based tariff.

1342 **Holloway, Ronald.** The problem of the ports. *Lloyds Bank Review* 99, January 1971, 13–26.

This article suggests that the sense of crisis in the port industry and serious labour difficulties are due to the scope and speed of technological change. What the ports have to cope with – three technological revolutions at once: (1) the port – cargo handling; (2) in ship size; (3) in the competition for other means of transport.

1343 **Hoyle, B.** Cities on the sea. *Maritime Policy and Management* 14(4), 1987, 345–6.

Brief report of symposium on harbours, port cities and coastal topography called 'Cities on the sea: Past and present', held in Haifa, September 1986.

1344 **Hoyle, B. S.** and **Hilling, D.** (eds) *Seaport systems and spatial change: Technology, industry and development strategies.* Chichester and New York (*et al.*): John Wiley & Sons, 1984. 481 pp.

Collection of papers concerned with the dynamic relationships between seaport systems and the process of spatial change, with special reference to the technologies, industrial development patterns and regional planning strategies that affect these relationships.

1345 **ICHCA.** *Port economics in developing countries and the appraisal of port warehouse extensions.* London: ICHCA, 1974.

1346 **Imakita, Junichi.** *A techno-economic analysis of the port transport system.* Farnborough, Hants: Saxon House, 1978. 206 pp.

Comprehensive analysis of technical operation of seaports. Main operations of port including ship guidance into berthing areas, cargo handling, warehousing, inland transport links. Resulting aggregate system used to find optimum patterns for rates of operation; employment of labour; equipment; investment strategies.

1347 **Jansson, Jan Owen** and **Shneerson, Dan.** *Port economics.* Cambridge (Massachusetts) and London: MIT Press, 1982. 183 pp.

Applies economic principles to salient issues of seaports. Covers historical development of port organization. Technology; production measures; short- and long-term cost functions; pricing; investment. Empirical testing of theory against port data.

1348 Kenyon, James B. Elements in inter-port competition in the United States. *Economic Geography* 46(1), January 1970, 1–24.

1349 Leontief, Wassily. The growth of maritime traffic and the future of world ports. In: Wassily Leontief. *Input–output economics.* 2nd edn. New York and Oxford: Oxford University Press, 1986, pp. 379–91.

1350 Mankabady, S. The concept of safe port. *Journal of Maritime Law and Commerce* 5(4), July 1974, 633–43.

Developments in concept of safe port. Problems arising in connection with concept of safe port. Fundamental principles of meaning of safe port.

1351 *Manual on port management. Part 2: Port Planning.* UNCTAD, 1976. 268 pp.

A collection of lectures by well-known experts in their fields. Planning and management. Technological changes. Requirements. Capacity.

1352 Marti, B. E. Shift-share analysis and port geography: A New England example. *Maritime Policy and Management* 9(4), 1982, 241–50.

Demonstrates validity of relative shift technique in aiding port geographer/planners in decision-making. Looks at New England's port facilities and applies the technique to Boston's port.

1353 Muller, E. J. Smaller ports gain amid load centers. *Distribution* 89(7), July 1990, 96–102.

The concentration of US cargo flows at megaports has not materialized. Small ports have developed value added services to attract shipping lines.

1354 Murphy, P. R., Daley, J. M. and **Dalenburg, D. R.** Port selection criteria: An application of a transportation research framework. *Logistics and Transportation Review* 28(3), September 1992, 237–55.

1355 Nettle, Stanley. *Port operations and shipping: A guide to ports and related aspects of the shipping industry.* London: Lloyd's of London Press, 1988. 152 pp.

Explains the basic principles of both the ports and the shipping industry. Emphasizes the interrelationship of the two industries and their importance to national economies, as well as their vulnerability to changes in world economic and political patterns.

1356 **Nonneman, W.** Benefits of dredging through reduced tidal waiting. *Maritime Policy and Management* 8(1), 1981, 17–20.

Using simple formulae to estimate benefits of dredging programmes to reduce tidal waiting. Important time savings can be made with larger, time-expensive vessels.

1357 **Oram, R. B.** and **Baker, C. C. R.** *The efficient port.* Oxford (*et al.*): Pergamon Press, 1971. 215 pp.

Sets out conditions in major ports at time of writing. Development of port and shipping industry. Suggestions for improvement of cargo handling. International consideration with some focus on British ports. Attempt to predict size and shape of future ports.

1357a **Pond, P. J.** Multi-user terminal: Manual documentation and control. In: National Ports Council. *Containers: Their handling and transport: A survey of current practice.* London: National Ports Council, 1978, pp. 124–36.

Examines the documentation and control systems used by port authorities in the wake of the development of purpose-built container berths in the UK in the 1960s. Reasons for implementing a flexible manual system. Manual documentation and control procedures. Import and export container procedure. Developments.

1358 **Quinn, Alonzo De F.** *Design and construction of ports and marine structures.* 2nd rev. edn. US: McGraw-Hill, 1972. 611 pp.

New edition incorporating developments in respect of ports and marine structures. Growth and regulation of ports. Wind, tides and waves. Harbour planning and construction. Breakwaters. Wharves, piers, bulkheads, dolphins and moorings. Port buildings. General cargo-handling equipment. Bulk cargo, container shipping terminals. Navigation aids. Marinas.

1359 **Reid, A. B., Dally, H. K.** and **Wellard, D. J.** Costs and charges. In: National Ports Council. *Containers: Their handling and transport: A survey of current practice.* London: National Ports Council, 1978, pp. 218–32.

Examines how ports price their services. Objectives and levels of container charges. Development and operating costs. Relationship between unit handling cost and annual throughput. Relationship between container rental and dwell time. Further considerations. Method of charging.

1360 **Schulze, G.** Investigation and review of periodical fluctuations in port turnovers. *Economic Computation and Economic Cybernetics Studies and Research* 12(3), 1978, 59–67.

1361 Stern, E. and **Hayuth, Y.** Developmental effects of geopolitically located ports. In: B. S. Hoyle and D. Hilling (eds) *Seaport systems and spatial change: Technology, industry and development strategies.* Chichester and New York (*et al.*): John Wiley & Sons, 1984, pp. 239–55.

Examines the effect of isolated, geopolitically located ports on the local and national development process. A schematic four-stage model is introduced to present the spatial developmental effects of such ports. Model tested, and trade and economic data.

1362 Taylor, L. G. *Ships and their ports: A basic study of their relationships.* Glasgow: Brown, Son and Ferguson, 1991.

A collection of guidelines, the interpretation of which should indicate the extent to which ships and ports liaise and how they endeavour to fit into the impact of the patterns of trade and the business of shipping.

1363 United Nations Conference on Trade and Development Secretariat. *Port development: A handbook for planners in developing countries.* 2nd rev. edn. New York: United Nations, 1985. 227 pp.

Provides guidance in formulation of national port development policy and preparing realistic programmes for extension and improvement of individual ports. General principles of port planning. Procedures to be applied for establishing programme of work, traffic forecasting, productivity. Methods of planning various port facilities.

1364 No entry.

1365 United Nations Conference on Trade and Development. *Port performance indicators.* New York: United Nations, 1975.

1365a United Nations Conference on Trade and Development Secretariat. *Port pricing.* New York: United Nations, 1975. 77 pp.

Calculation of port charges. Questions the necessity for charges to be directly related to the flow of disbursed costs as usually recorded by accountants. Argues that costs, utilization of installations and what traffic can bear are elements to be considered when building pricing system.

1366 United Nations Economic and Social Commission for Asia and the Pacific. *Port development and feeder services.* Bangkok: United Nations ESCAP, 1979.

1367 Verhoeff, J. M. Seaport competition: Some fundamental and political aspects. *Maritime Policy and Management* 8(1), 1981, 49–60.

Attempts to identify and classify complex nature of seaport competition. Considers role and policies of public authorities in port administration and management in Western European countries.

1368 **Viscencio-Brambila, Hector** and **Fuller, Stephen.** Estimated effect of deepened U.S. Gulf ports on export-grain flow patterns and logistics costs. *Logistics and Transportation Review* 23(2), June 1987, 139–53.

A spatial model is used to identify alterations in export grain flow patterns, by harbour deepening in various US Gulf ports.

1369 **Walters, A. S.** Marginal cost pricing in ports. *Logistics and Transportation Review* 11(4), 1975.

1370 **Waters, R. C.** Port economic impact studies: Practice and assessment. *Transportation Journal* 16, 1977, 14–18.

1371 **Wilder, Ronald P.** and **Pender, David R.** Economic behaviour of public ports in the United States. *Journal of Transport Economics and Policy* XIII(2), May 1979, 169–81.

Focuses on economic behaviour of public ports in the context of inter-port competition. Analyses the interactions of pricing behaviour of ports and carriers, port subsidies, cartelization.

Optimal cargo handling

1372 **Arlt, W. H.** Information requirements in strategic planning in the ports industry: Specification and management of a data base. *Maritime Policy and Management* 14(1), 1987, 49–61.

Describes way of analysing supply/demand balance faced by firms operating in port environment. Discusses what data is needed, how to use it and how to apply the results for planning.

1373 **Chappel, David.** Provision of optimal cargo handling facilities at a berth. *Maritime Policy and Management* 17(2), April–June 1990, 99–106.

Paper on optimal cargo-handling rates. Considers (i) under what circumstances optimal rate actually provided; (ii) how optimal rate affected as regards pre-booked time slots at berth, containerized cargoes, long queues.

1374 Clapp, I. C. The location, design, operation and economics of container terminals. Thesis submitted for PhD, University of Salford, 1984.

1375 Coopers and Lybrand Associates. *Container shipping and ports study for developing member countries.* Manila: Asia Development Bank Report, 1988.

1376 Cousen, B. and **Major, F.** A multi-user container terminal. In: National Ports Council. *Containers: Their handling and transport. A survey of current practice.* London: National Ports Council, 1978, pp. 233–61.

Defines the difference between a single or limited user and a common or multi-user terminal. Operating policy of a common-user terminal. The effects of external and internal delays on terminal operations. Computer control. Landslide handling systems. Straddle carrier operations. Groupage. Ro-ro and car handling operations. Security.

1377 Dally, H. K. and others. Terminal capacity. In: National Ports Council. *Containers: Their handling and transport. A survey of current practice.* London: National Ports Council, 1978, pp. 164–88.

Considers the relationships between dwell time and throughput capacity in respect of the length of time that containers remain on terminal. Influence of the shipping services. Data on container movements.

1378 Dawes, B. Limited user container terminals. In: National Ports Council. *Containers: Their handling and transport. A survey of current practice.* London: National Ports Council, 1978, pp. 263–82.

Considers the main advantage a limited-user terminal offers to a shipping line when compared with a common-user terminal. The original limited-user terminal concept. Operational procedures. Terminal interfaces. Centralized control system. System approach/computer-aided control. Operations department.

1379 De Weille, Jan and **Ray, Anandarup.** The optimum port capacity. *Journal of Transport Economics and Policy* VIII(3), September 1974, 244–59.

Deals with the question of what capacity a port should have. Possible use of operations research techniques for port planning within framework of conventional economic benefit–cost analysis.

1380 Dowd, T. J. and **Leschine, T. M.** Container terminal productivity:

A perspective. *Maritime Policy and Management* 17(2), April–June 1990, 107–12.

Problems and prospects of using quantitative measurements for container terminal productivity to estimate or compare productivity of terminals at ports.

1381 **El Noshokaty, S. M.** Optimum ship selection of ports, routes, and cargoes. Thesis submitted for PhD, UWIST, Cardiff, 1988.

The problem of ship selection of ports, routes and cargoes is solved by three mixed 0–1 fractional models. The first model is deterministic, the second is two-stage stochastic, and the third is chance-constrained stochastic. A computer programme is developed to solve the mixed 0–1 problem.

1382 **Gilman, S.** An investigation into some effects of new handling technologies on the structure of marine transport systems. *Maritime Studies and Management* 2(3), January 1975, 165–80.

Focuses on the capability of new handling technologies of providing for rapid ship turn-round with respect to deep-sea routes.

1383 **Goss, R. O.** Competition in cargo-handling: A response. *Maritime Policy and Management* 10(2), April–June 1983, 121–2.

A response to comments by J. A. Zerby, Vol. 9, No. 4 of same journal, pp. 295–301, made in respect of Goss's article 'Competition in cargo-handling: Some experience from Australia', Vol. 9, No. 1 in same journal. Agrees with views of Zerby, plus comments.

1384 **Griffiths, J. D.** Optimal handling capacity at a berth. *Maritime Studies and Management* 3(3), January 1976, 163–7.

Paper presents a model to show how decision of finding optimal cargo-handling capacity for a single berth is to be made. Two examples give approximate lower and upper bounds for the optimal capacity for many cases.

1385 **Griffiths, J. D.** Unloading facilities at an iron-ore terminal. *Maritime Studies and Management* 3(4), April 1976, 237–43.

Paper examines problem of installation of unloading facilities at a purpose-built iron ore terminal at a steelworks. Illustrates modifications possibly necessary to analysis given in 'Optimal handling capacity at a berth' by J. D. Griffiths in Vol. 3, No. 3 of same journal, pp. 163–7.

1386 **Grzelakowski, Andrzej S.** Natural equilibrium as optimum condition of port-services market. *International Journal of Transport Economics* 12(3), October 1985, 261–72.

1387 Hayuth, Y. Rationalization and deconcentration in the U.S. container port system. *Professional Geographer* 40, 1988, 279–87.

1388 Hayuth, Y. and **Wydra, H. N.** Determination of coal-discharging port-dimensions from a shipping point of view. *Maritime Policy and Management* 10(1), 1983, 41–51.

Considers optimal size of coal-discharging terminals, given huge expansion of world coal trade after oil crisis of 1973. Concludes that ports should cater for vessels of around 150,000 d.w.t.

1389 Ircha, M. C. Canadian container ports: Shipping trends and policy issues. *Canadian Public Administration* 31(1), 1988, 371–89.

1390 Jones, P. N. and **North, J.** Unit loads through Britain's ports: A further revolution? *Geography* 67, January 1982, 29–40.

1391 Macdiarmid, K. J. and **Chambers, J. D. C.** Inland ports: The U.K. containerbase system. In: National Ports Council. *Containers: Their handling and transport. A survey of current practice.* London: National Ports Council, 1978, pp. 295–310.

Considers the change in method of general cargo conveyance which follows the introduction of container shipping into a trade. The Full Container Load business as part of UK overseas trade. Consolidation and discharge services for unit loads. UK internal clearance depots.

1392 Maguire, F. J. Operational behaviour and performance. In: National Ports Council. *Containers: Their handling and transport. A survey of current practice.* London: National Ports Council, 1978, pp. 137–63.

Examines the operational performance of a container terminal. Continuous routine monitoring of the operation. A study of the operation in depth with a view to design or development.

1393 Marshall, J. K. Basic operational design of sea container terminals. In: National Ports Council. *Containers: Their handling and transport. A survey of current practice.* London: National Ports Council, 1978, pp. 1–29.

Highlights some of the factors which cause design and/or operational problems in the setting up of sea container terminals. Siting of sea terminal. Throughput and space. The quay-face approach. Operational and layout planning. Ship operator information requirements.

1394 Mitchell, C. Bradford. *Have served: The National Cargo Bureau's first quarter-century.* New York: National Cargo Bureau, Inc., 1977. 197 pp.

A review of the highlights of the century of cargo surveying and inspection to which National Cargo Bureau fell heir. Formation of the tripartite organization and aims for proper stowage and carriage of cargo. Achievements during first 25 years.

1395 National Ports Council. *Containers: Their handling and transport. A survey of current practice.* London: National Ports Council, 1978. 418 pp.

Brings together 21 contributions by people recognized as 'container experts', from their personal experience of the introduction and operation of container systems.

1396 *Port and cargo information study.* London: Simplification of International Trade Procedures, 1983. 99 pp.

A report about the feasibility of applying information technology to the control of cargo moving through British ports and inland clearance depots.

1397 Porter, Robert W. and **Swatton, Richard J.** *Intermediate bulk containers: An ICHCA survey.* London: International Cargo Handling Co-ordination Association, 1979. 52 pp.

A survey which is the product of sustained interest by ICHCA into matters of cargo handling, and intermediate bulk containers. The report concentrates on flexible bag types, which are most commonly used.

1398 Talley, W. K. Optimum throughput and performance evaluation of marine terminals. *Maritime Policy and Management* 15(4), 1988, 327–31.

In deregulated (competitive) environments, when natural hinterlands of ports are diffused, the economic approach for deciding optimum throughputs should be considered. Can they handle given amounts of cargo and can they compete?

1399 Taylor, Leslie Granville. *Cargo work: The care, handling and carriage of cargoes.* 11th edn. Glasgow: Brown, Son & Ferguson, 1985. 490 pp.

A treatise on the practicalities of ship cargo handling and the associated ports functions. Aimed at officers serving within a Merchant Navy parameter but also of interest to anyone in broader maritime transportation field. Current and envisaged developments which trade patterns impose upon ship husbandry.

1400 United Nations Conference on Trade and Development Secretariat. *Technological change in shipping and its effects on ports: The impact of unitization on port operations.* New York: United Nations, 1976. 65 pp.

Report which examines the spread of unitization. Growth of unitized ship types – present situation and future developments. New major developments in ports to cope with unitization. Practical problems experienced on unitized terminals in developed countries. The problems of transition for developing countries – multi-purpose terminal. Planning for efficiency.

Terminals and productivity

1401 Bendall, H. B. and **Stent, A. F.** On measuring cargo handling productivity. *Maritime Policy and Management* 14(4), 1987, 337–43.

With unitization of cargo handling in mind, paper develops technique for measuring productivity to aid policy analysts in assessing merits of different ship types.

1402 Bendall, Helen B. and **Stent, A. F.** Cargo handling productivity in liner trades. Part 1. The model. *Maritime Policy and Management* 15(1), January–March 1988, 19–33.

Investigates productivity differences in cargo-related operations between alternative ship types in liner trades. Study relates to one section of total production process in order to isolate and assess cargo-handling productivity.

1403 Bendall, Helen B. and **Stent, A. F.** Cargo handling productivity. Part 2. The findings. *Maritime Policy and Management* 15(2), 1988, 97–106.

Results produced by model developed in Part 1, designed to test a cargo-handling productivity measurement method with cross-sectional empirical data.

1404 Bennathan, Esra and **Walters, A. A.** *Port pricing and investment policies for developing countries.* New York and London: Oxford University Press (for the World Bank), 1979. 230 pp.

Analyses pricing policies for port services and use of prices to control distribution of benefits. Recommends pricing, investment policies designed to increase economic well-being of developing countries. Promotes concept of cost-based port tariffs.

1405 Dally, H. K. (ed.) *Container handling and transport: A manual* *of current practice.* 2nd edn. Worcester Park, Surrey: C. S. Publications, 1983. 238 pp.

A manual on container terminal practice consisting of papers contributed by 30 recognized experts in their respective fields. Divided into eight sections: Container trade growth; terminal design; terminal operation; terminal equipment; other operating factors; container service operating philosophy; the inland interface; developing countries.

1406 De Neufville, R. and **Tsunokawa, K.** Productivity and returns to scale of container ports. *Maritime Policy and Management* 8(2), 1981, 121–9.

Analyses performance of the five major container ports of East Coast of USA. Shows wide disparities in their productivity. Container ports give large returns to scale.

1407 Dowd, T. J. U.S. container terminal leasing and pricing. *Maritime Policy and Management* 11(4), 1984, 277–88.

Reviews methods used to lease container terminals and terminal facilities, leasing methodologies, pricing approaches used by US public port authorities, plus economic effects on ports.

1408 General Council of British Shipping and others. *Port performance comparison study: General cargo in conventional vessels.* London, 1978.

1409 International Chamber of Shipping. *Guide to contingency planning for the gas carrier alongside and within port limits.* London: Witherby & Co., 1987. 27 pp.

Addresses circumstances of accidents which might befall a gas carrier while laden within port limits or during handling of cargo while berthed alongside. Provides a reference for port authorities, operating managements of liquefied gas terminals and carriers.

1410 Laing, Edward Temperley. *Containers, pallets or lash: The economics of general cargo shipping.* London: Economist Intelligence Unit, 1973. 41 pp.

Types of cargo-handling methods. Examines whether container services more economic than conventional services. What are the comparative costs of competitive vessel types? How do handling costs differ between systems? Cost minimization in general cargo shipping. Non-cost considerations.

1411 Laing, E. T. and **Hecker, G.** A note on optimal waiting-to-service-time ratios in large ports. *Maritime Policy and Management* 16(1), 1989, 83–6.

A W/S ratio of 0.3 may be reasonable for small ports, but it should be much lower for large ports.

1412 **Marcus, H. S.** Improving port productivity: Research trends and issues. *Annual Proceedings of the Transportation Research Forum* 25(1), 1984, 347–51.

1413 Ports: The struggle to maintain traffic levels hits new investment. In: *EEC shipping: A guide to the sea transport industries of the European Economic Community.* Colchester: Seatrade Publications, pp. 55–61.

Problems in maintaining traffic levels in major EEC ports in light of years of economic gloom. Question arises of investing in new facilities and increased financial risk of such investment.

1414 **Suykens, F.** A few observations on productivity in seaports. *Maritime Policy and Management* 10(1), 1983, 17–40.

Discusses cargo-handling productivity in European seaports and factors influencing it. Looks at Port of Antwerp developments in mechanization, labour organization, port layout, etc.

1415 **Thomas, B. J.** Port productivity and costs, with particular reference to the handling of iron and steel products. Thesis submitted for PhD, UWIST, Cardiff, 1981.

1416 **Wright, A. J.** Potential port responses to new information systems on ships. *Maritime Policy and Management* 15(4), 1988, 291–7.

Information technology now makes it possible for ships to navigate extremely accurately, so ports have to decide whether and how they can meet shipping lines' requirements.

1417 **Yamajo, T.** A study of development cycle of container ports. *Journal of Municipal Problems* 35, 1983, 85–98.

Port purpose and congestion

1418 **Abrahamsson, B. J.** and **Vardi, Michael.** *Economic considerations with regard to port congestion surcharges.* Haifa: Israel Shipping Research Institute, 1973. 33 pp.

Provides guidelines for determination of: What conditions constitute port congestion; different degrees of congestion; conditions when surcharges are justified; levels of the surcharges; cost of congestion to shippers and carriers; suitable mechanism for the imposition, collection and cancellation of surcharges.

1419 Arnold, G. Systems analysis – Part 2: Enhancement of the physical/computer model. In: National Ports Council. *Containers: Their handling and transport. A survey of current practice.* London: National Ports Council, 1978, pp. 405–16.

Pursues the approach in previous chapter in same book, 'Systems Analysis – Part 1: Study of deep-sea multi-user berths', pp. 377–404 by H. K. Dally. Considers the simulation technique used to track and time container operations, handling and berth efficiency.

1420 Black, John N. Maplin: The case for a seaport. *Geographical Journal* 140, October 1974, 364–72.

1421 Buckley, D. J. and **Gooneratne, S. G.** Optimal scheduling of transport improvements to cater for traffic congestion. *Journal of Transport Economics and Policy* VIII(2), May 1974, 122–35.

Investigates some aspects of effect which construction year and characteristics of traffic growth have on certain types of investment. Suggests for certain purposes details of traffic growth not as important as might be suggested. Models applied, *inter alia*, to duplication of shipping berth.

1422 Chang, W. H. and others. Harbour expansion and harbour congestion charges. An application of nonlinear programming. *Journal of Transport Economics and Policy* IX(3), September 1975, 209–29.

Application of nonlinear programming techniques to plan future expansion in congested harbours of Taiwan. Aims are to secure optimal flows of goods through harbours and to consider interest of various inland regions.

1423 Couper, A. D. Environmental port management. *Maritime Policy and Management* 19(2), 1992, 165–70.

How can port management objectives relate to aims of saving planet through sustainable development and more integrated environmental planning and management?

1424 Dally, H. K. Systems analysis – Part 1: Study of deep-sea multi-user berths. In: National Ports Council. *Containers: Their handling and transport. A survey of current practice.* London: National Ports Council, 1978, pp. 377–404.

A study of the container berth system and its objective, which is to meet the demands of ships and land vehicles with minimum consumption of the berth's resources. Elucidates the possible effects of changes in the system.

1425 **Frankel, E. G.** Strategic planning applied to shipping and ports. *Maritime Policy and Management* 16(2), 1989, 123–32.

Changing economic, trade, technological and environmental conditions affect shipping and ports. Essential for them to update strategic decisions and objectives frequently.

1426 **Holt, J. A.** Some difficulties inherent in applying congestion theory to marine problems. *Maritime Studies and Management* 1(4), April 1974, 207–14.

Describes nature of congestion problem in a maritime context. Suggestions of solutions and areas which require more investigation.

1427 *Manual on port management. Part 3: Port operations.* Geneva: UNCTAD, 1976. 234 pp.

A collection of lectures by well-known experts in their fields. Management and operations of stevedoring operations. Palletization. Occupational problems. Berths.

1428 **Rol, Y.** and **Hayuth, Y.** Port performance comparison applying Data Envelopment Analysis (DEA). *Maritime Policy and Management* 20(2), April–June 1993, 153–62.

Management of port efficiency. Uses Data Envelopment Analysis to measure efficiency. The DEA approach demonstrated by a hypothetical numerical example where performances of twenty ports compared. Provides insight into port performance.

1429 **Scheaker, Eric** and **Brockel, Harry C.** (eds) *Port planning and development.* Cambridge (Maryland): Cornell Maritime Press, 1974. 327 pp.

1430 **Shim, Duke.** Port congestion at Kuching. *Transport Management* 11(3), March 1979.

1431 **Shneerson, Dan.** Short term planning for a ports system. *Maritime Policy and Management* 10(4), October–December 1983, 217–50.

Paper highlights short-term solutions that will increase port efficiency. Concerned with any system of ports that serve a country. Demonstrated by developing a model that minimizes total system costs of ports and inland networks. Nigeria is model's case study.

1432 **Stonham, P. E.** Cost–benefit of improving port entrance channels. *Maritime Studies and Management* 3(3), January 1976, 169–74.

Paper suggests advantages and ways of reducing port congestion. Focus is on reducing queue length by straightening, widening or deepening port channel. Provides a cost–benefit study of improving port approaches.

1433 **United Nations Conference on Trade and Development.** *Port congestion surcharges.* Geneva: United Nations, 1975.

1434 **United Nations Conference on Trade and Development.** *Port problems: The extent and causes of port cost increases.* Geneva: UNCTAD, 1977.

1435 **Williams, J. T.** The changing design of ports. *Royal Society of Arts Journal* 120, July 1972, 490–507.

The paper given and discussion following at the Thomas Gray Memorial Lecture, March 1972 to the Society. Considers designs in ports. Reviews the role that ports play in the changing pattern of trade.

Port and regional development

1436 **Bird, J. H.** Seaports are gateways. (Editorial.) *Maritime Studies and Management* 2(4), April 1975, 193–4.

Considers ports as historical gateways, as operational gateways and as regional or international gateways.

1437 **Blauwens, G.** and **Voorde, E. Van De.** The impact of port choice on inland transportation. *Maritime Policy and Management* 15(2), 1988, 127–40.

On impact of port choice on inland transportation in context of Belgian Government's intention to shift funding away from Antwerp towards Ghent and Zeebrugge. Disaggregated demand model constructed.

1438 **Bobrovitch, Dov.** Decentralised planning and competition in a national multi-port system. *Journal of Transport Economics and Policy* XVI(1), January 1982. 31–42.

Compares efficiency of centralized port planning with that of independent competitive ports by using simple mathematical model based on approach of Hotelling (1929) and considering ports as congestion-prone.

1439 Butler, Eamonn and **Pirie, Madsen** (eds) *Freeports.* London and Virginia: Adam Smith Institute, 1983. 79 pp.

Collection of papers based on discussions at seminar in London organized by the Adam Smith Institute. Papers look at the nature of freeports around the world and attempt to discover which ones work and why. Proposals ventured in Britain examined. Business advantages of freeports outlined.

1440 Carmichael, Arthur J. The World Bank and port development. *Journal of Maritime Law and Commerce* 2(1), October 1970, 115–30.

Criteria and policy for port development loans by World Bank Group – consisting of International Finance Corporation and International Development Association.

1441 Choo, Chung Kek. Port operations in less developed countries. In: Hans Ludwig Beth (ed.) *Liner shipping in the eighties: Report on the International Symposium, Bremen, October 24th–26th 1979.* Bremen: Institute of Shipping Economics, 1980, pp. 353–62.

Examines influences on port development: (i) the social, economic and political environment of the port's hinterland; (ii) the major portion of its users, the shipping lines, owned mainly by individuals/corporations in mature, developed economies.

1442 Clark, M. Fallow land in old docks: Why such a slow take-up of Britain's waterside redevelopment opportunities? *Maritime Policy and Management* 12(2), 1985, 157–67.

Process of renewal and reuse is slow and restricted. Successful schemes usually flexible, small-scale and piecemeal, and depend on good relations with wider social and economic environment.

1443 Dinwoodie, John. *Regional development impacts of transport schemes: A review of evaluation methods.* Plymouth Polytechnic, Department of Shipping and Transport, 1988. 50 pp.

1444 Dunbar-Nobes, A. C. Port problems and small-island economies: The case of the south west Pacific. In: B. S. Hoyle and D. Hilling (eds) *Seaport systems and spatial change: Technology, industry and development strategies.* Chichester and New York (*et al.*): John Wiley & Sons, 1984, pp. 81–97.

Examines some of the problems confronting island governments in their bid to maintain viable ocean ports. Pacific ports – central places in a marginal world. Ports, containerization and problems of small scale.

1445 **Frankel, Ernst G.** *Port planning and development.* New York: Wiley & Sons, 1987. 795 pp.

Provides a treatise on aspects of port planning, from the establishment of port requirements to port facility and operational planning. Also, port project financing and implementation, port operations and management. Relationships among the elements of port planning.

1446 **Goodman, Allen C.** Port planning and financing for bulk cargo ships: Theory and a North American example. *Journal of Transport Economics and Policy* XVIII(3), September 1984, 237–52.

Applies simple economic model to treatment of port-specific capital improvements relative to bulk cargo. Ship waiting time as real port cost. Implications of user financing of port improvements. Strategic behaviour of competing ports.

1447 **Grobey, Harold A.** Facilities planning: Management of LNG imports. *Journal of Maritime Law and Commerce* 5(1), October 1973, 157–63.

Selection of sites for LNG marine terminals for purpose of importing liquefied natural gas to north-eastern United States. Criteria for site selection. Problems encountered. Future development of LNG industry.

1448 **Hallet, G.** and **Randall, P.** *Maritime industry and port development in South Wales.* Cardiff: University College, 1970.

1449 **Hilling, D.** and **Hoyle, B. S.** Spatial approaches to port development. In: B. S. Hoyle and D. Hilling (eds) *Seaport systems and spatial change: Technology, industry and development strategies.* Chichester and New York (*et al.*): John Wiley & Sons, 1984, pp. 1–19.

Approaches to port geography. Seaports and development. Change in port systems. Port–hinterland links. Changing technology of cargo handling. Port planning.

1450 **Hoyle, B. S.** *Seaports and development.* Reading: Gordon and Breach, 1983. 272 pp.

1451 **Hoyle, B. S.** and **Pinder, A. D.** (eds) *Cityport industrialization and regional development: Spatial analysis and planning strategies.* Oxford: Pergamon Press, 1980.

1452 Hoyle, Brian. Waterfront redevelopment and the cityport economy. *Maritime Policy and Management* 15(3), July–September 1988, 255–6.

Discusses seminar held in Department of Geography, University of Southampton, 11–15 November 1987. Seminar discussed problems of port–city interface, redevelopment of urban waterfront zones and port–city economics.

1453 Leeper, John and **Tomassoni, Mark.** (eds) *Coal exports and port development.* 2nd edn. Centreville (Maryland): Cornell Maritime Press, 1982. 116 pp.

1454 Mather, Graham. The political background. In: Eamonn Butler and Madsen Pirie (eds) *Freeports.* London and Virginia: Adam Smith Institute, 1983, pp. 19–30.

Political history of freeports in Britain from 1979. Arguments raised for establishing freeports. The role of the Government Working Party which reported in preparation for the 1983 Budget announcement on freeports. The challenge.

1455 McConville, James. The southern challenge. (Editorial.) *Maritime Policy and Management* 15(1), January–March 1988, 1–3.

Developments in port industry. Suggests that opportunities for southern European ports will create challenge to supremacy of northern megaports.

1456 Meredith, B. Development problems of LDC ports. *Marine Policy* 5(4), 1981, 338–40.

1457 Nagorski, Bohdan. *Port problems in developing countries: Principles of port planning and organisation.* Tokyo: International Association of Ports and Harbors, 1972. 293 pp.

Provides a framework for decision-making in the establishment of efficient port facilities in developing areas. Analysis of economic, social and political factors affecting developmental policy.

1458 National Ports Council. *Financial objectives for the port industry.* London: National Ports Council, 1975. 16 pp.

Deals with issue of setting of realistic financial objectives for the nation's ports.

1459 National Ports Council. *Report to the Secretary of State for Transport on the ports industry.* London: National Ports Council, 1978. 74 pp.

Report in which National Ports Council were asked, whilst making a fundamental analysis of the present state of the ports industry, to consider how plans for its strategic future should be developed.

1460 **O'Brien, K. H.** and **Di Pilato, Vincent.** Setting up and operating freeports. In: Eamonn Butler and Madsen Pirie (eds) *Freeports.* London and Virginia: Adam Smith Institute, 1983, pp. 31–44.

History of American freeports. Lessons for the UK. Choosing the right customs concessions. Lessons in making a freeport work. Third generation freeports.

1461 **Pirie, Madsen.** Freeports: Their nature and success. In: Eamonn Butler and Madsen Pirie (eds) *Freeports.* London and Virginia: Adam Smith Institute, 1983, pp. 9–18.

What elements a freeport must have in order to work. What conditions are likely to make them succeed. The defining conditions. Advantages of freeports. Requirements from the locality. Shaping the freeports proposal.

1462 **Pollock, Eric E.** *Marketing, port development and the demand for port facilities.* London: British Transport Docks Board. 18 pp.

Paper which stresses relationship of port authority with development of a port's traffic and facility requirements.

1463 **Pollock, Eric E.** *Ports, port hinterlands, and regional development.* Paris: Organization for Economic Co-operation and Development, 1971.

1464 **Sinclair, Clive.** The attractions of freeports to businessmen. In: Eamonn Butler and Madsen Pirie (eds) *Freeports.* London and Virginia: Adam Smith Institute, 1983, pp. 57–60.

Proposes that right sort of freeport is of immense benefit to the country. Ensuring that a freeport isn't at a disadvantage with foreign freeports. Freeport is a focus – a focus for capital; a focus for buyers and for suppliers.

1465 **Smillie, Ian** and **Danter, John.** The potential sites: Air and sea ports. In: Eamonn Butler and Madsen Pirie (eds) *Freeports.* London and Virginia: Adam Smith Institute, 1983, pp. 61–73.

A discussion in respect of potential sites for free zones. Facilities needed – the environment; labour and shopping. Local effects – the need to adjust quickly. Goods handling and communications. Management aids.

1466 **Stanley, W. R.** Changing export patterns of Liberia's seaports. In

B. S. Hoyle and D. Hilling (eds) *Seaport systems and spatial change: Technology, industry and development strategies.* Chichester and New York (*et al.*): John Wiley & Sons, 1984, pp. 435–59.

Examines the relationship between modern seaports and hinterland economic and social development. Traces some of the more important aspects of its evolution from early 1920s and massive initial growth during Second World War.

1467–76 No entries.

1477 Tabernacle, J. B. European deep sea container terminals: Locational and operational perspectives. Thesis submitted for PhD, University of Liverpool, 1991.

This study examines the factors which contribute towards the survival of deep-sea container ports in an ever-changing environment. It argues a reduction in the number of ports of call, and an erosion of a port's natural hinterlands, has heightened the level of inter-port competition.

1478 Talley, W. K. The role of U.S. ocean ports in promoting an efficient ocean transportation system. *Maritime Policy and Management* 15(2), 1988, 147–55.

Since passing of Shipping Act of 1984, intending to promote efficiency, competition between US ocean ports has increased, shown by big reduction on port natural hinterlands.

1479 Tuppen, J. N. The port–industrial complex of FOS: A regional growth centre? In: B. S. Hoyle and D. Hilling (eds) *Seaport systems and spatial change: Technology, industry and development strategies.* Chichester and New York (*et al.*): John Wiley & Sons, 1984, pp. 303–25.

Examines the project of creation of port and industrial complex at FOS. Offers a general guide to the processes and problems inherent in development of such port–industrial areas in addition to an insight into the nature and impact of one specific undertaking.

1480 Ungar, A. Market areas for ocean shipping. *International Journal of Physical Distribution* 1973, 155–60.

This article argues that every port has an inland area tributary to it, a cargo space area on which it depends for its shipping markets. It suggests methods of determining the magnitude and location of the market requirements.

1481 Van Notten, Michael. The Belgian freeports proposal. In:

Eamonn Butler and Madsen Pirie (eds) *Freeports.* London and Virginia: Adam Smith Institute, 1983, pp. 53–6.

Using examples from abroad. The position in Belgium – features of Belgian free zones; easing the social security burden. The message for Britain.

1482 **Walters, Brian.** A case study: The Miami free zone. In: Eamonn Butler and Madsen Pirie (eds) *Freeports.* London and Virginia: Adam Smith Institute, 1983, pp. 45–52.

Physical location – of crucial importance; the host city. The work of the zone. Trade performance.

1483 **Wanhill, S. R. C.** A study in port planning: The example of Mina Zayed. *Maritime Studies and Management* 2(1), July 1974, 48–55.

Provides a model to solve problem of the correct number of berths that a port should provide.

1484 **Warf, B.** and **Cox, J.** The changing economic impacts of the port of New York. *Maritime Policy and Management* 16(1), 1989, 3–11.

Examines economic effects of maritime trade, especially changing cargo volumes and commodity mixes, through Port of New York on the New York metropolitan area.

1485 **Warf, B.** and **Kleyn, L.** Competitive status of US ports in the mid-1980's. *Maritime Policy and Management* 16(2), 1989, 157–72.

Paper measures shares of national waterborne imports and exports for eight largest US ports and how they changed between 1985 and 1987 – namely, decline of East Coast ports and rise of West Coast.

1486 **Jochum, Gilbert R.** and **Agarwal, Vinod B.** Economic impact of a port on a regional economy: Note. *Growth and change* 18(3), Summer 1987, 74–87.

Safety and pollution

1487 **Compton, M.** Terminal safety. In: National Ports Council. *Containers: Their handling and transport. A survey of current practice.* London: National Ports Council, 1978, pp. 189–208.

Changes brought about by the container revolution in respect of safety measures in the use of machinery. Considers safety measures which should be taken in the light of UK's Health and Safety Commission to guidelines on container berth safety and the International Labour Office's Code of Practice on Health and Safety in Dock Work.

1488 Kaijo Hoan Kokusai Kenkyukai. *Japan maritime safety laws and requisitions.* Tokyo: Kaijo Hoan Kyokai, 1980. 355 pp.

Examines laws and regulations relating to Japanese ports and harbours. Port and maritime safety organization. Marine traffic laws and regulations. Law relating to the prevention of marine pollution and maritime disaster. Customs, immigration and quarantine laws. Laws on the territorial sea and fisheries. Other laws.

1489 Marti, B. E. US port throughput of radioactive materials: A twenty-year study. *Maritime Policy and Management* 16(2), 1989, 147–56.

Half a million tons of radioactive materials, and waste, passed through US ports, 1965–84. Many ports used and are identified.

1490 Titz, M. A. Port state control versus marine environmental pollution. *Maritime Policy and Management* 16(3), 1989, 189–211.

Port state control found to be effective in fighting marine environmental pollution but with peripheral shortcomings that limit effectiveness.

1491 United Nations Conference on Trade and Development. *Port development: The role of UNCTAD.* Geneva: UNCTAD, 1990.

Finance and Investments

Finance

1492 Goss, Richard. Financial assistance for seaports. (Editorial.) *Maritime Policy and Management* 13(2), April–June 1986, 81–2.

Raises question of whether seaports should receive financial support at all.

1493 Grosdidier de Matons, Jean C. Economic and financial appraisal of port projects at the World Bank: A review of policy and practice. *Maritime Policy and Management* 13(4), October–December 1986, 259–75.

Paper examines two basic issues faced by the Bank when reviewing port projects for financing – those of economic and of financial appraisal.

1494 United Nations Conference on Trade and Development Secretariat. *Financial management of ports.* Geneva: United Nations, 1979. 103 pp.

A report on the accounting and financial functions in ports, written in non-technical language. Part One deals with general accounting matters. Part Two deals with management information services. Part Three contains two case studies: (i) preparing a balance sheet; (ii) Simportoe budgetary control system.

Subsidies

1495 Farrell, Sheila. The subsidization of seaports: An alternative approach. *Maritime Policy and Management* 13(2), April–June 1986, 177–84.

Examines seaport subsidies from theoretical point of view which considers why one might wish to have subsidies and type of subsidy which might result.

1496 Goss, R. O. Seaports should not be subsidized. *Maritime Policy and Management* 13(2), April–June 1986, 83–104.

Considers general case for subsidizing ports. Defines and gives examples of subsidy. Considers main arguments in favour of subsidies. Opposes such subsidies.

1497 Graillot, A. Public subsidy of seaports. *Maritime Policy and Management* 13(2), April–June 1986, 155–64.

Examines what lies behind the notion of government subsidy for ports. Draws conclusions from 20 years of French laws which have reformed status of French ports.

1498 Klaassen, Leo N. Subsidies to seaports. *Maritime Policy and Management* 13(2), April–June 1986, 169–76.

Examines port as part of infrastructure necessary to convey goods from producer to consumer. Essential for Europe to optimize trade and transport activities. Related arguments for port subsidies analysed.

1499 Robinson, Pamela. *A review of capital and operating costs for break-bulk and container berths.* Liverpool: Marine Transport Centre, University of Liverpool, 1977. 33 pp.

A review prepared as part of a wider comparison of general cargo systems. Examines the likely level of port sector costs for different systems of transporting general cargo and for different levels of usage within systems.

1500 Suykens, F. Ports should be efficient (even when this means that some of them are subsidized). *Maritime Policy and Management* 13(2), April–June 1986, 105–26.

Examines coexistence of management and financing of port infrastructure in relation to general system of government in which ports have to operate. Considers necessity of government regulations and subsidies.

1501 Tubielewicz, Andrzej. The economic principles of Polish seaports. *Maritime Policy and Management* 13(2), April–June 1986, 165–8.

Analysis of Polish ports' financial conditions in field of investment activity. Examines why Polish ports need special financial support from the state.

1502 Turnbull, P. and **Weston, S.** The British port transport industry. Part 1: Operational structure, investment and competition. *Maritime Policy and Management* 20(2), April–June 1993, 109–22.

Effect of abolition of National Dock Labour Scheme, July 1989. Consequent restructuring of port operations and new investment plans. Impact on industry. Customs, demands for lower charges.

1503 Vigarie, A. State aid for seaports: A Cartesian position in the French case. *Maritime Policy and Management* 13(2), April–June 1986, 127–37.

Examines extent to which it is possible to harmonize and achieve smoothing of fundamental differences in respect of French ports. Difficulties in treating French ports equally.

Investments

1504 Correa Chairez, R. The timing of civil engineering investments in ports. Thesis submitted for MSc, University of Manchester, 1979.

1505 Couper, Alastair D. *New cargo handling techniques: Implications for port employment and skills.* Geneva: I.L.O., 1986. 172 pp.

Examines impact on port employment of developments in cargo handling, including the growth of unitization and containerization and the introduction of computer-automated systems, information technology and robotics. Effect on social relationships, working conditions in ports. Progress of labour–management negotiations. Port worker training. International agencies.

1506 Fastenau, R. Investment in port facilities. In: Hans Ludwig Beth (ed.) *Liner shipping in the eighties: Report on the International Symposium, Bremen, October 24th–26th 1979.* Bremen: Institute of Shipping Economics, 1980, pp. 340–50.

Determination of port investments. Requirements of shipping lines. Immobility of port investments. Political influences. Education and training. Guideline for planning investment in port facilities – long-term master plan; medium-term project planning.

1507 Frankel, E. G. Port or shipping project appraisal under risk. *Maritime Policy and Management* 16(3), 1989, 213–21.

Discusses basic concepts of project appraisal under risk and effects of uncertainties on measures of project performance as they relate to ports and shipping.

1508 Garnett, H. C. Competition between ports and investment planning. *Scottish Journal of Political Economy* 17(3), November 1970, 411–24.

The basic theme of this paper is that the state of the technology for the transport of bulk and general cargoes is such that competition between ports for the business of British exporters and importers will result in higher than necessary overseas total distribution costs.

1509 Hawkins, J. E. Port investment appraisals for the 1990's. *Maritime Policy and Management* 18(3), 1991, 225–32.

Offers practical advice on port investment appraisal, covering three areas: Scope of appraisal, appraisal methods, and place of appraisal process in a port's overall strategy.

1510 La Saponara, Francesco. Seaports and public intervention. *Maritime Policy and Management* 13(2), April–June 1986, 139–54.

Examines arguments for and against subsidizing ports from public funds. Considers economic convenience of such subsidy. From Western Europe viewpoint.

1511 Mills, G. Investment planning for British ports. *Journal of Transport Economics and Policy* V(2), May 1971, 119–52.

Recent history of port investment planning in Britain. Describes, evaluates. Focus on mechanisms of investment appraisal. From Rochdale Committee 1961 to 1969 with postscript summary to 1971. Economic analysis; institutional and political setting.

1512 Shneerson, Dan. Investment in port systems: A case study of the

Nigerian ports. *Journal of Transport Economics and Policy* XV(3), September 1981, 201–16.

Considers questions whether, when and where investment in port systems should be made. Derives necessary port charges consistent with optimal port investment. Dynamic programming applied to planning Nigerian ports.

Manpower and Management

General

1513 Ahmed, A. E.-S. An analytical study of the organisation and administration of ports (with special reference to Scotland). Thesis submitted for PhD, University of Strathclyde, 1976.

1514 Bradshaw, M. R. Presentation of information to port traffic personnel. Thesis submitted for M.Phil, CNAA, 1979.

1515 Branch, Alan E. *Elements of port operation and management.* London and New York: Chapman and Hall, 1986. 265 pp.

Design, planning, management of ports, port operations. Salient commercial, operating, political aspects. Emphasis on economics and port management.

1516 Chu, David K. Y. *Modelling the process of the development of seaports in Great Britain and the formulation of a comprehensive future port planning policy.* London: London School of Economics and Political Sciences, Graduate Geography School. 8 pp.

Presents research on the growth and decline of British seaports with respect to non-fuel cargo tonnages. Explores pattern of major British seaports as anticipated up to 1985.

1517 Cook, T. C. C. The occupational group of ships' pilots: A bio-psychological study of their stresses and strains. Thesis submitted for PhD, University of London, 1983.

1518 Goss, R. O. *A comparative study of seaport management and administration. Vol. 1.* London: Government Economic Service, Department of Industry, Department of Trade, Department of Prices and Consumer Protection, 1979. 56 pp.

An international comparative study of the administration and management of major seaports. Comparison of ways in which certain selected countries administer major seaports. The ways in which those seaports are managed. Examination of the best practices.

1519 Goss, R. O. *A comparative study of seaport management and administration. Vol. 2. Appendices.* London: Government Economic Service, Department of Industry, Department of Trade, Department of Prices and Consumer Protection, 1979. 396 pp.

Appendices of material in Volume 1. Descriptions of general arrangements of seaport administration, and administration/management practices of individual seaports.

1520 Grosdidier de Matons, J. Ports as a public service: The evolution in France and in the United States. *Journal of Maritime Law and Commerce* 2(3), April 1971, 571–616.

Comparison of French and US port administration. Focus on (i) public character of port administration; (ii) establishment of specialized agencies in charge thereof.

1521 *Guide to safety and health in dock work.* Geneva: I.L.O., 1976. 288 pp.

1522 Hill, Stephen. Supervisory roles and the man in the middle: Dock foremen. *British Journal of Sociology* 24, June 1973, 205–21.

1523 Morton, Donald S. The Longshoremen's and Harbor Workers' Compensation Act: Coverage under the 1972 amendments. *Journal of Maritime Law and Commerce* 9(1), October 1977, 33–59.

Position of injured longshoremen before 1972 amendments. Major changes introduced by the amendments – coverage section and definition of 'employer' amended; term 'employee' redefined. Impact and jurisprudence discussion.

1524 Palmer, N. E. The stevedore's dilemma: Exemption clauses and third parties – I. *Journal of Business Law,* 1974, 101–18.

Examines background and implications of the Privy Council decision in *New Zealand Shipping Co. Ltd.* v. *A. M. Satterthwaite & Co. Ltd.*; The Eurymedon, 1974. Assesses its status as an authority and analyses generally legal position of those seeking protection of contracts to which they are not parties.

1525 Palmer, N. E. The stevedore's dilemma: Exemption clauses and third parties – II. *Journal of Business Law,* 1974, 220–24.

Follows on from Part I, same volume this journal. Summary of remaining ways (other than bilateral and unilateral devices set out in Part I) in which stevedore might secure immunity from suit at hands of consignor or consignee.

1526 **Pocock, Pamela.** Projects in the ports. *Personnel Management* 3, March 1971, 35–.

1527 **Ritchie, J. D.** An evaluation of a group development exercise carried out in the marine division of a large port organisation. Thesis submitted for M.Phil, University of London, 1978.

1528 *Safety and health in dock work: Code of practice.* I.L.O., 1977. 221 pp.

1529 **Scobie, W. E.** A case study of the liability of a port employer to his employees. Thesis submitted for MSc, UWIST, Cardiff, 1974.

1530 **United Nations.** *Transport modes and technologies for development.* New York: United Nations, 1970. 147 pp.

Publication which is one of a series undertaken under a resolution adopted by Economic and Social Council. Explores question of whether to invest in capital intensive processes or labour. Takes view that decision should be based on criteria of economic or cost effectiveness.

1531 **Vanplantinga, John R.** *Shipping for profit: A guide to stevedoring management.* London: Fairplay Publications, 1984. 57 pp.

Stevedoring management. Control of costs with ship alongside. Money saving, seeing how to cut across inefficient working practices. Improve dunnaging; stowage; prevention of damage; reduction of delay.

1532 **Walden, L.** Safety in the docks. *Occupational Safety and Health,* May 1984, 14.

1533 **Wormmeester, Ir. G. J.** and **Bosselaar, B. C.** Optimisation of manpower and employee relations. In: National Ports Council. *Containers: Their handling and transport. A survey of current practice.* London: National Ports Council, 1978, pp. 111–23.

Considers the optimization of manpower and employee relations from Dutch point of view. Considers cargo-handling companies in respect of this subject.

United Kingdom

1534 **Coghlan, F. P.** An evaluation of the training needs of shop stewards in a large port authority. Thesis submitted for M.Phil, University of London, 1980.

1535 **Collis-Squires, N.** An investigation of the National Dock Labour Scheme as an example of statutory joint-control of an industry. Thesis submitted for LLM., University of Warwick, 1977.

1536 **Connolly, D. J.** Social repercussions of new cargo handling methods in the Port of London. *International Labour Review* 105(6), June 1972, 543–68.

1537 **Creigh, S. W.** Work stoppages in the Humberside sub-region, 1960–1971. Thesis submitted for MScEcon, University of Hull, 1975.

1538 **Davis, D.** *Clear the decks: Abolish the National Dock Labour Scheme. (Policy Study No. 101.)* London: Centre for Policy Studies, 1988.

1539 **Fadem, J. A.** Technological change, worker attitudes and the employment relationship in the British docks industry. Thesis submitted for M.Litt, University of Oxford, 1981.

1540 **Fisher, J. P.** The Transport and General Workers' Union and the Devlin modernisation programme in the port transport industry, with particular reference to Southampton. Thesis submitted for PhD, University of Southampton, 1976.

1541 **Great Britain. Department of Employment.** *Employment in the ports: The Dock Labour Scheme, Cm. 664.* London: HMSO, 1989.

1542 **Hanson, Charles.** Time to end the National Dock Labour Scheme. *Economic Affairs* 8, June/July 1988, 34–6.

1543 **Hartwell, N. M.** Vehicle design and maintenance: Some human factors studies of container handling in British ports. Thesis submitted for M.Phil, University of London, 1978.

1544 Hill, S. *The dockers: Class and tradition in London.* London: Heinemann Educational Books, 1976. 252 pp.

1545 Hudson, Robert. The effects of dock strikes on U.K. international trade. *Applied Economics* 13(1), March 1981, 67–77.

Paper determines whether trade is lost or delayed due to a strike, whether there are differences between commodities, and any difference on the effect on imports and exports. Three dock strikes between 1967 and 1972 are used for the research.

1546 Jackson, Michael P. *Policy making in trade unions: The T. & G.W.U.'s policy on decasualisation of dock labour.* Aldershot: Avebury, 1991. 138 pp.

1547 Jensen, V. H. *Decasualization and modernization of dockwork in London.* Cornell, Ithaca: Cornell University Press, 1971.

1548 McCormick, B. J. Methods of wage payment and wages structures. *British Journal of Industrial Relations* 15, July 1977, 246–64.

This is a general article with a section on dockers. It looks at the Dock Labour Scheme, and argues that the 1967 reforms implied necessary changes in the wage system of the engagement of permanent labour and therefore the loss of flexibility of manpower deployment that was usual in the old system.

1549 Mellish, M. *The docks after Devlin.* London: Heinemann Educational Books, 1972. 146 pp.

1550 Milner, Simon and **Richards, Edward.** Determinants of union recognition and employee involvement: Evidence from London Docklands. *British Journal of Industrial Relations* 29(3), September 1991, 377–90.

1551 Phillips, Gordon and **Whiteside, Noel.** *Casual labour: The unemployment question in the port transport industry, 1880–1970.* Oxford: Oxford University Press, 1985. 324 pp.

Examines changing nature of work, structure, organization of workforce in British port transport industry since the late nineteenth century. Prevalence of casual employment, its influence on social attitudes, economic status, working practices of dockers. Industrial relations. Government policy.

1552 Powrie, P. J. Training in the British port industry. *Economisch en Sociaal Tijdschrift* 26(1), February 1972, 43–54.

1553 Rayner, Joe. How the port employers off-loaded the past. [Effects of abolition of National Dock Labour Scheme at Hartlepool Dock.] *Personnel Management* 65, June 1990, 42–5.

1554 Southwood, Peter. *British dockers: A survey of human rights in the former scheme ports.* London: Transport and General Workers' Union, 1992. 92 pp.

A report from a survey into what has happened in the former National Dock Labour Scheme ports since the abolition of the Scheme in 1989. The 1989 dock strike. Widespread and serious violations of human rights. Conclusions, recommendations.

1555 Stuart, Keith. Progress in Britain's seaports: Aspects of change in labour relations. *Royal Society of Arts Journal* 133, January 1985, 115–25.

The paper given and following discussion at the Thomas Gray Memorial Lecture, April 1984, to the Society. Aims to demonstrate the progress made in British seaports. UK ports industry in the national economy. Changes in port technology. Evolution of port industry labour, National Dock Labour Scheme. New directions.

1556 Thomas, B. J. The changing role of the UK port industry and its impact on stevedoring costs. *Maritime Policy and Management* 8(3), 1981, 141–61.

On the impact of changing maritime technology on the structure and development of the UK port transport industry.

1557 Turnbull, P. J. The docks after deregulation. *Maritime Policy and Management* 18(1), 1991, 15–27.

National Dock Labour Scheme abolition, strike of 1989, deregulation of employment and increase in competition brought job losses, new working practices and reports of better productivity.

1558 Turnbull, Peter and others. *Dock strike: Conflict and restructuring in Britain's ports.* Aldershot: Avebury, 1992. 280 pp.

1559 Turnbull, Peter and **Weston, Syd.** *Continuity and change in the British port transport industry: A study of the ports since abolition of the National Dock Labour Scheme.* Cardiff: Cardiff Business School, 1991. 48 pp.

Study forming part of a wider project funded by the Economic and Social

Research Council. An in-depth and systematic study of Britain's docks since the repeal of the scheme.

1560 **Turnbull, Peter** and **Weston, Syd.** Co-operation or control? Capital restructuring and labour relations on the docks. *British Journal of Industrial Relations* 31(1), March 1993, 115–34.

The abolition of the UK's National Dock Labour Scheme in July 1989 brought about a massive reorganization in Britain's ports. This saw decentralization, detachment, and disintegration. Dock workers' co-operatives enhance the port authority in terms of market position and commercial viability.

1561 **Turnbull, Peter J.** and **Sapsford, David.** Why did Devlin fail? Casualism and conflict in the docks. *British Journal of Industrial Relations* 29, June 1991, 237–57.

Devlin Report anticipated industrial peace. What emerged was unprecedented militancy late 1960s–early 1970s. Argues decasualization not sufficient condition for harmony in period of massive technological change, creating substantial insecurity.

1562 **West, A. C.** Introducing participation: An example from the British ports industry. *Journal of Occupational Psychology* 53, June 1980, 97–105.

1563 **Wilson, D. F.** *Dockers: The impact of industrial change.* London: Fontana, 1972.

1564 **Wilson, Peter.** A docker's view. *Industrial participation* 558, Spring 1976, 29–32.

North America

1565 **Bagwell, David Ashley.** Continuing problems after the supposed demise of Ryan Indemnity in U.S. Admiralty law. *Lloyd's Maritime and Commercial Law Quarterly,* November 1982, Part 4, 556–65.

Complexities of pre-1972 longshoreman's personal injury US litigation. Case law. 1972 amended US Statute. Post-1972 legal development. Remaining problems involving indemnity portion of Ryan case.

1566 **George, James A.** The content of the negligent action by longshoremen against shipowners under the 1972 amendments to the

Longshoremen and Harbor Workers' Compensation Act. *Journal of Maritime Law and Commerce* 11(1), October 1979, 1–41.

Addresses one of the significant changes wrought by the 1972 amendments – the content of the negligence action against the ship under S.905(b) with which longshoremen are left after abolition of seaworthiness action. Cases decided under S.905(b).

1567 George, James A. The content of the negligence action by longshoremen against shipowners under the 1972 amendments to the Longshoremen and Harbor Workers' Compensation Act: A postscript. *Journal of Maritime Law and Commerce* 11(2), January 1980, 253–6.

Examines case of *Santos* v. *Scindia Steam Navigation Co.* (1979) where Court of Appeals refused to follow previously used standard for determining negligence under the Act. Impact of standard of negligence adopted in Santos.

1568 Gorman, Francis J. The Longshoremen's and Harbor Workers' Compensation Act: After the 1972 amendments. *Journal of Maritime Law and Commerce* 6(1), October 1974, 1–29.

Examines important features of 1972 amendments to the US Longshoremen's and Harbor Workers' Compensation Act 1927. Analyses alterations in jurisdiction and third-party rights and discusses some of the practical aspects of settling cases and recovering liens under the new law.

1569 Ircha, M. C. and **Garey, D.** Labour/management relations in North American ports. *Maritime Policy and Management* 19(2), 1992, 139–56.

Survey set out to find whether North American ports are prepared to satisfy motivational needs of their port authority employees by improving labour/management relations.

1570 Kahn, L. M. Unions and internal labor markets: The case of the San Francisco long-shoremen. *Labor History* 21, Summer 1980, 369–91.

1571 Khan, E. Industrial relations in the Canadian longshoring industry: A case study of collective bargaining in the ports of Montreal and Vancouver. Thesis submitted for PhD, University of London, 1972.

1572 McTamaney, Robert A. The Longshoremen's and Harbor

Workers' Act of 1927: The twilight zone after Nacirema. *Journal of Maritime Law and Commerce* 1(3), April 1970, 443–56.

Examines controversy surrounding applicability of the Longshoremen's and Harbor Workers' Compensation Act 1927. Examined chronologically from its foundation, 1917 case of *Southern Pacific Company* v. *Jensen.*

1573 **Mills, Herb** and **Wellman, David.** Contractually sanctioned job action and workers' control: The case of San Francisco longshoremen. *Labor History* 28, Spring 1987, 167–95.

1574 **Mott, William.** The U.S. longshoremen's boycott. In: Curtis Cate (ed.) *The challenge of Soviet shipping.* London: Aims of Industry (in association with the National Strategy Information Center, New York), pp. 28–9.

1980 suspension of ILA workers in handling all Russian ships and all Russian cargoes in ports from Maine to Texas and Puerto Rico where ILA workers are employed. Results of this order.

1575 **Pilcher, William W.** Some concomitants of maritime employment: The longshore case. *Maritime Studies and Management* 2(4), April 1975, 195–201.

Points out some of the consequences of seeking a livelihood in a maritime occupation. A particular study of the longshoremen of Portland, Oregon, and their families.

1576 **Robertson, David W.** Negligence actions by longshoremen against shipowners under the 1972 amendments to the Longshoremen's and Harbor Workers' Compensation Act. *Journal of Maritime Law and Commerce* 7(3), April 1976, 447–85.

Impact of 1972 amendments. Focuses on the content of a longshoreman's negligence action against shipowners and the availability of any kind of credit, set-off, contribution or indemnity on behalf of the shipowner on account of the concurrent negligence of Compensation Act employer.

1577 **Ross, P.** Waterfront response to technological change: A tale of two unions. *Labour Law Journal* 21, 1970, 397–405.

1578 **Ryan, P.** The cost of job training. *British Journal of Industrial Relations* 18, November 1980, 334–52.

This is a general article which includes a case study approach to the problems of the size and distribution of investment in job skills for training programmes in arc welding at major United States shipyards.

1579 **Sher, Stanley O.** The Federal Maritime Commission and labor related matters: The aftermath of the Volkswagenwerk decision. *Journal of Maritime Law and Commerce* 3(4), July 1972, 647–81.

Examination of impact of US Supreme Court decision in the Volkswagenwerk case 1968. Subsequent litigation spawned by that decision. (Agreement among members of a Pacific Coast multi-employer association providing for fund to mitigate impact of technological unemployment upon longshoremen, subject to prior approval of FMC.)

1580 **Shorter, James Russell.** In the wake of the 1972 amendments to the L.& H.W.C.A.: The vessel's rights against the stevedore. *Journal of Maritime Law and Commerce* 7(3), July 1976, 671–96.

Effect of 1972 amendments to Longshoremen's and Harbor Workers' Compensation Act 1927 on the remedies of a negligent vessel against a concurrently negligent stevedore, where a judgment has been recovered from the former by an injured longshoreman in latter's company.

1581 **Steeg, Robert M.** The exclusivity of federal longshoremen's compensation after the L.H.W.C.A. amendments of 1972. *Journal of Maritime Law and Commerce* 10(3), April 1979, 395–409.

Analyses whether, after the 1972 amendments, the Longshoremen's and Harbor Workers' Compensation Act stands as the exclusive law governing the entire field of workmen's compensation for longshoremen.

1582 **Thuong, L. T.** The impact of US regulatory practices on the productivity of ocean liner services: A comment. *Maritime Policy and Management* 10(1), 1983, 61–5.

Comments on article by G. K. Sletmo, with above title, in *Maritime Policy and Management* 9(3), 1982, pp. 189–205, and points out misconceptions regarding liner cost, pricing and competition.

Other countries

1583 **Connolly, John.** A European charter for port workers. *European Labour Forum* 6, Winter 1991–92, 31–5.

1584 EEC ports: 1992 and onwards. *The dock and harbour authority* 71(818), May 1990, 3–13.

1585 **Hemson, D.** Class consciousness and migrant workers: Dock workers of Durban. Thesis submitted for PhD, University of Warwick, 1979.

1586 Korsakov, E. Technological progress and its effect on Soviet dockworkers. *International Labour Review* 105(6), June 1972, 531–42.

1587 Morris, R. The Melbourne Dockers War. *Maritime Policy and Management* 10(1), 1983, 53–60.

Considers so-called Melbourne Dockers War of 1971–3; analyses origins of unprecedented level of violence in trade union elections and the implications for Australian politics and law.

1588 Tongzon, J. A model for forecasting future supply of shipping services at Australian ports. *Maritime Policy and Management* 18(1), 1991, 55–68.

Develops an integrated model for forecasting both number of ship visits and their characteristics in medium and long term, with reference to port of Melbourne.

1589 Yamamoto, Hiromasa. Decasualization of dock labour-experiences at the Port of Kobe. *Kobe Economic and Business Review* 17, 1970, 43–52.

Article which points out the merits and demerits of the Dock Labour Decasualization Act 1965 in relation to the enforcement of the Act at the Port of Kobe. Additionally, makes a proposal for remedies of the Act.

Management

1590 Beth, Hans Ludwig (ed.) *Port management textbook.* Bremen: Institute of Shipping Economics, 1978. 393 pp.

Collection of papers from several port management seminars organized in developing countries by the Institute of Shipping Economics, Bremen. Problems of port organization and planning, labour management, education of workforce.

1591 Beth, H. L. (ed.) *Port management textbook: Containerization.* Bremen: Institute of Shipping Economics and Logistics, 1985. 409 pp.

A collection of papers from researchers of the Institute and representatives from industry. General impact of containerization, terminal descriptions, operations issues, the respective documentation and information systems, the container and cargo aspects, aspects of hinterland connection. Planning procedures.

1592 De Monie, G. *The challenges facing port management in developing countries.* Geneva: UNCTAD, 1979.

1593 Great Britain. Department of Trade and Industry. *Management of port maintenance: A review of current problems and practices.* London: HMSO, 1989. 247 pp.

1594 Harte, M. Commercial management in the Royal dockyards. *Public Administration* 66(3), Autumn 1988, 319–.

1595 Hershman, Marc J. (ed.) *Urban ports and harbor management: Responding to change along U.S. waterfronts.* Bristol (Pennsylvania): Taylor & Francis, 1988. 366 pp.

1596 Klein, A. and **Verbeke, A.** The design of an optimal short term forecasting system for sea port management: An application to the port of Antwerp. *International Journal of Transport Economics* 14(1), 1987, 57–70.

1597 *Manual on port management. Part 4: Modern management techniques.* Geneva: UNCTAD, 1976. 212 pp.

A collection of lectures by well-known experts in their fields. Management statistics. Accounting; costing; port congestion; pricing, etc. UNCTAD's work.

1598 Stuchtey, Rolf W. (ed.) *Port management textbook. Vol. 1: General aspects of port management.* Bremen: Institute of Shipping Economics and Logistics, 1990. 274 pp.

Collection of papers examining aspects of port management. Volume 1 of three, which deals mainly with the general aspects of port management.

1599 Stuchtey, Rolf W. (ed.) *Port management textbook. Vol. 2: Intermodal transport systems.* Bremen: Institute of Shipping Economics and Logistics, 1991. 191 pp.

Collection of papers examining aspects of port management. Volume 2 of three, which deals with aspects of intermodal transport.

1600 Stuchtey, Rolf W. (ed.) *Port management textbook. Vol. 3: Port* *marketing.* Bremen: Institute of Shipping Economics and Logistics, 1991. 248 pp.

Collection of papers examining aspects of management. Volume 3 of three, which deals mainly with port communications, marketing, competition, port selection pricing, management accounting.

1601 **Suykens, Fernand.** Administration and management at the port of Antwerp. *Maritime Policy and Management* 12(3), July–September 1985, 181–94.

Paper describes administration and management of port of Antwerp from perspective of improving efficiency. Involves collaboration between the private and public sectors.

1602 **Thomas, B. J.** Port management development: A strategy for the provision of a training capability in developing countries. *Maritime Policy and Management* 8(3), 1981, 179–90.

Examines nature and scale of management-training requirements of ports of developing countries. Looks at UNCTAD's strategy in management training field for 1980s.

1603 **Thomas, B. J.** and **Roach, D. K.** *Management of port maintenance: A review of current problems and practices.* London: HMSO, 1989. 247 pp.

1604 **United Nations Conference on Trade and Development. Swedish International Development Authority.** *Manual on port management. Part 1: Transport economics, port administration.* Geneva: UNCTAD, 1976. 220 pp.

A collection of lectures by well-known experts in their fields. The role of maritime transport and ports in economic development.

9 Sea Canals and Waterways

Panama

1605 **Black, Jan Knippers.** Panama in crisis. *Contemporary Review* 253, August 1988, 72–4.

1606 **Duran, Oyden Ortega.** The Panama Canal and Panama's national interests. *Contemporary Review* 249, August 1986, 74–7.

1607 **Elton, C.** Studies of alternatives to the Panama Canal. *Maritime Policy and Management* 14(4), 1987, 289–99.

Panama Canal Alternatives Study Commission was set up in 1986 and from 1989 had five years to recommend best alternative to existing canal and develop plans for that alternative.

1608 **Fuller, S., Makus, L.** and **Gallimore, W.** Effects of increasing Panama Canal toll rates on U.S. grain exports. *Southern Journal of Agricultural Economics* 16(2), December 1984, 9–19.

1609 **Gibbs, Stephen R.** The economic value of the Panama Canal. *Water Resources Research* 14(2), April 1978, 185–9.

1610 **Habeeb, William Mark** and **Zartman, I. William.** *Panama Canal negotiations.* Baltimore (Maryland): Johns Hopkins Foreign Policy Institute, School of Advanced International Studies, 1988. 72 pp.

1611 **La Feber, Walter.** *The Panama Canal: The crisis in historical perspective.* New York: Oxford University Press, 1990. 270 pp.

1612 **Lael, R. L.** Dilemma over Panama: Negotiation of the Thomson–Urrutia Treaty. *Mid-America* 61, January 1979, 35–45.

1613 Morris, Charles. 75 years of security history: The Panama Canal. *Security Management* 33(9), September 1989, 68–74.

Security of the Canal is essential to promote its use.

1614 Padelford, Norman J. Ocean commerce and the Panama Canal. *Journal of Maritime Law and Commerce* 4(3), April 1973, 397–423.

Examines procedure relating to transiting the Canal and relating law. Compliance with relevant federal laws and regulations governing navigation, safety and behaviour in the Canal waters.

1615 Padelford, Norman J. and **Gibbs, Stephen R.** *Maritime commerce and the future of the Panama Canal.* Cambridge (Maryland): Cornell Maritime Press, 1975. 206 pp.

Examines the operation of the Panama Canal at time of writing. Trends in cargo movement and shipbuilding. Limits of capacity of the Canal. Fundamentals of toll policy. Policy issues – modernizing treaty relationships with Republic of Panama. Changes in maritime commerce. Scientific and technological innovations.

1616 Rubin, R. Panama Canal retrospective. *Contemporary Review* 234, February 1979, 80–84.

1617 Summ, G. H. and **Kelly, T.** (eds) *Good neighbours: America, Panama and the 1977 Canal Treaties.* Athens (Ohio): Ohio University, Centre for International Studies, 1988. 135 pp.

1618 Van Alstyne, R. W. Panama Canal: A classical case of an imperial hangover. *Journal of Contemporary History* 15, April 1980, 299–316.

1619 Zimbalist, Andrew and **Weeks, John.** *Panama at the crossroads: Economic development and political change in the twentieth century.* Berkeley: University of California Press, 1991. 219 pp.

Suez

1620 Abou-el-Hassan, A. A. A study of Suez Canal traffic and an analysis of the canal tariff level and structure. Thesis submitted for PhD, UWIST, Cardiff, 1976.

1621 Abu-El-Hassan, Aly. Suez Canal dues: Past and future. *Maritime Studies and Management* 2(1), July 1974, 32–7.

Suggests rational basis for pricing policy and implementation of re-opening Suez Canal.

1622 Drewry Shipping Consultants. *The Suez Canal and its impact on tanker trades and economics.* London: H. P. Drewry, 1978. 56 pp.

Report – part of series covering world's bulk shipping markets. The Canal up to 1967. The Canal since 1975 and plans. Economics of Suez Canal transits. Suez Canal development plans – transportation costings.

1623 Ezzat, H. The Suez Canal toll regime. Thesis submitted for M.Phil, University of Salford, 1989.

This seeks to establish that the Suez Canal Authority, following re-opening in 1975, could and should introduce a new tariff regime which, while increasing its revenue, would not discourage traffic.

1624 Hansen, Bent and **Tourk, Khairy.** The Suez Canal project to accommodate super-tankers. *Journal of Transport Economics and Policy* VIII(2), May 1974, 103–21.

Appraises Suez Canal project to accommodate supertankers, by Canal authorities. From economic viewpoint, not political aspects. From both national Egyptian and Canal users' viewpoint. Concludes that deepening, widening of Canal beneficial to both.

1625 Hassan, E. M. A study of shipping operations in the Suez Canal. Thesis submitted for PhD, UWIST, Cardiff, 1978.

1626 Mountjoy, Alan B. New life for the Suez Canal? *Contemporary Review* 219, November 1971, 263–8.

1627 Suez Canal Authority. *The Suez Canal: Present conditions and future prospects.* Ismailia, Egypt: Suez Canal Authority Press, 1975. 38 pp.

Examines features of and prospect for Suez Canal. Characteristics of present canal. System of traffic in the Suez Canal. Impact on world trade of interruption of navigation in the Suez Canal. Impact of resumption of navigation. The 1967 aggression and its effects on Suez Canal. Development since nationalization.

1628 Suez Canal Authority. *The Suez Canal development project.* Ismailia, Egypt: Suez Canal Authority Press, 1975. 8 pp.

Study of the project for the development of the Suez Canal.

1629 United Nations Conference on Trade and Development Secretariat. *Economic effects of the closure of the Suez Canal.* New York: United Nations, 1973. 45 pp.

Examines the importance of the Suez Canal to world shipping, as an international sea route. Effects of the Suez Canal closure on shipping services, on tanker shipping costs, on dry cargo shipping costs, on other shipping costs, on international trade, on ports. Other economic consequences.

1630 Wassermann, Ursula. Closure of the Suez Canal: Its effect on shipping and trade patterns. *Journal of World Trade Law* 7(6), November/December 1973, 701–10.

Tables showing volume of cargo moving through Suez Canal and world international trade prior to closure in 1967. Closure effects – on shipping services; tanker shipping costs; dry cargo shipping; on ports.

Barge systems

1631 Chang, Semoon. Estimating short-run costs of a barge company. *Maritime Policy and Management* 15(1), January–March 1988, 67–76.

Paper estimates firm-specific short-run costs of a barge company by extending a methodology developed by US Army Corps of Engineers. Questionnaire survey corroborates findings of paper.

1632 Diehl, Niel N. and **Philip, Craig E.** Critical issues facing the U.S. barge industry in the 1980's. *Transportation Journal* 25(1), Fall 1985, 5–10.

Sea and inland waterways

1633 Baxter, Richard and others. *International navigable waterways: Financial and legal aspects of their improvement and maintenance.* New York: UNITAR, 1975. 253 pp.

Report on the Symposium held at Buenos Aires 30 November–4 December 1970. Evaluation of financial aspects; definition of costs, benefits; distribution of costs; reimbursement of costs; tolls and other dues; legal aspects; operating entities; special problems.

1634 **Belcher, G. L., Jones, J. R.** and **Lindeborg, K. H.** Inland/ocean waterborne transportation innovations and port charges. *Southern Journal of Agricultural Economics* 12(1), July 1980, 179–88.

1635 **Bierman, Don Edward.** *The Oder river: Transport and economic development.* Evanston (Illinois): The Transportation Center, Northwestern University, 1973. 247 pp.

A study of the function of the Oder River as an artery of transportation within the western provinces of Poland. Study based on their requirements for existing traffic: (i) functioning mode of transportation; (ii) a demanding market with terminal facilities; (iii) accessible supplies of economic goods.

1636 **Case, Leland S.** and **Lave, Lester B.** Cost functions for inland waterways transport in the United States. *Journal of Transport Economics and Policy* IV(2), May 1970, 181–91.

Waterway cost functions. Main hypothesis that long-run average cost of £1m declines as size of firm expands. Firm's average cost falls in short run as output expands.

1637 **Collins, J. W. F.** Inland waterway freight carrying. *Industrial Management and Data Systems*, May/June 1984, 3–8.

1638 **Collins, J. W. F.** Inland waterway freight transport. *International Journal of Physical Distribution and Materials Management* 14(3), 1984, 45–54.

This shows the extent of the underestimation of water transport for inland trade carrying and outlines the potential for future developments and the problems which will have to be overcome to realize that potential.

1639 **De Jong, J.** *The Rhine–Main–Danube connection and its economic implications for Europe.* Rotterdam: Netherlands Maritime Institute, 1976. 15 pp.

1640 **European Communities.** *Carriage of goods, 1989: Inland waterways.* London: HMSO, 1991. 170 pp.

1641 **European Conference of Ministers of Transport.** *Competitive position and future of inland waterway transport. Round Table. No. 49.* Paris: ECMT, 1980. 97 pp.

1642 **European Conference of Ministers of Transport.** *Inland water-*

way transport in ECMT countries to the year 2000: A new dimension. London: HMSO, 1990. 123 pp.

1643 **Hauser, R. J., Beaulieu, J.** and **Baumel, C. P.** Impacts of inland waterway user fees on grain transportation and implied barge rate elasticities. *Logistics and Transportation Review* 21(1), March 1985, 37–55.

1644 **Howell, Paul** (ed.) *Jonglei Canal: Impact and opportunity.* Cambridge: Cambridge University Press, 1987. 568 pp.

1645 Inland waterways: Cinderella of the land transport systems reaches fulfilment. In: *EEC shipping: A guide to the sea transport industries of the European Economic Community.* Colchester: Seatrade Publications, pp. 41–3.

Development of inland waterways and impact on economic fortunes of major European ports.

1646 **Martin, Michael V.** and **Clement, David A.** An analysis of port specific international grain freight rates: The case of the Lower Columbia River Port area. *Transportation Journal* 22(1), Fall 1982, 18–26.

1647 **Montgomery, T. M.** The future of the Manchester Ship Canal: Implications of land use and water quality in the upper reaches. Thesis submitted for MSc, University of Manchester, 1988.

1648 Waterways. *Export Direction,* December 1983, 14–.

The Great Lakes

1649 **Gemmel, A. W., Uhm, I. H.** and **Shaw, G. C.** Economics of Canadian water carriers on the Great Lakes and St. Lawrence seaway system. *Journal of Transport Economics and Policy* XVII(2), May 1983, 191–210.

Examines extent to which economies of scale prevail in shipping in the Great Lakes/St Lawrence system, considering constraints of, *inter alia,* short navigation season; short, narrow, shallow locks. Prospects for greater vessel use by developing vessels for Great Lakes plus oceans.

10 Human Resources in Shipping

General

1650 **Fricke, Peter H.** (ed.) *Seafarer and community: Towards a social understanding of seafaring.* London: Croom Helm; New Jersey: Rowman & Littlefield, 1973. 164 pp.

Collection of papers prepared for 'Symposium on Seafarer and Community', at University of Wales, Institute of Science and Technology, May 1972.

1651 **Grey, Michael.** *Changing course: A second career for the seafarers within the maritime industry.* London: Fairplay Publications, 1980. 116 pp.

Gives practical advice to the mariner in respect of a second career ashore in the marine industry. Looking ahead – planning a future ashore/qualifications/making the move ashore. Adjustment to life ashore. Shipping companies. Port industry. Commerce. Insurance and the law. Safety and technical. Nautical education. Miscellaneous careers.

1652 **Wright, Joseph P.** Drug and alcohol testing of maritime personnel. *Journal of Maritime Law and Commerce* 20(4), October 1989, 541–57.

Impact of Supreme Court's consideration for the first time of the constitutionality of on-the-job drug testing. Impact on regulations promulgated by Coast Guard. In *Skinner* v. *Railway Labor Executives Association* (1989) and *National Treasury Employees Union* v. *Von Raab* (1989), drug testing programmes upheld.

General; cost and wages; contracts

1653 **Argiroffo, E.** Recruitment of seamen in Asia. *International Labour Review* 95, 1967.

1654 Barnes, L. *Evolution and scope of mercantile marine law relating to seamen in India.* New Delhi: Maritime Law Association of India, 1983. 464 pp.

1655 British Shipping Federation. *Fleet and manpower inquiry: 1972–1976.* London: British Shipping Federation, 1972. 17 pp.

A report presenting results from Inquiry and showing consistency of the replies involving UK registered trading vessels of 200 g.r.t. or more. Provides a breakdown of the UK registered fleet at the time and its manning. Discovers trends in the size, composition and manning of the UK fleet.

1656 Chapman, Paul K. *Trouble on board: The plight of international seafarers.* New York: ILR Press, 1992. 176 pp.

1657 *Crew size and maritime safety.* Washington (DC): National Academic Press, 1991. 184 pp.

1658 Forsyth, Craig J. Sea daddy: An excursus into an endangered social species. *Maritime Policy and Management* 13(1), January–March 1986, 53–60.

Examines stereotypic image of American merchant seaman. A typology constructed on basis of (i) primary identification either with seamen subculture or shore-side activity; (ii) response to stigma of occupation.

1659 Franco, G. Robert. The wage subsidy index in the Merchant Marine Act of 1970. *Journal of Transport Economics and Policy* VII(3), September 1973, 283–90.

Analyses forces that led to creation of wage index included into Merchant Marine Act 1970. Computation and implementation of index; impact of wage trends of seafarers.

1660 Fricke, Peter H. Family and community: The environment of the ship's officer. In: Peter H. Fricke (ed.) *Seafarer and community: Towards a social understanding of seafaring.* London: Croom Helm; New Jersey: Rowman & Littlefield, 1973, pp. 132–50.

Problems of seafarer and his family and subsequent effect on attitudes to seafaring.

1661 General Council of British Shipping. *Fleet and manpower inquiry: 1976–1981.* London: GCBS (Annual).

1662 General Council of British Shipping. *A master's guide to the* *law concerning unfair dismissal as it applies to seafarers.* London: GCBS, 1978. 19 pp.

A guide to assist Masters in handling any situations that occur on board where discharge would be considered appropriate so as to be fair to the individuals concerned, preserve good employee relations and avoid as far as possible incurring financial penalties through a breach of acceptable practices.

1663 Great Britain. Department of Transport. *Human element in shipping casualties.* London: HMSO, 1991. 29 pp.

1664 Ha, T. M. *The seamen's rights to wages under contracts of employment.* Malta: LLM-I.M.O. International Maritime Law Institute, Malta.

Presents and analyses the rights of seamen to wages under the contract of employment set out in the British legal system.

1665 Hodges, S. L. H. (ed.) *Legal rights of seafarers.* London: Nautical Institute, 1988.

1666 Hodgin, W. R. *Merchant Shipping Act, 1970: Crew employment.* London: Lloyd's of London Press, 1987.

1667 Hope, R. *Seamen's world.* London: Marine Society, 1982.

1668 International Labour Office. *Accident prevention on board ship at sea and in port.* Geneva: I.L.O., 1984.

1669 International Labour Office. *Inspection of labour conditions on board ship: Guidelines for procedure.* Geneva: I.L.O., 1990. 76 pp.

1670 International Labour Office. *Labour standards on merchant ships.* Geneva: I.L.O., 1992. 187 pp.

1671 International Shipping Federation. *ISF Guide to international maritime labour supply.* London: Lloyd's of London Press, 1990. 212 pp.

Examines the international alternatives available on the maritime labour market. Country by country explanation of prices, availability and other essential information required by shipowner for manning decisions.

1672 Marine Technology Society. *The ocean: An international work-*

place. Proceedings of a conference. Halifax, Canada: I.E.E.E. Ocean Engineering Society, 1987.

1673 **McCowen, Peter** and **Barry, Michael.** *Central manpower supply to the Merchant Navy.* Sealife, 1978. 39 pp.

The second Sealife report on central manpower supply. This report asks further questions about the performance of the present system of manpower supply through the MNE being used by employers and seafarers.

1674 **Moreby, D. H.** Crew costs. *Maritime Policy and Management* 12(1), 1985, 55–60.

What factors are involved in calculating costs of crews? Different views are taken by owners, unions, etc. Essential to check what items have been included and seamen's conditions of employment.

1675 **Moreby, David H.** *The human element in shipping.* Colchester: Seatrade Publications, 1975.

Complexity of personnel management within shipping industry. Investigation of shipping managerial philosophy. Areas of potential difficulty in development of manning policies. Problems of technological change and increasing capital intensity.

1676 **Muldowney, T.** An Irish experiment in worker participation: B & I Line (Dublin) Works Council. *Industrial Participation* 558, Spring 1976, 22–7.

1677 **Nautical Institute.** *Legal rights of seafarers.* London: Nautical Institute, 1985.

1678 **Norris, Martin J.** *The law of maritime personal injuries.* (2 vols.) 4th edn. New York: Clark Boardman Callaghan, 1990.

1679 **Norris, Martin J.** *The law of seamen.* (3 vols.) 4th edn. Rochester (New York): Clark Boardman Callaghan, 1985.

1680 **NUMAST.** *Only human: The human element in safe shipping.* London: NUMAST, 1991. 14 pp.

Examines the human factors present in accidents at sea.

1681 **Roggema, J.** and **Smith, M. H.** Organizational change in the shipping industry: Issues in the transformation of basic assumptions. *Human Relations* 36, August 1983, 765–90.

1682 Sea change: Safety at sea. *Occupational Safety and Health* 18(6), 1988, 12.

1683 Sea change: Safety at sea. *Occupational Safety and Health* 18(7), July 1988, 37–.

1684 Sherar, Mariam G. *Shipping out: A sociological study of the American merchant seaman.* Cambridge (Maryland): Cornell Maritime Press, 1973. 96 pp.

1685 Shneerson, D. *The effect of crew cost on Israeli shipping competitiveness.* Haifa: Israel Shipping and Aviation Research Institute, 1985.

1686 Social Surveys (Gallup Poll) Limited. *The attitude of seafarers to their employment.* London: Board of Trade, 1970. 469 pp.

A report submitted to the Committee of Inquiry into Shipping by Social Surveys (Gallup Poll) Ltd. Volume I – commentary. Volume II – statistical tables of officers and men currently serving at sea. Volume III – statistical tables of former seafarers.

1687 Söhmen, Helmut. *The human element in shipping.* Singapore: Chua Chor Teck Memorial Fund Trustee Committee, [1990]. 11 pp.

Paper given at lecture 19 January 1990. The mystique of shipping. Traditions and current concerns. Social attitudes. Casual labour as seafaring career. Shipboard environment. The shipowner's dilemma. Shipping and innovation. Maritime safety. Discipline. Opportunities. Peculiarities of shipping.

1688 Strevens, Peter. *Seafaring. Vol. 1.* London: Collier Mac and Cassell, 1972. 144 pp.

1689 Van Voorden, W. *The shipping industry in the Netherlands.* Geneva: International Institute for Labour Studies, 1976. 17 pp.

Part of the Research Series of the International Institute for Labour Studies, considering strategic factors in industrial relations systems. Industrial relations in the Dutch shipping industry. Employers' organizations. Workers' organizations. Role of government. Relationships between the parties. The industrial relations system in action.

1690 Westinform Service. *Manning of merchant fleets of the world.* London: Westinform, 1973. 21 pp.

1691 Yamamoto, Hiromasa. Redundancy of seafarers and the practice of permanent employment. *Kobe Economic and Business review* 25, 1979, 1–13.

Explains the reason why the practice of permanent employment has become established in the deep-sea Japanese shipping industry since the Second World War. Considers how this practice was evaluated by management and the seamen's union before the depressed years.

Officers

1692 Barnes, B. I. *Mental health of officer personnel in the Indian merchant marine.* London: ITF, 1989.

1693 Hershey, R. The primacy of the Master and its consequences. *Maritime Policy and Management* 15(2), 1988, 141–6.

Is there a cause–effect connection between authoritarian Masters and maritime casualty rates? Evidence from airline industry suggests so. Could maritime industry benefit?

1694 Watson, Mary-Anne L. The interactional effects of lower level participant power and performance in attributional leadership: Ship model. Thesis submitted for PhD, University of South Carolina, 1988.

Ratings

1695 Dooley, G. In search of a miracle: The future of professional American seamen. *The Lookout* 74(3), October–November 1982, 2–4.

1696 Doxley, J. M. The seamen's strike of 1960. Thesis submitted for MA (Econ), University of Manchester, 1963.

Education and training

1697 Al-Hassan, O. S. A. A critical review of maritime educational facilities, with special reference to the Arab world. Thesis submitted for MSc, UWIST, Cardiff, 1979.

1698 Bassis, M. and **Rosengren, W.** Socialization for occupational disengagement: Vocational education in the merchant marine. *Sociology of Work and Occupations* 2(2), 1975, 133–49.

1699 Burger, W. and **Corbett, A. G.** Training in maritime technology. *Maritime Studies and Management* 3(2), October 1975, 81–6.

Considers formal training of deck officers for the merchant service. Views marine technology as topic which should be seen as part of syllabus with subjects interconnecting and not as separate issues.

1700 Erez, Miriam. Expectancy theory prediction of willingness to be retrained: The case of ratings' advancement in the Israeli Merchant Navy. *Journal of Occupational Psychology* 52, 1979, 35–40.

1701 Gold, E. Canadian marine training: Time to change course? A comparative comment. *Maritime Policy and Management* 17(3), 1990, 177–88.

Examines problems facing marine training in Canada, compared with systems developed in UK, Australia, China and at World Maritime University in Sweden.

1702 Gunton, G. Paul. *Guide to marine training.* London: Lloyd's of London Press, 1991. 64 pp.

1703 Hlaing, Maung Tin. A new approach to education and training of mariners, with special reference to South East Asian countries. Thesis submitted for PhD, UWIST, Cardiff, 1984.

1704 Holder, L. A. *Training on board.* London: Witherby & Co., 1991. 76 pp.

This is a guide to managers and seafarers on how best to take the opportunities for on-board ship training.

1705 Hope, Ronald. *Introduction to the Merchant Navy.* 4th edn. London: Seafarers Educational Service, 1973. 80 pp.

1706 Hopwood, Warren H. Some problems associated with the selection and training of deck and engineer cadets in the British Merchant Navy. In: Peter H. Fricke (ed.) *Seafarer and community: Towards a social understanding of seafaring.* London: Croom Helm; New Jersey: Rowman & Littlefield, 1973, pp. 97–116.

Examines career structures of grades of personnel; patterns of initial training; certain problems associated with adaptation to ship environment.

1707 Jeans, C. A critical study of the education and training of masters and mates in British merchant ships. Thesis submitted for M.Tech, Brunel University, Uxbridge, 1971.

1708 Moore, Michael. *COS: The College of the Sea 1938–1988.* London: Marine Society, 1988. 32 pp.

Collection of essays to commemorate the 50th anniversary of the College.

1709 Moreby, David. Comment. *Maritime Studies and Management* 2(3), January 1975, 164.

Critical comment in respect of research carried out by Rosengren and Bassis for their article in the same volume, 'Ship contingencies and nautical education'. Essence of comment is that research is from men who had not yet been to sea.

1710 Moreby, David H. On the management of nautical colleges. (Editorial.) *Maritime Studies and Management* 2(3), January 1975, 129–31.

Considers issue of how nautical colleges must decide on their diversification strategies.

1711 Morrison, W. S. G. Technology upgradation and human resources development in the maritime industry. In: Uddesh Kohli and Vinayshil Gautam (eds) *Human resources development and the planning process in India.* New Delhi: Vikas, 1988, pp. 104–12.

1712 Moukhtar, G. A. The development and role of the Arab Maritime Transport Academy. *Maritime Studies and Management* 2(2), October 1974, 110–11.

Consideration of ultimate objective of United Nations Development Programme's authorized project – the establishment of Maritime Training Institute for Arab countries.

1713 Rosengren, William R. and **Bassis, Michael S.** External control and organisational adaptability: American, British and Spanish merchant marine academies. In: Peter H. Fricke (ed.) *Seafarer and community: Towards a social understanding of seafaring.* London: Croom Helm; New Jersey: Rowman & Littlefield, 1973, pp. 117–31.

General description and explanation of differences in the organization of merchant marine training schools in the USA, Great Britain and Spain.

1714 Rosengren, William R. and **Bassis, Michael S.** Ship contingencies and nautical education: A problem of level and scope of certification. *Maritime Studies and Management* 2(3), January 1975, 154–63.

Reports a preliminary and partial analysis of a study of merchant navy cadets in maritime schools in the USA, Great Britain and Spain. Highlights consequences for work attitudes in respect of differing training systems.

1715 *Ship manning present and future: Need for innovation in management design and training within the shipping industry.* London: Honourable Company of Master Mariners, Nautical Institute, 1983.

1716 Singh, Nagendra. *International maritime law conventions – Vol. 3. Training, employment and welfare; environment.* 3rd edn. London: Stevens & Sons, 1988. 744 pp.

Training of seafarers – IMCO Maritime Training Convention and Recommendations; employment, welfare and status of seamen – I.L.O. Maritime Conventions and Recommendations; environment – prevention and control of marine pollution, sanitary conventions.

1717 *Training and manning, IMAS73. International Marine and Shipping Conference. Institute of Marine Engineers.* London: Institute of Marine Engineers, 1973.

1718 Waters, D. M. Nautical education within the national education system. Thesis submitted for MSc, UWIST, Cardiff, 1971.

On-board systems

1719 Cadwallader, F. The seaman's charter. *Current Legal Problems* 23, 1970, 156–86.

1720 Committee on effect of smaller crews (for National Research

Council) *Crew size and maritime safety.* Washington (DC): National Academy Press, 1990. 162 pp.

Identifies economic and technological changes which cause manning reductions. Forecasts evolution of ship operating technology and associated changes in crew organization and training. How smaller crews are affected by changes in manning patterns, and implications for training, skill levels, crewing patterns and vessel operating technology. Safety.

1721 **Coveney, A.** Olsen's Democracy. *Industrial Society* November/December 1975, p. 11.

1722 **Fricke, P. H.** The social structure of the crews of British dry cargo merchant ships. Thesis submitted for PhD, University of Durham, 1974.

1723 **Fricke, Peter H.** Of ships and social scientists. (Editorial.) *Maritime Studies and Management* 3(3), January 1976, 129–30.

Views who is considering problems of management and life on board. Questions why hierarchical structure of ship life still exists.

1724 **Fricke, Peter H.** *The social structure of the crews of British dry cargo merchant ships: A study of the organization and environment of an occupation.* Cardiff: UWIST, 1974. 458 pp.

A sociological study of seafarers on board ship. Study undertaken from 1967 to 1969 of the crews of 28 British merchant ships. Concerned with the analysis of the organizational variables of shipboard life in relation to attitudes and expectations of seafarers.

1725 **Holder, L. A.** *Technology and manning for safe ship operation in the 1990's. Vol. 1. Management summary. Vol. 2. Main report.* Ilford: Kelvin Hughes, 1989.

A report on manning and training for UK shipping in the next decade, for a range of modern vessels.

1726 **Hughes, C. N.** *Shipboard operational problems.* London: Lloyd's of London Press, 1989. 238 pp.

1727 **Jackson, John** and **Wilkie, Roy.** General purpose manning: A case study of organizational innovation. Part I. *Maritime Studies and Management* 2(3), January 1975, 132–7.

Examines the introduction of a general-purpose manning scheme within one particular company.

1728 **Jackson, John** and **Wilkie, Roy.** General purpose manning: A case study of organizational innovation. Part II. *Maritime Studies and Management* 2(4), April 1975, 215–20.

How general-purpose manning was operated within (fictitiously named) Armstrong Shipping Company and the difficulties involved for those at sea.

1729 **Jackson, John** and **Wilkie, Roy.** General purpose manning: A case study of organizational innovation. Part III. *Maritime Studies and Management* 3(1), July 1975, 21–6.

Third article of series concerned with the introduction of general-purpose-manning scheme into (fictitiously named) Armstrong Shipping Company. Part III – impact of the new work system and difficulties it met.

1730 **Koishi, Y.** Work simplification through the transformation of work on board. *Maritime Policy and Management* 8(4), 1981, 261–6.

Research on Japanese ships since 1961 showed that relationship between crew number and resulting structural change of work on board ship can be formulated by a linear regression.

1731 **Nolan, B.** Social life in a merchant ship: A study of attitude change during assimilation into a total institution. Thesis submitted for PhD, University of London, 1975.

1732 **Nolan, Bryan.** A possible perspective on deprivation. In: Peter H. Fricke (ed.) *Seafarer and community: Towards a social understanding of seafaring.* London: Croom Helm; New Jersey: Rowman & Littlefield, 1973, pp. 85–96.

Theoretical approach to certain aspects of seafaring life. The ship as a total institution. Deprivations experienced by seamen. Social interaction aboard merchant ships.

1733 **Perry, Nick** and **Wilkie, Roy.** Social theory and shipboard structure – II. Models, metaphors and ships. *Maritime Studies and Management* 1(3), January 1974, 136–46.

Overall assessment of concept of the total institution for analysis of life on board ship. What lessons can be drawn *re* social life on board – organizational behaviour in general.

1734 **Schrank, Robert** (ed.) *Industrial democracy at sea: Authority and democracy on a Norwegian freighter.* Cambridge (Massachusetts): MIT Press, 1983. 220 pp.

Presents the accounts of three American merchant seamen who were taken on as crew members for the course of a voyage from the Pacific Northwest to Europe. Their reactions and adjustments form the basis of a case study of an experimental democratization of working life on ship. Includes contributed essays.

1735 *Ship organization and personnel.* Annapolis (Maryland): US Naval Institute, 1972. 140 pp.

1736 Smith, M. H. and **Roggema, J.** *Crew stability and shipboard management: Technical paper.* London: Sealife, 1978. 20 pp.

Report which examines concept of continuity of crew. Examines conditions underlying crew stability. Considers the kind of organizational features required to cope with lack of continuity and the ensuing advantages and disadvantages of lack of crew stability. Suggested definitions. Consequences of continuity.

1737 *Technology and manning for safe ship operations in the 1990's: Project sponsored by Department of Transport.* (2 vols). 1986. Vol. 1 40 pp., vol. 2 549 pp.

Project carried out by joint team supervised by senior Polytechnic staff with relevant industrial and research experience. Examines status of shipping industry with respect to manning, automation, safety from which to extrapolate information concerning manpower, fleet and trade mix, new technology, etc. Requirements of shipboard personnel in 1990s. Training. (Vol. 2: Main report.)

1738 Wada, Keiji. *The progress and future prospects of modernization of the manning system in Japan.* Tokyo: Japan Maritime Research Institute, 1986. 30 pp.

Booklet, part of series covering shipping activities. The establishment of the Committee on the modernization of the manning system. Progress of experiments in respect of the system. Recent developments and future prospects.

1739 Wan, Hua. The economic responsibility system on board Chinese ships. *Maritime Policy and Management* 15(1), January–March 1988, 77–82.

Paper examines background, contents and results of experiment on economic responsibility system on board Chinese deep-sea ships. Discusses problems and future directions.

1740 Yamamoto, Hiromasa. Available jobs and employment in

Japanese foreign-going shipping. *Kobe Economic and Business Review* 20(20), 1973, 51–9.

Reduction of the manning scale on Japanese foreign-going vessels. Trends in merchant fleet, available jobs and productivity of labour. Employment of seafarers on Japanese foreign-going vessels. Changes of composition of seafaring labour by age – the impact of technological development on board ships.

Trade Unions; Professional Associations; Collective Bargaining

1741 Albury, P. F. Settlement of maritime disputes: The New Zealand way. *Maritime Policy and Management* 10(4), October–December 1983, 275–84.

Paper describes and discusses the formal process whereby maritime industrial disputes are solved in New Zealand.

1742 British Shipping Federation. *Commission on IR: Approved closed shop agreement.* 30th edn. London: HMSO, 1972.

1743 Donn, C. B. *Concession bargaining in the ocean-going maritime industry.* Syracuse (New York): Le Moyne College Institute of Industrial Relations, 1987.

1744 Donn, C. B. *Foreign competition, technological change and the decline in maritime employment.* Syracuse (New York): Le Moyne College Institute of Industrial Relations, 1987.

1745 Donn, C. B., Richard, M. and **Phelan, G.** *Reforms of work practices and industrial relations procedures in the maritime industry: An Australia–United States comparison.* Syracuse (New York): Le Moyne College Institute of Industrial Relations, 1990.

1746 Donn, Clifford B. Concession bargaining in the ocean-going maritime industry. *Industrial and Labour Relations Review* 42(2), January 1989, 189–200.

Examines the concessionary phase of US maritime collective bargaining – concessions in staffing, work rules, wages and benefits. Argues that these concessions come in both standard multi-employer agreements and in departures from those agreements on an employer-by-employer and vessel-by-vessel basis. Impact on industry's decline.

1747 Driesen, George B. Arbitration and seamen's rights under collective agreements in the wake of Arguelles. *Journal of Maritime Law and Commerce* 3(3), April 1972, 429–44.

Overtime and section 596 (46 U.S.C. 8596) late payment penalty claim. Adjudication by federal court. Contractual claims; statutory claims. Examination of 1871 case *United States Bulk Carriers* v. *Arguelles.*

1748 Fletcher, B. The government were determined to make the men as strong as the masters . . . The experience of the Shipping Federation, 1906 to 1910. *Maritime Policy and Management* 11(4), 1984, 261–8.

Considers extent of changes in 'balance of power' between employers and labour which took place between 1906 and 1914, especially conflict between NSFU and the Shipping Federation.

1749 Forsyth, C. J. Expansionism: New forms of old conflicts for US maritime labour unions. *Maritime Policy and Management* 15(4), 1988, 279–82.

On fresh conflicts among US maritime labour unions as they move into other job areas. New form of union, that of the 'top to bottom' union, also discussed.

1750 Forsyth, C. J. and **Cullison, A.** Intraunion conflict: The changing marine engineer. *Maritime Policy and Management* 14(4), 1987, 279–87.

Describes shift in membership of Marine Engineers' Beneficial Association, the major US marine engineer labour union, and problems associated with this change.

1751 Forsyth, C. J. and **Cullison, A. C.** Conflict in the US maritime industry: An account of historical cycles. *Maritime Policy and Management* 11(4), 1984, 251–9.

Different configurations of US maritime labour unions and shipping companies have led to distinct periods of conflict in industry. Analyses basis and continuation of problem.

1752 Hearn, George H. Recent law of the Federal Maritime Commission and collective bargaining. *Journal of Maritime Law and Commerce* 6(1), October 1974, 31–46.

Examines the law which places Federal Maritime Commission in position to play a part with other agencies in association with management and labour relations. Importance of ocean commerce.

1753 Heaton, P. *A trade union view of change in the shipping industry.* Edinburgh: Scottish Academic Press, 1984.

1754 Marsh, Arthur Ivor and **Ryan, Victoria.** *The seamen: History of the National Union of Seamen.* Oxford: Malthouse Publications, 1988. 248 pp.

1755 McConville, J. Industrial relations in the U.K. shipping industry since the Second World War. Thesis submitted for PhD, University of Warwick, 1982.

Examines the decline in the UK industry and its labour requirement. Analyses the impact this had on industrial relations institutions – in particular on collective bargaining.

1756 McConville, James. *The shipping industry in the U.K.* Geneva: International Institute for Labour Studies, 1977. 87 pp.

A study, part of series for the International Institute for Labour Studies. Focus of this monograph is the maritime industry's collective bargaining structure, the National Maritime Board, particularly post-Second World War.

1757 Morris, R. Shipping's new industrial relations: The international context and the Australian experience. *Journal of Industrial Relations* 32(3), 1990, 17–33.

1758 Rowan, R. L., Northrup, H. R. and **Immediata, M. J.** International enforcement of union standards in ocean transport. *British Journal of Industrial Relations* 15, November 1977, 338–55.

This examines the controversial work of the International Transport Workers' Federation (ITF) in their union campaign to control flags and crews of convenience.

1759 Rubenowitz, Sigvardi and **Gleerup, Alf.** *The shipping industry in Sweden.* Geneva: International Institute for Labour Studies, 1977. 31 pp.

Fourth of a number of national studies dealing with the shipping industry in Sweden. Part of series covering international shipping.

1760 Sasaki, H. *The shipping industry in Japan.* Geneva: International Institute for Labour Studies, 1976. 47 pp.

Second in a number of national studies. Part of series covering international shipping.

1761 Underwood, David. Industrial relations in the shipping industry. *Personnel Management* 10, March 1978, 22–6.

1762 Urhoma, I. K. Strategic factors in industrial relations systems: Their impact on bargaining outcomes in the Nigerian shipping industry. Thesis submitted for PhD, UWIST, Cardiff, 1991.

An explanation of the Nigerian industrial relations phenomenon – especially the wage behaviour, which owes much to the traditional British model. It holds that a new interpretation is possible if the notion of strategic factors and factors strategies are introduced into the analysis.

1763 Wainwright, David. Launching into participation: Two exercises in participation at sea. *Personnel Management* 10, December 1978, 28.

1764 Yamamoto, Hiromasa. Productivity bargaining and the change of labour management relations in the Japanese foreign-going shipping: 1965–1975. *Kobe Economic and Business Review* 22, 1976, 17–25.

An article which traces the main issues of collective bargaining of the shipping industry in Japan and reveals the process and reasons why the labour–management relations changed, culminating in the 1972 three-month strike of the seamen's union.

Theory

General

1765 Bryant, D., Debievre, A. and **Dyer-Smith, M.** *Investigating human factors in marine casualties.* London: Tavistock Institute of Human Relations, 1987.

1766 Burton, V. C. The work and home life of seafarers, with special reference to the port of Southampton. Thesis submitted for PhD, University of London, 1989.

1767 Chandler, E. J. Sailors' wives and husband absence. Thesis submitted for PhD, CNAA, Plymouth Polytechnic, 1987.

1768 Fleming, D. K. Seascapes. *Maritime Policy and Management* 9(1), 1982, 29–33.

Literary descriptions of oceans and seas of the world: Mediterranean, Caribbean, N. Atlantic, Southern Ocean, and various sea conditions, storm and calm. Is the sea a home for the mariner?

1769 **Forsyth, C.** and **Gramling, R.** Merchant seamen: Historical stereotypes and current status dilemmas. *Free Inquiry in Creative Sociology* 13(2), November 1985, 148–50.

1770 **Forsyth, C. J.** The creation and maintenance of a stigmatised occupation: An historical analysis of the American merchant marine. *Maritime Policy and Management* 14(2), 1987, 99–108.

Historical background to stigmatization of US merchant seaman. Stigma persists despite better wages and working conditions. Improved technology may bring higher status.

1771 **Forsyth, C. J.** The future of the American seaman. *Maritime Policy and Management* 17(4), 1990, 249–55.

Analyses occupational future of American merchant seaman. Reasons for decline cited and strategies for resurgence discussed.

1772 **Forsyth, C. J.** and **Bankston, W.** The merchant seaman as a social type: A marginal life-style. *Free Inquiry in Creative Sociology* II(1), May 1983, 8–12.

1773 **Forsyth, Craig J.** *The American merchant seaman and his industry: Struggle and stigma.* New York: Taylor and Francis, 1989. 125 pp.

1774 **Goldberg, J.** The maritime industry. *Labour Law Journal* 21(8), 1970, 505–13.

1775 **Goldberg, Joseph P.** Seamen and modernization of merchant shipping. *Monthly Labor Review* 94(2), February 1971, 49–54.

1776 **Gordon, Paul** and **Reilly, Danny.** Guestworkers of the sea: Racism in British shipping. *Race and Class* 28, Autumn 1986, 73–82.

1777 **Haralambides, Hercules E.** An econometric analysis of the seagoing labour market. *Logistics and Transportation Review* 17(1), March 1991, 15–31.

Analysis of factors that affect the demand for seafarers by a country's shipping industry; the supply of seamen and the reasons why people go to sea; the determination of the seamen's wage.

1778 **Lysenko, Vladil.** How Soviet seamen really live. In: Curtis Cate (ed.) *The challenge of Soviet shipping.* London: Aims of Industry (in association with the National Strategy Information Center, New York), pp. 10–16.

An account of how Soviet seamen live based on author's own experiences. Wages and calculation of wages. Problem of food allocation at sea. Political control. The role that the union plays in a seaman's life. Retirement benefits and disability pensions.

1779 **Metwally, H. I.** Management and human relations problems and policies in merchant ships, with special emphasis on the Egyptian mercantile fleet. Thesis submitted for PhD, UWIST, Cardiff, 1978.

1780 **Moreby, D. H.** Communication problems inherent in a cross-cultural manning environment. *Maritime Policy and Management* 17(3), 1990, 199–205.

Paper explains causes and consequences of cultural differences, with reference to power differentials and rule orientation. Dangers of stereotypes are highlighted and communication problems discussed.

1781 **Nordvik, Helge N.** Norwegian sailors in the Canadian Merchant Marine 1863–1914: A preliminary survey. In: Einar Hope (ed.) *Studies in shipping economics: In honour of Professor Arnljot Stromme Svendsen.* Oslo: Bedriftsokonomens Forlag A/S, 1981, 8 pp.

Role that Norwegian seamen played in maritime activities of other major shipping nations. Quantitative historical information. Maritime labour force.

1782 **Punekar, S. D.** *The socio-economic conditions of Indian seamen and their welfare.* Bombay: Tata Institute of Social Services, 1977.

1783 **Robertson, David W.** The law of seamen clarified. *Journal of Maritime Law and Commerce* 23(1), January 1992, 1–28.

Determining a worker's status as a seaman within American maritime law. Effects of the decision of Supreme Court in *McDermott International, Inc.* v. *Wilander* (1991).

1784 **Robinson, M. D.** A study of industrial injuries in British shipping. Thesis submitted for MSc, UWIST, Cardiff, 1974.

1785 Roggema, J. and others. *The acceptance by seafarers of future technology and organizational changes in shipping.* London: Tavistock Institute, 1970.

1786 Salmans, S. Management moves from shore to ship. *International Management*, March 1978, 18–.

1787 Sharpe, Thomas. Morflot and the KGB. In: Curtis Cate (ed.) *The challenge of Soviet shipping.* London: Aims of Industry (in association with National Strategy Information Center, New York), pp. 42–5.

Control and surveillance of merchant fleet by KGB. Selection process for career in merchant fleet. On-board KGB surveillance. Additional 'offensive' aspects of this control.

1788 Tresselt, D. *The controversy over the division of labour in international seaborne transport.* Bergen: Institute of Shipping Research, 1970.

1789 Wakeford, G. W. Manning merchant ships: Problems and remedies. *Royal Society of Arts Journal* 126, March 1978, 182–92.

The paper given and discussion following at the Thomas Gray Memorial Lecture, November 1977, to the Society. Suggests there are four points to the issue of manning merchant ships: (a) what is the problem? (b) how has the problem changed over the previous twenty years? (c) what needs to be done? (d) how can it be achieved?

Communication

1790 Bernard, H. R. On the social structure of an ocean going research vessel and other important things. *Social Science Research* 2, 1973.

1791 Bernard, H. Russell and **Killworth, Peter D.** Scientists and crew: A case study in communication at sea. *Maritime Studies and Management* 2(2), October 1974, 112–25.

A study on scientist/crew relations aboard research vessels at sea. (A case study in a long-range project dealing with the relations of natural scientists in the USA to various segments of lay population.)

1792 Cadwallader, F. J. J. Discipline in British ships. *Journal of Maritime Law and Commerce* 2(1), October 1970, 147–64.

Court of Inquiry set up to report upon causes and circumstances of May 1966 strike by British seamen. Pearson Report. Merchant Shipping Act 1970. Provisions in relation to discipline. Effect of Inquiry.

1793 **Fricke, Peter H.** Seafarer and community. In: Peter H. Fricke (ed.) *Seafarer and community: Towards a social understanding of seafaring.* London: Croom Helm; New Jersey: Rowman & Littlefield, 1973, pp. 1–7.

Views life of seafarer as form of occupational community. Impact of technological and social change upon this community.

1794 **Perry, Nick.** Conflict on board ship: An interpretation. *Sociological Review* 22, November 1974, 557–80.

1795 **Poole, M.** Maritime sociology: Towards a delimitation of themes and analytical frameworks. *Maritime Policy and Management* 8(4), 1981, 207–22.

Evaluates the main areas of focus within maritime sociology and links various general theoretical and conceptual approaches.

Job satisfaction

1796 **Alexander, L.** and **Sharpe, P.** The merchant shipping industry's employee retention scheme. *International Journal of Manpower* 1(2), 1979, 8–.

1797 **Forsyth, C. J.** and **Bankston, W. B.** The social psychological consequences of a life at sea: A causal model. *Maritime Policy and Management* 11(2), 1984, 123–34.

Influence of seafaring occupation on alienation and social integration of its members; 293 officers and crew surveyed.

1798 **Forsyth, Craig J.** Factors influencing job satisfaction among merchant seamen. *Maritime Policy and Management* 17(2), April–June 1990, 141–6.

Explores occupational and extra-occupational sources of job satisfaction in merchant seamen. From sample of 302 seamen, officers and crew. Variance in job satisfaction examined using multiple regression technique.

1799 **Goss, R. O.** The revival of Australian shipping: A review article. *Maritime Policy and Management* 10(2), April–June 1983, 133–7.

Review of Australian shipping industry. Fall of seafaring employment and consequent efforts of Minister for Transport to form committee to revitalize Australian shipping.

1800 Hill, J. M. M. *The seafaring career: A study of the forces affecting joining, serving and leaving the Merchant Navy.* London: Centre for Applied Social Research, Tavistock Institute of Human Relations, 1972. 158 pp.

Research commissioned by National Maritime Board into the causes of wastage from the British Shipping Industry. Careers of seafarers; why certain organizations more likely to retain entrants than others; extent and main features of wastage.

1801 Lane, Tony. *Grey dawn breaking: British merchant seafarers in the late twentieth century.* Manchester: Manchester University Press, 1986. 198 pp.

Aims to capture the changes and continuities of the seafaring life; to show how seafarers construct and inherit a particular image of themselves. Portrays the social relations of a cramped and hierarchical community. Describes and explains the significance of the daily routine of shipboard life and the nature of hazards and voyages.

1802 Lawrence, D. P. Skills analysis aboard ships. Thesis submitted for PhD, CNAA, Plymouth Polytechnic, 1988.

1803 Mannheim, Bilha F. and **Rosenstein, Eliezer.** Occupational behaviour and commitment of seamen. – I. Activity and its correlates. *Maritime Studies and Management* 3(1), July 1975, 41–51.

First of two papers which report the result of a study of seamen of the Israeli Merchant Marine. This paper examines extent to which social and occupational characteristics and job attitudes of seamen in operating departments relate to differential occupational behaviour.

1804 Mannheim, Bilha F. and **Rosenstein, Eliezer.** Occupational behaviour and commitment of seamen. – II. Staying or leaving the occupation. *Maritime Studies and Management* 3(1), July 1975, 53–65.

Second of two papers which report the result of a study of seamen in the Israeli Merchant Marine. Focuses on factors related to the commitment of seamen to their occupation.

Productivity

1805 Bendall, H. B. and **Stent, A. F.** Labour productivity of Australian registered ships, 1948–1978. *Maritime Policy and Management* 15(4), 1988, 269–73.

Follows previous study on period 1839 to 1919. Present study looks at continuing advances in maritime technology and differentiates between specialized carrying classes.

1806 Goss, R. O. The measurement of productivity in shipping. *Maritime Policy and Management* 9(1), 1982, 73–6.

Looks at forms of measurement of labour productivity applicable industry-wide, to a company, to a ship-type or to a specific ship.

1807 Goss, R. O. 'Social' costs, transfer payments and international competition in shipping. *Maritime Policy and Management* 12(2), 1985, 135–43.

Crew costs consist of wages levels and productivity. Must also consider costs of employment not appearing in take-home pay. Discusses ratio between a seaman's net pay and overall cost of employing him.

1808 Katz, H., McCaul, J. R. and **Fettner, L.** Productivity in the maritime industry. *Journal of Maritime Law and Commerce* 6(3), April 1975, 449–59.

Research study to determine the pattern of productivity growth in the maritime industry, and for strategic sectors within the industry. Productivity and theory behind its calculation. Results of calculations of productivity for entire maritime industry. Application of concept in containerized section of industry.

1809 Saxena, K. B. C. and **Joshi, P. B.** Motivating ship managers to use management support technology. *Maritime Policy and Management* 19(1), 1992, 55–62.

Suggests five principles for development of 'user seductive' ship management support systems, and illustrates successful use with case histories.

Differential subsidies

1810 Franco, G. R. The operating differential subsidy in the 1970 Merchant Marine Act. *Transportation Journal*, 1973.

11 Management

General

1811 Anders, Przemyslaw. The ship's agent (maritime agent) and shipbroker in Poland. *Lloyd's Maritime and Commercial Law Quarterly*, August 1991, Part 3, 408–12.

Development of basic civil law and the agency contract. The maritime agent and the Maritime Code of 1961. The shipbroker or maritime broker. Present day practice.

1812 Branch, Alan E. *Import/export documentation.* London: Chapman and Hall, 1990. 274 pp.

Explains the most important documents concerned in import/export shipping practice. Air freight, chartering, commercial certificates, customs, dangerous cargo, finance, insurance, liner conferences, road transport and shipping.

1813 Frankel, Ernst G. Promoting efficiency in shipping. *Marine Policy* 14(5), September 1990, 438–48.

Examines the main areas of technological change in the last two decades and examines the implications of these developments on shipping and past management. Details economic impact of crew reductions, fuel savings, automation, ship operations and ship-condition management.

1814 Franklin, P. *Computer programming ships' business.* Centreville (Maryland): Cornell Maritime Press, 1988.

1815 Hughes, C. N. *Ship performance: Some technical and commercial aspects.* London: Lloyd's of London Press, 1987. 112 pp.

Practical guide to analysing the operational potential of deep-sea vessels. Ability of a merchant vessel to perform well under the conditions faced on ocean voyages – what variables to consider?

1816 Marcus, Henry S. *Marine transportation management.* London, Sydney, and Massachusetts: Croom Helm, 1987. 322 pp.

Vessel technology issues faced by shipping companies in recent years. Specific technologies with economic, regulatory, political aspects. Dealt with as case studies on management situations.

1817 Marsh, Allan G. *Introduction to quality assurance in shipping.* London: Nautical Institute, 1991. 34 pp.

Origins of quality assurance. Certification and accreditation. The process of initiating quality assurance. How companies should prepare for quality assurance in the maritime industry. Case study. Sources of information.

1818 Melis, D. Application of marketing planning in smaller sized Greek shipping companies. Thesis submitted for MSc, University of Strathclyde, 1983.

1819 Nakagawa, Yoshiaki. *How the Japanese shipping companies and related transport operators utilize computer software programs in daily business activities.* Tokyo: Japan Maritime Research Institute, 1986. 54 pp.

Applications using large-sized computer, medium-sized computer, microcomputer.

1820 Nautical Institute. *Maritime commercial practice.* London: Nautical Institute, 1978.

1821 Nicholson, T. A. J. and **Pullen, R. D.** Dynamic programming applied to ship fleet management. *Operational Research Quarterly,* September 1971, 211.

1822 Ronen, David. Collecting information in the shipping industry. *Information and Management,* July 1981, 127–32.

An article which discusses the potential for legal commercial intelligence in the ocean shipping industry through the analysis of the information in the industry; it also describes an information system for processing such industry data.

1823 Walz, C. J. The impact of international standards for seafarers on maritime management. Thesis submitted for MSc, UWIST, Cardiff, 1979.

1824 **Wentworth, F.** and **Christopher, M.** (eds) *Managing international distribution.* Farnborough: Gower Press, 1979. 272 pp.

Ship Management

1825 **Benford, H.** A simple approach to fleet deployment. *Maritime Policy and Management* 8(4), 1981, 223–8.

How a fleet manager can select the best mix of ships to provide a given transport service, in non-liner trades.

1826 **Claessens, E. M.** Optimization procedures in maritime fleet management. *Maritime Policy and Management* 14(1), 1987, 27–48.

Addresses a number of well-known problems in shipping logistics by means of the 'assignment problem' case in linear programming. Uses realistic cost-figures and operations data.

1827 **Cowap, G. R.** Alternative computer solutions to general cargo management. Thesis submitted for M.Phil, CNAA, 1979.

1828 **Frankel, E. G.** Ship management systems development. *Maritime Policy and Management* 9(2), 1982, 135–43.

Considers complex array of decisions required to ensure effective operation and performance of a ship. How is new technology changing the nature of ship management?

1829 **Frankel, Ernst G.** *Management and operations of American shipping.* Boston (Massachusetts): Auburn Ho. Publishing Co., 1982. 200 pp.

1830 **General Council of British Shipping.** *A manual on the law concerning unfair dismissal as it applies to seafarers.* 2nd rev. edn. 28 pp.

A manual designed primarily to assist shore-based staff departments with responsibility for employment and dismissal practices and procedures relating to seafarers.

1831 **Gorton, Lars.** Ship management agreements. *Journal of Business Law* 1991, 562–74.

Examines shipowner's role in management of vessel and relationship with independent ship manager. Legal problems which may occur in relation to duties and functions of ship manager; his authority and liability in different relations to third parties, cargo owners, passengers, crew, officers, etc.

1832 **Hudson, G. F., Johnson, R. A.** and **Lawrence, D. P.** *Integrated ship management and instrumentation.* London: Witherby & Co., 1988. 112 pp.

Examines the technological changes, modern instruments and integrated systems which aid the seafarer in marine transportation. Three areas highlighted: Fuel management, man-power management, minimizing equipment downtime. (Book supports series of four videotapes of same title.)

1833 **Laurence, C. A.** *Vessel operating economies.* London: Fairplay Publications, 1984. 60 pp.

Considers overall theme of operating economies in respect of operation of ships. Considers whole philosophy of performance monitoring and motivation of people. The total approach of all units of activity in ship management. An undepartmental approach.

1834 **Lavery, H. I.** *Shipboard operations.* 2nd rev. edn. Oxford: Heinemann Professional, 1990. 306 pp.

Covers basic routines and procedures, and the many regulations governing their use, for the safe and efficient day-to-day operation of merchant ships. In particular, SOLAS Chapter III; MARPOL 73/78; Annex II and Annex V; 1988 Occupational Health and Safety Regulations; Ro-Ro Ferry Regulations.

1835 *Leading developments in ship management.* London: Lloyd's of London Press, 1992. 100 pp.

Selection of papers from the 2nd International Lloyd's Ship Manager Ship Management Conference 1991.

1836 **Lorange, Peter** and **Norman, Victor D.** (eds) *Shipping management: Papers presented at an International Research Seminar jointly sponsored by the Maritime Research Centre, The Hague, and the Institute for Shipping Research, Bergen, August 23–26 1972.* Bergen: Institute of Shipping Research; The Hague: Maritime Economic Research Centre, 1973.

1837 **Society of Naval Architects and Marine Engineers.** *Ship Operations, Management and Economics Symposium.* London: SNAME, 1990.

1838 **Society of Naval Architects and Marine Engineers, American**

Society of Naval Engineers, Society of Marine Port Engineers of New York. *1987 Ship Operations, Management and Economics: International Symposium* (in two parts). London: SNAME, 1987.

1839 **Spruyt, John.** *Ship management.* London: Lloyd's of London Press, 1990. 222 pp.

Up-to-date work of reference which describes history of ship management and analyses the participants, markets and legal responsibilities. Examines the difficulties and opportunities facing ship managers, and areas of future growth.

Education and training

1840 **Bassis, Michael S.** and **Rosengren, William R.** The organization of nautical training and executive leadership. *Maritime Studies and Management* 3(2), October 1975, 87–94.

Indicates some of the distinctive problems of executive leadership which surround the control of authority in merchant navy training schools in the US, Great Britain and Spain.

1841 **Brown, Robert Henry.** *Training notes for brokers: Insurance of merchant ships and shipowners' interests. Bk. 1. Bk. 2. Bk. 3.: Supplementary notes.* London: Witherby & Co., 1990.

1842 **Couper, A. D.** and **Couroux, M. B.** *Maritime management and training in developing countries.* Geneva: United Nations Conference on Trade and Development, 1979.

1843 **Ircha, M. C.** and **Thomas, B. J.** Systematic training for maritime transport managers. *Maritime Policy and Management* 20(2), April—June 1993, 163–72.

Focuses on transport training initiatives in developing countries, particularly within the marine mode. Examines the two parallel training programmes initiated by UNCTAD in 1980s: TRINMAR and IPP.

1844 **Menon, P. K.** The World Maritime University. *Maritime Policy and Management* 14(2), 1987, 157–61.

From birth of IMO in 1958 it has been concerned with lack of trained maritime personnel. Technical assistance programme was tried, finally UN and Sweden helped to create WMU at Malmö in 1983.

1845 **Molina, E. L.** Changes in shipping management and implications for shipping management training. Thesis submitted for MSc, University of Bath, 1974.

1846 **United Nations. Economic Commission for Asia and the Far East.** *Training course on the management of tramps, tankers and purpose-built tonnage, 15th–26th May, 1972, Singapore.* New York: United Nations, 1973. 221 pp.

Export and freight forwarding

1847 **Bird, James** and **Bland, Gail.** Freight forwarders speak: The perception of route competition via seaports in the European Communities Research Project. Part 1. *Maritime Policy and Management* 15(1), January–March 1988, 35–55.

Part 1 of two papers. To discover views about seaport development and port routeing by group of 'landward customers' – freight forwarders. Seventy-two interviews in 11 European countries.

1848 **Bird, J.** Freight forwarders speak: The perception of route competition via seaports in the European Communities Research project. Part 2. *Maritime Policy and Management* 15(2), 1988, 107–25.

Analyses 72 interviews to show how freight forwarders see their jobs. Emphasis on increasing symbiosis of land and sea operations; vital topics of costs and time; frustration with Customs, etc.

1849 **Branch, Alan E.** *Elements of export marketing and management.* 2nd rev. edn. London and New York: Chapman and Hall, 1990. 339 pp.

Practical overall understanding of techniques of export marketing. Role of export management. Export marketing strategy. 1992 Single Market Entity. Export selling/negotiation; market research; licensing; pricing; joint ventures; overseas business visits; market selection.

1850 **Branch, Alan E.** *Elements of export practice.* 2nd rev. edn. London and New York: Chapman and Hall, 1985. 450 pp.

Practical treatment of subject. Theory of export. Technique of exporting goods – profitable, competitive contract. Role of export management. Export documentation. Customs; finance; cargo insurance; freight rates; packing; export facilitation organizations.

1851 Branch, Alan E. *Elements of import practice.* London and New York: Chapman and Hall, 1990. 289 pp.

Companion to *Elements of Export Practice* (2nd edn) by same author. This book deals with the theory and practice of importing on a cost-effective and profitable basis. The shipping office. Buying. Physical distribution. Freight tariff. Customs. Finance. Cargo insurance. Documentation. Incoterms 1990. The import order. Facilitation organizations. Overseas business trips. Transport.

1852 Bugg, Ralph and **Whitehead, Geoffrey.** *Elements of transportation and documentation.* London: Woodhead-Faulkner, 1984. 290 pp.

Covers procedures and methods which need to be followed when transporting freight. Examination of export documentation. Function and structure of export office. Consignment delivery terms for transport. Modes of transport and rates, charges, procedures for special cargoes. Customs practice. Costing and accounting. Postal services.

1853 Denniston, Thomas. Freight forwarders face the 1990's. In: *Leading developments in international marine insurance: An industry report.* London: Lloyd's of London Press, 1991. 10 pp.

The economic regulation of freight forwarders trading in the US; liabilities; need for liability insurance; new efficiencies for 1990s; real value of damages.

1854 *Freight forwarders' liability.* Oyez Longman, Oyez Science and Technical Services, 1978. 28 pp.

1855 *Freight forwarding: A market sector overview.* 7th edn. London: Key Note, 1993.

1856 Gooding, K. Freight forwarders. *Export Direction*, January 1978, 12.

1857 Grimsley, R. H. Forwarding: A costing formula. *Freight Management*, November 1972, 56.

1858 Hill, D. J. *Freight forwarders.* London: Stevens, 1972. 376 pp.

Places law within commercial framework. Common law countries of Commonwealth treated as one jurisdiction to establish pattern of relationships. Relevant American law and practice. Some reference to Continental civil law, in respect of status and secured rights over goods for non-payment.

1859 Jacobs, Jon. SOS! Save on shipping. *International Business* 5(9), September 1992, 27–31.

The reduction of shipping costs becoming an increasingly bigger factor in US companies business.

1860 John, M. and **Wright, D.** Impact of information technology on the methods and working systems of the freight forwarding industry. *Maritime Policy and Management* 18(2), 1991, 113–22.

Reviews background to freight forwarding industry and analyses reasons for poor implementation of successful information technology systems.

1861 Matear, S. and **Gray, R.** Factors influencing freight service choice. *International Journal of Physical Distribution and Logistics Management* 23(2), 1993, 25–36.

This asks the question as to whether shippers and freight forwarders apply different criteria in the selection of freight transport and their unseen transport services; coming to fairly broad conclusions and suggesting further research.

1862 Newborne, M. J. *A guide to freight consolidation for shippers.* Andover: International Thomson Transport Press, 1977.

1863 O'Kane, Gerry. Those in peril on the sea. *Asian Business* 28(9), September 1992, 50–56.

The problems a company can face when shipping goods, by sea, land or air. The solutions to save time and money depend on the company's size.

1864 Slack, Brian. Containerization, inter-port competition, and port selection. *Maritime Policy and Management* 12(4), October–December 1985, 293–303.

Paper explores criteria shippers employ in the port selection process. Focuses on containerization traffic between North American Midwest and Western Europe. Examines factors considered by exporters and freight forwarders.

1865 Stapleton, J. *Elements of export management.* Cambridge: Woodhead-Faulkner Ltd, 1984. 170 pp.

1866 Ullman, Gerald H. Ocean freight forwarders in the United States. *Journal of Maritime Law and Commerce* 7(4), July 1976.

Considers role of US freight forwarder and variety of services which begin before goods leave export plant. Distinction between US and European freight forwarders. Supervision by Federal Maritime Commission.

1867 Ullman, Gerald H. The role of the American ocean freight forwarder of intermodal, containerized transportation. *Journal of Maritime Law and Commerce* 2(3), April 1971, 625–43.

Traditional role of ocean freight forwarder. Effects of containerization on forwarder's role. Government proposals for authorization of through transport. Meeting the challenge.

1868 Walker, A. G. *Export practice and documentation.* 3rd edn. Guildford: Butterworths-Heinemann, 1987. 272 pp.

1869 Whitehead, Geoffrey. *Elements of export law.* Cambridge: Woodhead-Faulkner Ltd, 1983. 160 pp.

Shipbrokers

1870 Gorton, L., Ihre, R. and **Sandevarn, A.** *Shipbroking and chartering practice.* 3rd edn. London: Lloyd's of London Press, 1990. 297 pp.

Provides a reference about the practice and commerce of shipping and sets out the legal framework which regulates the shipping industry. Cases and arbitration decisions.

1871 Institute of Chartered Shipbrokers. *The Shipbroker's Manual, Vol. 1.* London: Lloyd's of London Press, 1983. 167 pp.

Part one of series of shipping manuals. Aimed as compendium for students of Institute's professional examination. Law and practice. Contract issues. P & I clubs. Admiralty jurisdiction. Limitation of liability. Arbitration. Shipowner. Master. Conflict of laws. Damages. Ownership/registration of British ships.

1872 *Shipbroking in the '80's.* Hamburg: German Shipbrokers' Association, 1980.

1873 United Nations Conference on Trade and Development. *Merchant fleet development: Ways and means of establishing shipbroking centres in developing countries on a regional basis.* Geneva: UNCTAD, 1986. 26 pp.

Preliminary report prepared pursuant to resolution on merchant fleet development adopted by Committee, 11th Session. Existing shipbroking centres and international institutions. Ways and means of establishing shipbroking centres in developing countries on a regional basis.

12 Law

General

General

1874 **Abbott, Charles** and others. *A treatise of the law relative to merchant ships and seamen.* 14th edn. Littleton (Colorado): Fred B. Rothman & Co., 1993. 1356 pp.

1875 **Abecassis, David W.** and others. *Oil pollution from ships: International, United Kingdom and United States law and practice.* 2nd edn. London: Stevens & Sons, 1985. 619 pp.

Outlines structure in detail; highlights points of particular interest or importance. Comprehensive survey of relevant legal principles.

1876 **Ahnish, Faraj Abdullah.** *International law of maritime boundaries and the practice of states in the Mediterranean sea. Oxford monographs in international law.* Oxford: Oxford University Press, 1993. 430 pp.

1877 **Anderson, Andrew W.** Jurisdiction over stateless vessels on the high seas: An appraisal under domestic and international law. *Journal of Maritime Law and Commerce* 13(3), April 1982, 323–42.

Examines issue of whether US courts may give effect to authority claimed by Congress over vessels without nationality. Fundamental areas of constitutional and international law.

1878 **Astle, W. E.** *Legal developments in maritime commerce: Matters of general interest and importance arising out of selected legal issues.* London: Fairplay Publications, 1983. 125 pp.

Examines selection of important developments arising out of the governing legislation in maritime commerce. Includes arbitration procedure, uniformity in application of international rules, radar, potential of fire from spillage of oil. Trade Union and Labour Relations Act, Warsaw Convention, documentary credits, Mareva injunction.

1879 **Banda, C. A.** The maritime status of landlocked states in international law. Thesis submitted for M.Phil, University of Kent at Canterbury, 1979.

1880 **Baylis, Arthur E.** The standard shipping format: A guide to simplified documents. *Journal of Maritime Law and Commerce* 2(2), January 1971, 399–412.

Study made jointly with Dept of Transport. Aims to make recommendations to eliminate, simplify and otherwise reduce international trade documents and procedures that cause them.

1881 **Briggs, Adrian.** Wider still and wider: The bounds of Australian exorbitant jurisdiction. *Lloyd's Maritime and Commercial Law Quarterly*, May 1989, Part 2, 216–24.

Considers doctrine of *forum non conveniens* in applications for leave to serve out of jurisdiction under R.S.C. Ord. 11. The English judiciary's acceptance of the Spiliada (1987) approach. The High Court of Australia's decline to follow the Spiliada, in *Oceanic Sun Line Special Shipping Co. Inc.* v. *Fay* (1988).

1882 **Charity, David E.** 'Mareva' injunctions: A lesson in judicial acrobatics. *Journal of Maritime Law and Commerce* 12(3), April 1981, 349–62.

Development of this remedy. Statutory basis. Case law. Effect of the injunction. Consequential relief. Prospective application.

1883 **Cole, Stamford Darley.** *Merchant Shipping Acts.* 8th edn. 1960.

1884 **Comité Maritime International.** *Apportionment of risk in maritime law.* Paris: CMI, 1976.

1885 **Corbet, A. G.** Development of vessel traffic services: Legal considerations. *Maritime Policy and Management* 16(4), 1989, 277–92.

Legal changes, rather than more technological advances, are needed in order to extend development of vessel traffic services.

1886 Curzon, L. B. (ed.) *Chance's principles of mercantile law.* 22nd edn. London: Cassell Ltd, 1980. 362 pp.

A textbook based on the general syllabus requirements in mercantile law of the professional examining bodies in commerce, accountancy and secretarial studies. References to changes effected by the introduction of statutes from 1974 to 1979.

1887 Davison, Richard and **Snelson, Anthony.** *The law of towage.* London: Lloyd's of London Press, 1990. 266 pp.

Comprehensive guide to legal principles. Detailed discussion of standard form agreements in most common use, e.g. UK Standard Conditions. Provisions affecting employment, health, safety, and insurance. Limitation Convention, LOF 90.

1888 Evans, Malcolm D. *Relevant circumstances and maritime delimitation.* Oxford: Oxford University Press, 1989. 255 pp.

1889 Fogarty, Aengus. *Merchant shipping legislation.* London: Lloyd's of London Press, 1992. 1376 pp.

1890 Gandhi, D. S. Shipping and the new law of the sea. Thesis submitted for MSc, UWIST, Cardiff, 1978.

1891 Gold, Edgar. International maritime law in transition: New challenges for education and training. *Marine Policy* 13(3), July 1989, 178–92.

Examines the developing base of modern maritime law and recommends that maritime education and training must today include an increased portion of legal and policy knowledge in order to ensure the overall competency of those who operate vessels in an ever more complex maritime era.

1892 Gold, Edgar. *Maritime transport: The evolution of international marine policy and shipping law.* Lexington (Massachusetts): Lexington Books, 1981. 425 pp.

Traces evolution of private and public maritime law. International marine policy and shipping laws. Traditions of maritime states. Unites divergences among various marine law and policy aspects affecting international maritime transport.

1893 Gough, Neville. *Institute cargo and related trade clauses: A detailed analysis.* London: Witherby & Co., 1988.

1894 Harris, Nigel and others. *Use of offshore jurisdictions.* London: Longman Law, Tax and Finance, 1991. 840 pp.

1895 Hazelwood, Steven J. Barratry: The scuttler's easy route to the 'Golden Prize'. *Lloyd's Maritime and Commercial Law Quarterly,* August 1982, Part 3, 383–400.

Establishing 'perils of the sea' and 'barratry' in order to claim insurance on vessel. Case law development. Burden of proof upon shipowner to prove claim. Burden of proof upon underwriter to show scuttling.

1896 Hoffman, Stanley. *Modern legal forms for shippers.* Mamaroneck (New York): Transport Law Research, Inc., 1970. 508 pp.

Contains a number of forms which have been developed and refined by/in conjunction with author's associates. Trackage and property forms. Transportation equipment. Transportation service. Storage forms. Freight consolidation. Private carriage. Loss, damage, overcharge. Vendor–vendee. Practice.

1897 Hopkins, F. J. (revised by G. G. Watkins) *Business and law.* 7th rev. edn. Glasgow: Brown, Son and Ferguson, 1989. 900 pp.

Reference for many aspects of maritime law and shipping business. Registry, classification, crew matters, health, safety, carriage of passengers and goods, customs formalities, towage, pilotage, marine insurance, limitation of liability.

1898 Hoyle, Mark D. W. *Mareva injunction and related orders.* 2nd edn. London: Lloyd's of London Press, 1989. 240 pp.

Provides a practical and detailed examination of the Mareva injunction. All relevant aspects of the Mareva injunction, demonstrating how it works in practice and how it is likely to develop. The position in the EEC countries given. Implications of the Civil Jurisdiction and Judgments Act 1982.

1899 Jackson, D. C. *Enforcement of maritime claims.* London: Lloyd's of London Press, 1985. 483 pp.

Based on enquiries made by potential claimant in a maritime cause of action in English court. Jurisdiction and mode of enforceability. Interim relief. Security on the merits – the lien concept. The foreign element. Remedies.

1900 Jackson, David. *Civil jurisdiction and judgments: Maritime claims.* London and New York (*et al.*): Lloyd's of London Press, 1987. 356 pp.

Role of national laws. Framework for recognition and enforcement of judgments introduced by the Civil Jurisdiction and Judgments Act 1982. Examination of Scottish framework and extended framework for intra-UK disputes.

Provisions of 1968 Convention and interpretation by European Court. Relevance of defendant's domicile.

1901 **Lee, D. C.** *C.I.I. marine law.* London: Witherby & Co., 1986.

1902 *Lloyd's maritime law yearbook.* London: Lloyd's of London Press.

1903 **Lowe, A. V.** Jurisdiction over foreign merchant ships in marginal waters. Thesis submitted for PhD, UWIST, Cardiff, 1980.

1904 **Malek, Ali** and **Lewis, Caroline.** Worldwide Mareva injunctions: The position of international banks. *Lloyd's Maritime and Commercial Law Quarterly,* February 1990, Part 1, 88–98.

Examines effect on banks of the judicial widening of the Mareva jurisdiction to encompass assets abroad. Suggests that the courts have left the position of banks carrying on business in different jurisdictions in an uncertain and unacceptable state.

1905 **Manca, Plinio.** *International maritime law.* Vol. I. Brussels: European Transport Law, 1970. 325 pp.

First of two volumes analysing international conventions, projects of international conventions, maritime rules, international law issues – conflict of laws, territorial waters, ports, ships, shipowners and liabilities, agents, crew, employment contracts.

1906 **Manca, Plinio.** *International maritime law.* Vol. II. Brussels: European Transport Law, 1970. 461 pp.

Second of two volumes analysing international conventions, projects of international conventions, maritime rules, international law issues – contracts for employment of ships, carriage of passengers/luggage, stowaways, carriage of goods by sea, international combined transport.

1907 **Mangone, Gerard J.** *Law for the world ocean.* London: Stevens & Sons, 1981. 313 pp.

Develops concepts of law that have been applied to the coastal margins and high seas from ancient times through age of mercantilism to middle of twentieth century. Public international plus private maritime law issues – regulation of world merchant marine; legal protection of marine environment.

1908 **Martinez, Robert D.** Marine radar: A proposal for mandatory usage. *Journal of Maritime Law and Commerce* 11(1), October 1979, 109–19.

Use of marine radar aboard commercial vessels. Examination of legal status. The need for a revised statutory status.

1909 **Massey, Eugene A.** Prospects for a new intermodal legal regime: A critical look at the TCM. *Journal of Maritime Law and Commerce* 3(4), July 1972, 725–57.

Examines questions and issues raised by TCM. Background to TRC Convention and aims of Combined Transport of Goods. IMCO/ECE Draft. Scope and applicability of the Convention.

1910 **Mathews, John.** Law of the sea. *New Left Review* 95, January/February 1976, 55–62.

1911 **Meeson, Nigel.** *Ship and aircraft mortgages.* London: Lloyd's of London Press, 1989. 217 pp.

Considers position of both mortgagor and mortgagee under modern form of statutory mortgage and under an equitable mortgage. Rights and obligations of mortgagor and mortgagee. Priorities; collateral deeds; transfers; discharges.

1912 **Ngantcha, Francis.** *The right of innocent passage and the evolution of the international law of the sea: The current regime of free navigation in coastal waters of third states.* London: Pinter, 1990. 250 pp.

1913 **O'Connell, D. P.** *The international law of the sea.* Vol. II. Oxford: Clarendon Press, 1984. 566 pp.

Volume II is concerned with public international law, and transcends traditional divisions and includes relevant principles of private maritime law. Shipping law issues *re* nationality, registration. International Carriage by Sea. Jurisdiction over navigation and port areas. Arrest. Salvage. Collision.

1914 **O'Connell, D. P.** The law of the sea. *Royal Society of Arts Journal* 124, June 1976, 367–79.

The paper given and discussion following at the Thomas Gray Memorial Lecture, February 1976, to the Society. Reviews problems of fixing limits of coastal jurisdiction. The Law of the Sea Conference. Impact on British mercantile marine.

1915 **Pineus, Kaj.** *Time-barred actions.* London: Lloyd's of London Press, 1984. 157 pp.

Questionnaire response studied. Terminology. Rationale of time bars. Practical impact. Conventions and national law. How to stop time bar running out. Interruption and suspension.

1916 Powles, David G. The Mareva injunction. *Journal of Business Law*, 1978, 11–18.

Development in 1975 by Court of Appeal of *ex parte* injunction preventing a defendant from removing assets from jurisdiction where judgment not entered against him. Historical grounds and grounds of commercial expediency.

1917 Powles, David G. The Mareva injunction expanded. *Journal of Business Law*, 1981, 415–31.

Widening of application of the remedy. Relationship between Mareva injunction and other actions and remedies. Limits of its application.

1918 Powles, David G. Mareva round-up – I. *Journal of Business Law*, 1982, 383–90.

Developments in respect of application of remedy. Basis of jurisdiction. Nature of the order. Policy. Guidelines.

1919 Powles, David G. Mareva round-up – II. *Journal of Business Law*, 1982, 489–95.

Continues from Part I in same volume, pp. 383–90. The Mareva injunction in other common law jurisdictions. Future developments.

1920 Purvis, Rodney N. and **Darvas, Robert.** *The law and practice of commercial letters of credit shipping documents and termination of disputes in international trade.* Sydney: Butterworths, 1975. 247 pp.

Letters of credit; Uniform Customs and Practice for Documentary Credits; strict compliance; bill of lading; marine insurance; contract and terms; discharge of contract; settlement of disputes and arbitration; conflict of law.

1921 Rogers, Andrew. The extra-territorial reach of the Mareva injunction. *Lloyd's Maritime and Commercial Law Quarterly*, May 1991, Part 2, 231–46.

Examines whether (i) Mareva injunctions can be granted against defendants who are out of jurisdiction of the court and in respect of foreign assets; (ii) should such orders be made and if so in what form?

1922 Rogers, David N. The action in rem and 'Mareva' injunction:

The need for a coherent whole. *Journal of Maritime Law and Commerce* 14(4), October 1983, 513–36.

Examines why claimant might lose security for its claim within jurisdiction of Canada. Why Mareva injunction might be dissolved. How might problem be overcome. Development of proceeding in rem, examination of 'The Beldis' rule. How to overcome problems in the rule – statutory reform; common law concept of Mareva.

1923 **Simmonds, Kenneth R.** (ed.) *Cases on law of the sea.* 4 vols. New York: Oceana Publications, 1984.

1924 **Singh, Nagendra.** *International maritime law conventions – Vol. 2. Safety.* 3rd edn. London: Stevens & Sons, 1988. 897 pp.

Load lines; safety of life at sea; passenger ships; safe containers; fishing vessels; maritime search and rescue; tonnage measurement of ships.

1925 **Singh, Nagendra.** *International maritime law conventions – Vol. 4. Maritime law.* 3rd edn. London: Stevens & Sons, 1988. 680 pp.

Law of the sea, landlocked states, ports, canals and straits. Freedom of transit and navigable waterways. Private maritime law, miscellaneous conventions and international organizations.

1926 **Stanton, L. F. H.** *Guide to the Merchant Shipping Acts.* 2 vols. Glasgow: Brown, Son and Ferguson Ltd, 1987. 526 pp.

1927 **Sturley, Michael F.** The future of the COGSA fair opportunity requirement: Is there life after Carman Tool and Chan? *Journal of Maritime Law and Commerce* 20(4), October 1989, 559–69.

Examines *Carman Tool and Abrasives Inc.* v. *Evergreen Lines* (1989) in light of fair opportunity doctrine under S. 4(5) COGSA. Also, *Chan* v. *Korean Air Lines* (1989) and guidance for future cases. Significance of court's opinion and speculation on future of fair opportunity requirement.

1928 **Tetley, William.** *Marine cargo claims.* 3rd edn. Montreal: International Shipping Publications, 1988. 1305 pp.

Examines the law relating to marine cargo claims. Preliminaries to marine cargo claims and their defence. What the cargo claimant must prove. What the carrier must prove. Counter proof by the claimant. Related arguments available to both parties. National summaries.

1929 **Tettenborn, A. M.** Transferable and negotiable documents of title: A redefinition? *Lloyd's Maritime and Commercial Law Quarterly,* November 1991, Part 4, 538–42.

Considers concepts of 'negotiability' and 'transferability'. Problems caused by lack of universally agreed definition. Comparison of different theories.

1930 **Trakman, Leon E.** The evolution of the law merchant: Our commercial heritage. *Journal of Maritime Law and Commerce* 12(1), October 1980, 1–24.

First of two parts. Examines ancient and medieval law merchant. Development of ancient law merchant. Thesis is that modern international law and regulatory institutions have developed from ancient custom and usage.

1931 **Trakman, Leon E.** The evolution of the law merchant: Our commercial heritage. *Journal of Maritime Law and Commerce* 12(2), January 1981, 153–82.

Second of two parts. Examines the modern law merchant. Attitude of nineteenth- and twentieth-century common law judges. American cases. A 'new' law merchant.

1932 **Trappe, Johannes.** Too remote. *Lloyd's Maritime and Commercial Law Quarterly,* August 1982, Part 3, 427–32.

Comparison of various national laws – considers whether different answers given internationally to some questions raised in different places. Examples of such situations.

1933 *United Nations Commission on international trade law: International legislation on shipping.* New York: United Nations, 1970.

1934 **Verzijl, J. H. W., Heere, W. P.** and **Offerhaus, J. P. S.** (eds) *International law in historical perspective. Vol. II: Law of maritime prize.* The Hague: Nijhoff, 1991. 696 pp.

1935 **Whitehurst, C. H.** Is the legal definition of a passenger ship obsolete? *Journal of Maritime Law and Commerce* 11(4), July 1980, 527–35.

Asks whether current definition of passenger ship obsolete in light of present technology. If applicable law less restrictive, what options open to shipping firms?

United Kingdom

1936 Ademuni-Odeke. U.K. Merchant Shipping Act, 1988: Protection of shipping and trading interests. *Marine Policy* 15(4), July 1991, 272–88.

Examines impact of the 1988 Act. Suggests that despite noble objectives of providing for the protection of both shipping and trading interest, the Act has probably reintroduced flag discrimination, maritime subsidies, cabotage restrictions and protectionism against developing countries and some of the UK's EEC and OECD partners.

1937 Cadwallader, F. J. J. English shipping cases: 1970. *Journal of Maritime Law and Commerce* 2(4), July 1971, 815–35.

Survey of English shipping law 1970: Admiralty practice; collisions; salvage; charterparties; bills of lading; stevedores; marine insurance.

1938 Cadwallader, F. J. J. English shipping cases: 1971. *Journal of Maritime Law and Commerce* 4(2), January 1973, 255–94.

Survey of English shipping cases for 1971. Admiralty practice; collisions; salvage; towage; charterparties; bills of lading.

1939 Cadwallader, F. J. J. English shipping cases: 1972. *Journal of Maritime Law and Commerce* 5(3), April 1974, 407–61.

Survey of English shipping cases for 1972. Admiralty practice; collisions; pilotage; charterparties; bills of lading; marine insurance.

1940 Cadwallader, F. J. J. English shipping cases: 1973. *Journal of Maritime Law and Commerce* 6(3), April 1975, 331–93.

Survey of English shipping cases for 1973. Admiralty practice; collisions; charterparties; bills of lading; marine insurance; harbours.

1941 Davies, M. B. Legal measures for the encouragement of British shipping and shipbuilding, 1485–1977. Thesis submitted for PhD, London, University College, 1981.

1942 Harper, J. C. Statutes: The Merchant Shipping Act, 1970. *Modern Law Review* 34, January 1971, 55–74.

The Merchant Shipping Act 1970 tracks the employment conditions of merchant seamen and fishermen kicking and screaming into the nineteenth century. Essentially an enabling measure – it reaches over a number of issues but is basically concerned with terms and conditions of employment and replacement.

1943 Lewis, Charles. The case of the Cuban sugar. *Lloyd's Maritime and Commercial Law Quarterly,* February 1982, Part 1, 96–103.

Modern development of English law in respect of state immunity in commercial transactions. House of Lords decision in Primero Congreso del Partido (1981). Restrictive doctrine of immunity applied to actions in respect of state-owned trading vessels.

1944 Lewis, N. Merchant Shipping Act 1970 status. *Modern Law Review* 34, 1971, 55–61.

1945 Singh, Nagendra and **Colinvaux, Raoul.** *Shipowners. British shipping laws.* Vol. 13. London: Stevens & Sons, 1967. 403 pp.

Ownership and acquisition of tonnage; English law of sale; operation of merchant shipping; organizations aiding shipowners; insurance in London; ports of UK agents; bills of lading; settlement of disputes in London; flag discrimination.

1946 Thomas, D. R. *Maritime liens. British Shipping Laws.* Vol. 14. London: Stevens & Sons, 1980. 397 pp.

Analyses the concept and nature of a maritime lien and examines individually the principal maritime liens recognized in English law. Examines various associated issues such as right in rem enjoyed against the Crown and foreign sovereigns. Priorities. The assignment and extinction of maritime liens. Conflict of law. International unification.

1947 Thomas, Michael and **Steel, David.** *Temperley: The Merchant Shipping Acts. British Shipping Laws.* Vol 2. 7th rev. edn. London: Stevens & Sons, 1976. 1001 pp.

Examination of English shipping related statutes. From Merchant Shipping Act 1894 to Merchant Shipping Act 1974. Case law. Rules and Orders.

Western Europe and Scandinavia

1948 Humphreys, Gordon and **Higgs, Andrew.** Waybills: A case of common law laissez faire in European commerce. *Journal of Business Law,* September 1992, 453–80.

Examines some of the problems which waybills present when used in international trade, the extent to which these issues are likely to be remedied by pending legislation, and the solutions which the shipping industry might introduce for itself to plug remaining loopholes.

1949 Karatzas, Theodoros and **Ready, Nigel.** *Greece: The Greek Code* *of Private Maritime Law.* The Hague and London: Nijhoff, 1982. 98 pp.

Examines the provisions of the Greek Code of Private Maritime Law.

1950 Vella, Edwin A. Malta: The Merchant Shipping Act, 1973 – tax concessions. *Bulletin for International Fiscal Documentation* 40(8/9), August–September 1986, 413–16.

1951 Yi, Seong Deog and **Choi, Chong Ju.** The Community's unfair pricing practices in the maritime transport sector. *European Law Review* 16(4), August 1991, 279–94.

The unfair pricing practices regulations in maritime transport were adapted in 1986. This article analyses The Hyundai to which the regulation was first applied.

Central and Eastern Europe and Middle East

1952 Chakradaran, C. and **Bashir, Ahmed.** Marine cargo claims under the laws of the United Arab Emirates. *Lloyd's Maritime and Commercial Law Quarterly,* November 1992, Part 4, 538–48.

The Provisional Constitution on which legal system in UAE based. Western jurisprudence and influence. Maritime law – Articles 256–87 of the Maritime Law. Similarity to Hague Rules.

1953 Haberbeck and Galloway. *Saudi shipping law.* London: Fairplay Publications.

1954 Makovskaia, A. A. Development of contractual relations in maritime transport: 1991 Soviet civil legislation. *Marine Policy* 16(2), March 1992, 15–17.

Examines the significance of the 1991 Fundamental Principles of Civil Legislation of the USSR and the Republics for maritime transport. Analyses the new regime of contractual relations using carriage of goods as an example.

1955 Price, Richard. *Maritime laws of the Arabian Gulf Co-operation Council States.* 2 vols. London: Graham and Trotman, 1986. 400 pp.

1956 Vafai, G. H. Commercial laws of Iran: Iran. In: Allen P. Keesee and Gamal Badr. *Commercial laws of the Middle East.* 10 vols. New York: Oceana Publications, 1980.

North America

1957 Bauer, R. Glenn. What has become of the U.S. fire statute? *Lloyd's Maritime and Commercial Law Quarterly*, February 1982, Part 1, 1–4.

Developments in US law in respect of cargo damage by fire.

1958 Bowman, Roy G. The Merchant Marine Act of 1970. *Journal of Maritime Law and Commerce* 2(4), July 1971, 715–34.

Examines problems and explains philosophy leading to the 1970 Act. Operating subsidy under 1936 Act. Index system. Buy American provisions – Title VI; Title V. Eligibility. Other provisions. Tax deferred funds. Construction – differential subsidy. Procurement procedures. Productivity goals.

1959 Buchwald, Todd. Federal Maritime Commission jurisdiction over terminal operators. *Journal of Maritime Law and Commerce* 12(2), January 1981, 209–32.

'Other persons' within regulatory power of Federal Maritime Commission under the Shipping Act 1916. Includes terminal operators. Development of concept of 'other persons'. Definition of 'other persons' – to allow comprehensive regulation of maritime commerce.

1960 Cervone, John B. The manner in which state law affects federal maritime law. *Journal of Maritime Law and Commerce* 8(4), July 1977, 455–88.

When US state law affects admiralty and maritime law (as connoted by Article III Section 2 US Constitution) and varying degrees of those effects. Criteria used by courts to determine when state law will be annexed or infused into maritime law.

1961 Dillon, Stephen D. Pre-emption and the commerce clause revisited: The 1975 Washington tanker law. *Natural Resources Journal* 17(4), October 1977, 691–702.

1962 Engerrand, Kenneth G. and **Brann, Scott R.** Troubled waters for seamen's wrongful death actions. *Journal of Maritime Law and Commerce* 12(3), April 1981, 327–48.

Analysis of area of litigation in field of maritime wrongful death: The conflict between the application of the Jones Act 1977 and *Morague* v. *States Marine Lines Inc.* (1970) to an action for wrongful death of a seaman in state territorial waters. Damages available.

1963 **Fernandez, M.** and **Burke, Christopher.** *Annotated Canada Shipping Act.* Canada: Butterworth, 1987.

1964 **Finc, Mark J.** and **Bureau, Lisa C.** The redundant Foreign Shipping Practices Act of 1988: Now it's up to the Federal Maritime Commission. *Law and Policy in International Business* 20(3), 1989, 555–69.

1965 **Goldman, Steven E.** Non-pecuniary damages in a seaman's wrongful death action. *Journal of Maritime Law and Commerce* 13(1), October 1981, 53–70.

Impact of Supreme Court's decision. *Mobil Oil Corpn.* v. *Higginbotham* (1978) – death outside US territorial waters leads to pecuniary loss only. Examines reverse side of this – jurisprudence, case-law development of principle that death within territorial waters leads to non-pecuniary damages.

1966 **Howard, Alex T.** Personification of the vessel: Fact or fiction? *Journal of Maritime Law and Commerce* 21(3), July 1990, 319–29.

Case-law development of American doctrine of personification of the vessel.

1967 **Kindred, H. M.** and **Brooks, M. R.** The effects of marine cargo delays in law and commerce. *Maritime Policy and Management* 17(3), 1990, 189–97.

Considers incidence of delay in marine transportation, its legal and commercial consequences. Discusses nature of legal obligation and its adequacy based on survey of Canadian shippers etc.

1968 **Lowe, A. V.** The extraterritorial application of United States shipping law. Thesis submitted for LL M, UWIST, Cardiff, 1978.

1969 **McConnell, John W.** (Jr) *The U.S. Shipping Act of 1984.* London and New York: Lloyd's of London Press, 1985. 339 pp.

The 1984 Act almost a complete rewrite of Shipping Act 1916. Review of legislative history of 1984 Act to interpret its provisions in overall legislative scheme. Objectives and scope of new Act.

1970 **McDorman, T. L.** Shipping policy as a British export product: The Canadian case. *Maritime Policy and Management* 11(1), 1984, 1–13.

Outlines development of Canadian shipping law and policy in last fifty years, highlighting impact of 1931 Merchant Shipping Agreement seen as impeding development of truly Canadian shipping policy.

1971 **Robertson, David.** *Admiralty and federalism: History and analysis of problems of federal-state relations in the maritime law of the United States.* Mineola (New York): Foundation Press, 1970.

1972 **Shashikumar, N.** *Shipping Act of 1984: A scrutiny of controversial provisions.* Maine: Maine Maritime Press, 1987. 104 pp.

1973 **Shashikumar, N.** U.S. maritime regulation: History and evolution of the 1984 Act. *Maritime Policy and Management* 15(3), July–September 1988, 241–6.

Background and development; relevant case reference to evolution of Shipping Act 1984, US maritime regulation.

1974 **Sweeney, Joseph C.** Happy birthday Harter: A reappraisal of the Harter Act on its 100th anniversary. *Journal of Maritime Law and Commerce* 24(1), January 1993, 1–42.

Historical overview. Interpretation of the Harter Act. The effect of COGSA on the Harter Act. Vitality of the Harter Act: Its interaction with COGSA.

1975 **Tetley, William.** Canadian maritime law decisions 1987–88. *Lloyd's Maritime and Commercial Law Quarterly,* February 1989, Part 1, 89–103.

General average. Jurisdiction. Definition of a ship. Administrative law. Arbitration. Carriage of goods. Ship mortgages. Limitation of liability. Marine insurance. Collision. Set-off. Pilotage. Labour. Fishing. Seizure. Agency.

1976 **Tetley, William.** Canadian maritime law decisions 1988–1989. *Lloyd's Maritime and Commercial Law Quarterly,* February 1990, Part 1, 99–106.

Collisions and limitation. Judicial sale. Jurisdiction. Charterparties. Carriage of goods. Fishing offences. Marine insurance. Shipbuilding.

1977 **Tetley, William.** Canadian maritime law decisions 1989–1990. *Lloyd's Maritime and Commercial Law Quarterly,* May 1991, Part 2, 264–79.

Jurisdiction. Agency. Arbitration. Carriage of goods. Collisions. Conflict of law. Economic loss. Judicial sale. Limitation. Marine insurance. Mortgages and liens. Procedure. Tug and tow.

1978 **Tetley, William.** Canadian maritime law decisions 1990–1991. *Lloyd's Maritime and Commercial Law Quarterly*, February 1992, Part 1, 115–23.

Jurisdiction. Maritime liens. Carriage of goods by sea. Sale of ship. Arbitration. Collision. Ship construction and repair, modification and judgment.

1979 **Tetley, William.** Canadian maritime law judgments in 1981. *Lloyd's Maritime and Commercial Law Quarterly*, November 1982, Part 4, 542–55.

Jurisdiction of Federal Court of Canada and its Admiralty side. Carriage of goods. Collision and limitations. Labour law. Procedure. Injury to passenger. Tug and tow. Administrative law.

1980 **Tetley, William.** Canadian maritime law judgments in 1982. *Lloyd's Maritime and Commercial Law Quarterly*, November 1983, Part 4, 603–14.

Jurisdiction of the Federal Court. *Forum non conveniens.* Carriage of goods. Practice and procedure. Liens. Administrative law. Labour law. Marine insurance. Crown liability.

1981 **United States Department of Justice.** *Analysis of the impact of the Shipping Act of 1984.* Washington (DC): US Government Printing Office, 1990, pp. 31–5.

1982 **Villareal, Dewey R.** The concept of due diligence in maritime law. *Journal of Maritime Law and Commerce* 2(4), July 1971, 763–77.

Meaning of 'due diligence' (as opposed to results of shipowner's failure to exercise due diligence) considered under the Harter Act and COGSA.

Socialist Asia

1983 **Feng Li Qui.** *Maritime legal proceedings in China.* University of Southampton, 1989. 28 pp.

Africa

1984 **Mbanefo, L. N.** *Nigerian shipping cases.* Professional Books, Butterworth, 1987. 650 pp.

1985 **Mbanefo, L. N.** *Nigerian shipping laws.* Oxford: Professional Books, Butterworth, 1983. 1500 pp.

1986 **Ramsden, W. A.** Maritime law in South Africa: Recent developments. *Journal of Maritime Law and Commerce* 20(2), April 1989, 191–203.

Enactment of Admiralty Jurisdiction Act 105 of 1983. Impact and consequent maritime developments.

1987 **Rembe, N. S.** Africa and the international law of the sea: A study of the legal problems of the African States and of their contribution to the Third United Nations Conference on the Law of the Sea. Thesis submitted for PhD, UWIST, Cardiff, 1977.

Central and South America

1988 **Arria, Luis Coja.** The Venezuelan legislative treatment of maritime law. *Journal of Maritime Law and Commerce* 21(2), April 1990, 251–72.

Examines treatment of maritime law in Venezuelan legislation in order to demonstrate the need to update the relevant legal provisions by enactment of further rules.

Asia, Australasia and Oceania

1989 **Calder, Korella J.** The Commonwealth of Australia Admiralty Act of 1988. *Journal of Maritime Law and Commerce* 20(2), April 1989, 169–90.

Enactment of Act, 1988. Impact and side effects. Examines application of the Act. Changes effected by the Act.

1990 **Chen, C. V.** *Taiwan.* Volume 6. In: Kenneth R. Simmonds (ed.) *Commercial business and trade laws* (21 vols). New York: Oceana Publications, 1981.

1991 **Davies, Martin.** Australian maritime law decisions 1990. *Lloyd's Maritime and Commercial Law Quarterly,* August 1991, Part 3, 326–41.

Carriage of goods by sea. Marine insurance. Procedure. Collisions and liability for damage and personal injury. Salvage. Property and ownership. Navigation. Conflict of laws.

1992 **Davies, Martin.** Australian maritime law decisions 1991. *Lloyd's Maritime and Commercial Law Quarterly,* August 1992, Part 3, 351–75.

Legislative developments. Carriage of goods by sea. Marine insurance. Procedure. Collisions and liability for damage and personal injury. Salvage. Pollution.

1993 **Hornick, Robert N.** Indonesian maritime law. *Journal of Maritime Law and Commerce* 8(1), October 1976, 73–85.

Introduction to more important aspects of Indonesian maritime law including vessel ownership requirements, ship registration procedures, ship mortgaging and the regulation of foreign vessel operations in Indonesian waters.

1994 **Smith, M. F.** Conflict management and maritime strategy in the Indian and western Pacific Oceans: Aspects of international law. Thesis submitted for PhD, University of London, 1984.

1995 **Tangsubkul, Phiphat.** *ASEAN and the laws of the sea.* Singapore: Institute of Southeast Asian Studies, 1982. 149 pp.

1996 **White, M. W.** (ed.) *Australian maritime law.* William W. Gaunt and Sons, 1991. 356 pp.

Carriage of Goods

Contract

1997 **Adams, John** and **Brownsword, Roger.** 'The Aliakmon' and the Hague Rules. *Journal of Business Law,* 1990, 23–35.

Argues that an assault on the privity in contract rule can be justified in the special context of bailments. Decision by House of Lords in *The Aliakmon,* 1986.

1998 **Appleby, George** and **Meisel, Frank.** 'Hongkong Fir' revisited: The House of Lords and certainty in contracts. *Lloyd's Maritime and Commercial Law Quarterly,* November 1992, Part 4, 592–610.

A reconsideration of construction of contractual terms over preceding few years and, in particular, two cases of House of Lords: (i) *Bunge Corporation* v. *Tradax Export S.A.* (1981) and (ii) *The Chikuma* (1981). Considers question of time clauses in contracts and how they should be construed.

1999 **Astle, W. E.** *Hague Rules law digest.* London: Fairplay Publications, 1981. 90 pp.

A digest of leading British law cases in respect of issues arising from the Rules. Liability for cargo loss/damage.

2000 **Astle, W. E.** *The Hamburg Rules.* London: Fairplay Publications, 1981. 163 pp.

An appreciation of the cause and effect of the amendments to the Hague Rules and the Hague–Visby Rules. Presents Hamburg Rules in their final form. Reference to relative Hague Rule and explanation of nature and manner of amendment to Rule concerned. Reviews findings of UNCTAD on this area.

2001 **Astle, W. E.** *Shipping and the law: A practical review of the Hague Rules as amended by the Brussels Protocol, 1968.* London: Fairplay Publications, 1980. 260 pp.

Examines history and origin of the Hague/Hague–Visby Rules and the introduction of Carriage of Goods by Sea Acts 1924, 1971. Application and limitations on application of Rules. Nuclear Damage Convention not affected by Rules.

2002 **Bauer, R. Glenn.** Conflicting liability regimes: Hague Visby v. Hamburg Rules – a case by case analysis. *Journal of Maritime Law and Commerce* 24(1), January 1993, 53–74.

Examines areas of controversy between Hague–Visby regime and the Hamburg Rules in the context of cases and considers in what degree results might differ under these conflicting regimes.

2003 **Beatson, J.** and **Cooper, J. J.** Right of suit in respect of carriage of goods by sea. *Lloyd's Maritime and Commercial Law Quarterly,* May 1991, Part 2, 196–208.

Examines why the Law Commission's 1991 Report on Rights of Suit in Respect of Carriage of Goods by Sea was necessary. Main principles on which proposed reform was based. Summary of 1991 Report. Consideration of Carriage of Goods by Sea Bill.

2004 **Betz, John.** Anchors aweigh for the Hamburg Rules? *Distribution* 91(2), February 1992, 48–54.

2005 **Cashmore, Chris.** Who are consignors and consignees for the purposes of a contract of carriage? *Journal of Business Law*, 1990, 377–93.

Examines who are 'consignors' and 'consignees' for the purposes of a contract of carriage and explains importance of these issues.

2006 Cashmore, Christopher. *Parties to a contract of carriage – or who can sue on a contract of carriage of goods?* London: Lloyd's of London Press, 1990. 252 pp.

Comprehensive and practical analysis of the three-party contract. Chronology of contract – assesses, identifies parties involved; discovery of owner of lost, damaged, delayed goods; legal nature of resulting actions; problems at each stage. Comparative look at all major common-law jurisdictions.

2007 Chandler, George F. After reaching a century of the Harter Act: Where should we go from here? *Journal of Maritime Law and Commerce* 24(1), January 1993, 43–52.

Considers choices of law questions in respect of cargo claims. Why such resistance to Hamburg Rules? Impact of Hamburg Rules. What changes to COGSA preferred by various interests? Changing COGSA by commercial compromise.

2008 Clarke, Malcolm A. *Aspects of the Hague Rules: A Comparative study in English and French Law.* The Hague: Nijhoff, 1976. 253 pp.

2009 Coffee Trade Federation. *Standard contract forms, conditions of sale and arbitration rules.* London: Coffee Trade Federation, 1963. 40 pp.

Sets out standard contract forms, conditions of sale and arbitration rules for f.o.b.; C & F and c.i.f. contracts.

2010 Colinvaux, Raoul. *Carriage by sea. British shipping laws.* 2 vols. 13th rev. edn. London: Stevens & Sons, 1982.

Covers the law on carriage of goods by sea. Updates Carver's work on the approach of its centenary. Volume One deals with the obligations implied and expressly undertaken in a contract for carriage of goods. Volume Two concludes contractual issues and covers the voyage, the delivery, remedies.

2011 D'Arcy, Leo. *Ridley's law of the carriage of goods by land, sea and air.* 7th rev. edn. Crayford: Shaw and Sons, 1992. 368 pp.

Law of carriage of goods by road, rail, inland waterways, sea, and air. Carriage of goods in passengers' luggage. Issues of law having aspect in connection with carriage of goods rather than contract *per se* or maritime law.

2012 DeGurse, John L. The 'Container Clause' in Article 4(5) of the 1968 protocol to the Hague Rules. *Journal of Maritime Law and Commerce* 2(1), October 1970, 131–46.

Limitation of liability of the ocean carrier for carriage of goods. Interpretation of Article 4(5) of the Protocol. Particular portions of Article 4(5) to be considered: 'unless clause', 'container clause'.

2013 Dockray, Martin. *Cases and materials on the carriage of goods by sea.* Abingdon, Oxon.: Professional Books, 1987. 804 pp.

Selection of materials on the English law of carriage of goods by sea from variety of primary and secondary sources. Standard form contracts; law reports; statutes; conference reports; texts. Examines policies on which legal rules and interpretations based.

2014 Feltham, J. D. The appropriation to a C.I.F. contract of goods lost or damaged at sea. *Journal of Business Law,* 1975, 273–80.

Considers issue of risk of loss or damage to goods during carriage under c.i.f. contract. Examines liability and whether the fact that loss or damage at sea not covered by insurance does not prevent risk passing to buyer. Case law. International sale of goods, uniform law.

2015 Finnie, Alasdair. Short form shipping documents. *Journal of Maritime Law and Commerce* 7(4), July 1976, 697–707.

Considers increased popularity of short-form shipping document. Advantages. Problems. Highlights need for careful consideration of consequences of this procedure.

2016 Ganado, Max and **Kindred, Hugh.** *Marine cargo delays: The law of delay in the carriage of general cargoes by sea.* London: Lloyd's of London Press, 1990. 175 pp.

A practical and thorough examination of the law of delay in the carriage of goods by sea under bills of lading. Covers all aspects of delay from loss by damage and delay to the carrier's responsibilities for despatch and the possible frustration of the contract of carriage.

2017 Glass, David A. and **Cashmore, Chris.** *Introduction to the law of carriage of goods.* London: Sweet & Maxwell, 1989. 336 pp.

Principles of law of United Kingdom, supplemented by European, Commonwealth and US authorities. Law *re* carriage by air; inland carriage; freight forwarding; carriage by water; international carriage by road, by rail; containerization and combined transport. Summary *re* passengers and luggage.

2018 Goldie, C. W. H. Documentation: The writing on the bill. Articles 15–18. In: Samir Mankabady (ed.) *The Hamburg Rules on the carriage of goods by sea.* Leiden and Boston: Sijthoff, 1978, pp. 209–20.

Critical look at Arts 15–18. Of view that many particulars required by Art. 15 to be in bill of lading unnecessary. Although compliance not mandatory, author questions whether some countries to impose penalties for non-compliance Art. 15.

2019 Gorman, Francis J. The choice between proportionate fault or Ryan indemnity in maritime property damage cases. *Journal of Maritime Law and Commerce* 10(3), April 1979, 325–48.

Case law and development of 'Ryan' indemnity following *Ryan Stevedoring Co.* v. *Pan Atlantic Steamship Corporation.* Impact of Longshoremen's and Harbor Workers' Compensation Act 1972 Amendments. Ryan warranty and Ryan indemnity in light of these developments. Application of Ryan indemnity to other than personal injury/death cases.

2020 Gorton, Lars. *The concept of the common carrier in Anglo-American law.* Gothenburg: Akademiforlaget, 1971. 271 pp.

Mainly a comparison between English and American law with some references to Scandinavian law. How is common carrier defined? Important ingredients? Implications on legislation? Law of contract. Conditions concomitant with business.

2021 Gorton, Lars. Escalation and currency clauses in shipping contracts. *Journal of World Trade Law* 12(4), July–August 1978, 319–41.

2022 Gorton, Lars and **Ihre, Rolf.** *A practical guide to contracts of affreightment and hybrid contracts.* 2nd rev. edn. London: Lloyd's of London Press, 1990. 125 pp.

A basic introduction to the contract of affreightment and hybrid contracts. Presents a practical survey to the commercial use of such contracts. Example of a private contract of affreightment involving a dispute is annexed, together with comments on the disputed items.

2023 Graham, Thomas and **Chrispeels, Erik.** Revision of the Hague Rule. *Journal of World Trade Law* 7(2), March/April 1973, 252–65.

Third and Fourth (special) sessions of UNCITRAL Working Group on International Legislation on Shipping 1972. To examine rules and practices concerning bills of lading, including Hague Rules as amended by 1968 Protocol.

2024 Gram, Per. and **Bonnick, Stanley.** *Gram on chartering documents.* 2nd rev. edn. London: Lloyd's of London Press, 1988. 220 pp.

Commentary on specimen standard forms, complemented by list of abbreviations. Contracts for sea transport discussed in detail. References to important cases. Hague–Visby Rules. Tankervoy 87.

2025 Gronfors, Kurt. Non-contractual claims under the Hamburg Rules. In: Samir Mankabady (ed.) *The Hamburg Rules on the carriage of goods by sea.* Leiden and Boston: Sijthoff, 1978, pp. 187–97.

Considers Art. 7, 'Application to non-contractual claims'. Traces background of Article; analysis of Himalaya clause. Refers to various clauses in bills of lading under Hague Rules.

2026 Healy, Nicholas J. Commentary in 1981 revision of the New York Produce Exchange form time charter. *Journal of Maritime Law and Commerce* 13(4), July–October 1982, 521–6.

Background and Revised NYPE Form 1981. Text of the Revision. Utilization of the 1981 Revision.

2027 Hill, C. J. *Introduction to the law of carriage of goods by sea.* New York: Beekman Publishers, 1974. 80 pp.

2028 Honan, William J. Defective bunkers: What is a shipowner to do? *Journal of Maritime Law and Commerce* 13(1), October 1981, 21–33.

Impact of enactment of the Uniform Commercial Code which is a body of statutory law governing commercial transactions, including sale of goods. A shipowner's rights under UCC; theory of strict liability; time charters case law.

2029 Honnold, John O. Ocean carriers and cargo: Clarity and fairness – Hague or Hamburg? *Journal of Maritime Law and Commerce* 24(1), January 1993, 75–110.

Compares and evaluates the clarity, efficiency and fairness of the competing regimes of Hague Rules and Hamburg Rules. Legal and commercial background. Contributions of Hague and Hamburg to clarity and international uniformity. Fairness and efficiency of risk allocation under the two regimes.

2030 Honour, J. P. The P & I Clubs and the new United Nations Convention on the Carriage of Goods by Sea 1978. In: Samir Mankabady (ed.) *The Hamburg Rules on the carriage of goods by sea.* Leiden and Boston: Sijthoff, 1978, pp. 239–50.

Outlines consequences of new Convention on P & I Clubs. Central article in this paper is Art. 5, which omits the negligence clause. Consideration of Arts 6–11. Concludes that transfer of some of risks from cargo to ship will lead to increase in shipowner's P & I insurance.

2031 **Howard, Tim.** The Carriage of Goods by Sea Act 1992. *Journal of Maritime Law and Commerce* 24(1), January 1993, 181–90.

Review of modernization of law relating to bills of lading. Shortcomings of Bills of Lading Act 1855. Changes made by 1992 Act.

2032 **Humphreys, Gordon** and **Higgs, Andrew.** An overview of the implications of the Carriage of Goods by Sea Act 1992. *Journal of Business Law,* January 1993, 61–6.

Considers two major areas of cargo claims upon which 1992 Act has effect: Privity of contract; the rule in *Grant* v. *Norway.*

2033 **Ivamy, E. R. Hardy.** *Casebook on carriage by sea.* 6th rev. edn. London: Lloyd's of London Press, 1985. 221 pp.

Cases *re* charterparties; bills of lading; the Hague Rules; exclusion and limitation of shipowner's liability; general average; demurrage; freight; third-party rights under contract; giving of an indemnity; liens. Highlights from judgments.

2034 **Ivamy, E. R. Hardy.** *Encyclopaedia of carriage law sources.* 2nd edn. London: Lloyd's of London Press, 1987. 556 pp.

2035 **Jackson, D. C.** The Hamburg Rules and conflict of laws. In: Samir Mankabady (ed.) *The Hamburg Rules on the carriage of goods by sea.* Leiden and Boston: Sijthoff, 1978, pp. 221–38.

Conflict of laws in light of Arts 2, 21 and 22. Observes that Convention concentrates on one narrow point – freedom to choose forum. Concludes that key problem whether national courts accept Convention as a code.

2036 **Kimball, John D.** Shipowner's liability and the proposed revision of the Hague Rules. *Journal of Maritime Law and Commerce* 7(1), October 1975, 217–52.

Content of and historical forces leading to rules governing carriage of goods by sea. The Harter Act. The Hague Rules. Limitation of Liability Act. Development of Hague Rules. Impact and application of the Rules.

2037 Kindred, Hugh M. *Kindred: Modern shipping contracts.* Oxford: Butterworth, 1992.

2038 Kwang, Peter K. (ed.) *Carriage of goods by sea.* Oxford: Butterworth, 1986. 216 pp.

2039 Luddeke, C. F. and **Johnson, A.** *Hamburg Rules: Introduction and comments.* London: Lloyd's of London Press, 1991. 96 pp.

2040 Mankabady, Samir. Comments on the Hamburg Rules. In: Samir Mankabady (ed.) *The Hamburg Rules on the carriage of goods by sea.* Leiden and Boston: Sijthoff, 1978, pp. 27–116.

Analyses provisions of the Convention leading to Rules. Comparison of situation under Hague Rules with situation under Hamburg Rules. English law central to approach. Some reference to American, French, Belgian law.

2041 Mankabady, Samir (ed.) *The Hamburg Rules on the carriage of goods by sea.* Leiden and Boston: Sijthoff, 1978. 358 pp.

Book of articles dealing with various issues of the Hamburg Rules and carriage of goods by sea.

2042 Mendelsohn, Allan I. Why the U.S. did not ratify the Visby Amendments. *Journal of Maritime Law and Commerce* 23(1), January 1992, 29–54.

Non-revisionist history of why the Visby Amendments to the Hague Rules were not promptly ratified by the United States after being adopted in Brussels (and signed there by the US Delegation) on 23 February 1968.

2043 Meng, Tan L. *The law in Singapore on carriage of goods by sea.* Oxford: Butterworth, 1986. 509 pp.

2044 Mills, C. P. The future of deviation in the law of the carriage of goods. *Lloyd's Maritime and Commercial Law Quarterly*, November 1983, Part 4, 587–96.

The history of deviation. Development of fundamental breach. The demise of fundamental breach. The present status of deviation.

2045 Mitra, B. C. *Carriage by sea.* (Tagore law lectures.) Calcutta: Eastern Law House, 1972. 261 pp.

2046 Moore, John C. The Hamburg Rules. *Journal of Maritime Law and Commerce* 10(1), October 1978, 1–11.

Outline of international uniformity. The Hague Rules and Visby Amendments. Changes incorporated into Hamburg Rules.

2047 **Morris, J. H. C.** The scope of the Carriage of Goods by Sea Act 1971. *Law Quarterly Review* 95, January 1979, 59–67.

2048 **Nicoll, C. C.** Do the Hamburg Rules suit a shipper-dominated society? *Journal of Maritime Law and Commerce* 24(1), January 1993, 151–80.

Explores question of whether New Zealand should adopt the Hamburg Convention. Alterations in the core of liability to be assumed by carrier; changes in applicable rules of evidence; application, security and privity.

2049 **Payne, William** and **Ivamy, E. R. Hardy.** *Carriage of goods by sea.* 13th rev. edn. London: Butterworth & Co., 1989. 395 pp.

Thirteenth edition updated in light of changes wrought by enactment of Merchant Shipping Act 1979, plus case law. Commercial practice plus explanations of common shipping terms.

2050 **Peacock, J. Hoke.** Deviation and the package limitation in the Hague Rules and the Carriage of Goods by Sea Act: An alternative approach to the interpretation of international uniform acts. *Texas Law Review* 68(5), April 1990, 977–1002.

2051 **Peyrefitte, Léopold.** The period of maritime transport: Comments on Article 4. In: Samir Mankabady (ed.) *The Hamburg Rules on the carriage of goods by sea.* Leiden and Boston: Sijthoff, 1978, pp. 125–36.

Pinpoints difference between English and French texts of the Convention in respect of Art. 4 on the period of maritime transport. Concept of taking charge, its application in transportation of bulk cargo, lighterage operations, carriage of containers and barges.

2052 **Powles, David G.** The Carriage of Goods by Sea Act 1971. *Journal of Business Law,* 1975, 141–6.

Aims and provisions of the Act. Application of the Hague–Visby Rules and other matters dealt with in the body of the Act.

2053 **Ramberg, Jan.** *Cancellation of contracts of affreightment on account of war and similar circumstances.* Gothenburg: Gothenburg School of Economics and Business Administration, 1970. 476 pp.

2054 Richardson, John. *A guide to the Hague and Hague Visby Rules.* 2nd rev. edn. London: Lloyds of London Press, 1989. 82 pp.

Broad explanation of operation of the Rules. Interpretation of the Rules in English courts. Both Hague and Hague–Visby Rule set out in full. References to interpretation of certain foreign courts of some of Rules' provisions.

2055 Sassoon, David M. A footnote to the Mendelsohn article on the Visby Amendments. *Journal of Maritime Law and Commerce* 23(3), July 1992, 485–6.

Examines article by Mendelsohn in *Journal of Maritime Law and Commerce* 23(1), January 1992, pp. 29–54, titled 'Why the U.S. did not ratify the Visby Amendments'. Controversy *re* container clause and 'Diplock explanation'.

2056 Sassoon, David M. and **Cunningham, John C.** Unjustifiable deviation and the Hamburg Rules. In: Samir Mankabady (ed.) *The Hamburg Rules on the carriage of goods by sea.* Leiden and Boston: Sijthoff, 1978, pp. 167–86.

Compares deviation under Hague Rules with that under the new Rules. Causation. Limitations on amount of damages on cargo. Limitations on time for suit.

2057 Schmitthoff, Clive M. Carriage of Goods by Sea Act 1971. *Journal of Business Law*, 1971, 191–6.

Reviews main new provisions of 1971 Act: Extension of scope of Rules; coasting trade; documents of title; conclusive evidence; limitation; the container clause; Himalaya overruled; time limit.

2058 Shachar, Y. Containers in the law of carriage of goods. Thesis submitted for D.Phil, University of Oxford, 1976.

2059 Shah, M. J. The revision of the Hague Rules on bills of lading within the U.N. system: Key issues. In: Samir Mankabady (ed.) *The Hamburg Rules on the carriage of goods by sea.* Leiden and Boston: Sijthoff, 1978, pp. 1–26.

Historical background; basic concerns expressed by UNCTAD, UNCITRAL; key issues for concern – basic rule of liability, sustaining carrier liability throughout contract, monetary limitation of liability, invalid clauses, jurisdiction clauses.

2060 Silard, Stephen A. Carriage of the SDR by sea: The unit of account of the Hamburg Rules. *Journal of Maritime Law and Commerce* 10(1), October 1978, 13–38.

Examines unit of account as defined in Art. 26 of the Hamburg Rules. Uniform limits of liability. Historical evolution of SDR as a unit of account. Formulation of unit of account provision of Hamburg Rules.

2061 Simmonds, K. R. The interpretation of the Hamburg Convention: A note on Article 3. In: Samir Mankabady (ed.) *The Hamburg Rules on the carriage of goods by sea.* Leiden and Boston: Sijthoff, 1978, pp. 117–24.

Examines Art. 3 on the interpretation of the Convention leading to the Hamburg Rules. Consideration of various international aspects and application of Hamburg Rules by national courts.

2062 Simon, Seymour. Admiralty jurisdiction and the liability of wharfingers, *Journal of Maritime Law and Commerce* 3(3), April 1972, 513–36.

Examination of findings of the Court of Appeals in *Leather's Best* v. *S.S. Mormaclynx.* Question: Is a 40-foot container a package within COGSA limitation of $500 per package? Also, unlitigated issues regarding jurisdiction of the court and corresponding substantive law.

2063 Singh, Avtar. *Law of carriage.* Eastern (Iowa), 1988.

2064 Sturley, Michael F. Bill of lading choice of forum clauses: Comparisons between United States and English law. *Lloyd's Maritime and Commercial Law Quarterly*, May 1992, Part 2, 248–60.

Considers application of law in both US and England. When will a court decline to exercise its own jurisdiction on the grounds that a court in the chosen forum should govern? Do Hague Rules prohibit a carrier from including an enforceable choice of forum clause in bill of lading?

2065 Sturley, Michael F. (ed.) *The legislative history of the Carriage of Goods by Sea Act and the Travaux Préparatoires of the Hague Rules* (3 vols). Littleton (Colorado): Fred B. Rothman & Co., 1990. 680 pp.

2066 Sturley, Michael F. and **Grover, Steven F.** Ad valorem rates under Article 4(5) of the Hague Rules: A response to Mendelsohn (and De Gurse). *Journal of Maritime Law and Commerce* 23(4), October 1992, 621–30.

Comment on article by Allan Mendelsohn, 'Why the U.S. did not ratify the Visby Amendments', 23(1), January 1992, pp. 29–54, in same journal. Takes view that Hague 'unless clause' does not prohibit *ad valorem* rates.

2067 Sweeny, J. C. Article 6 of the Hamburg Rules. In: Samir Mankabady (ed.) *The Hamburg Rules on the carriage of goods by sea.* Leiden and Boston: Sijthoff, 1978, pp. 151–66.

Comprehensive study on Arts 6 and 26 on the unit of account. Preliminary drafting in UNCITRAL; container clause; elaboration in UNCITRAL; Hamburg Conference; package deal of Hamburg.

2068 Tetley, William. Acceptance of higher Visby liability limits by U.S. courts. *Journal of Maritime Law and Commerce* 23(1), January 1991, 55–72.

Suggests how judges of the Southern District Court of New York in six recent decisions might have decided question of whether US court may apply higher Hague–Visby Rules package, and kilo limitation in a shipment from a Visby Rules nation to US, by applying conflict of law theory and practice.

2069 Tetley, William. Articles 9 to 13 of the Hamburg Rules. In: Samir Mankabady (ed.) *The Hamburg Rules on the carriage of goods by sea.* Leiden and Boston: Sijthoff, 1978, pp. 197–208.

States new Rules do not favour shippers sufficiently. Examines Art. 9 – deck carriage; Arts 10, 11 – liability carrier, actual carrier; Arts 12, 13 – liability of shipper. Lists other Articles imposing liability on shipper.

2070 Tetley, William. The Hamburg Rules: A commentary. *Lloyd's Maritime and Commercial Law Quarterly,* February 1979, 1–20.

2071 Tetley, William. An update on the per package limitation and national intentions regarding future carriage of goods by sea legislation. *Journal of Maritime Law and Commerce* 14(3), July 1983, 301–46.

Package limitations and their currencies; carriage of goods by sea legislation; from perspective of national jurisdictions involved.

2072 Tetley, William. Waybills: The modern contract of carriage of goods by sea. *Journal of Maritime Law and Commerce* 14(4), October 1983, 465–511.

First of two parts (Part II, vol. 15, no. 1. p. 41). Carriage of goods by sea under ocean waybills. Definition. Examples. Problems. Hague Rules; Hamburg Rules. UK position. US position. Canadian position. Italian position. Who may sue on a waybill? Case law.

2073 Timmermans, W. A. *Carriage of goods by sea in the practice of* *the USSR Maritime Arbitration Commission.* The Hague: Nijhoff, 1990. 368 pp.

2074 Todd, Paul. *Contracts for the carriage of goods by sea.* Oxford: BSP Professional, 1988. 270 pp.

Covers all varieties of charterparties and bills of lading. Includes standard dry-cargo charterparties. INTERTANKO tanker forms, development of out-turn loss and cargo retention clauses, bunker fuel quality clauses, the problems of delivery without production of bills of lading, 'reachable on arrival' clause. Himalaya clauses, fraud. Tanker charterparties.

2075 UNCITRAL. UNCITRAL: Revision of the Hague Rules. *Journal of World Trade Law* 5(5), September/October 1971, 577–90.

UNCITRAL – working group April 1971 – study revision of Hague Rules on bills of lading international shipping. The need to revise the rules, per UNCTAD, in favour of developing countries.

2076 *United Nations Convention on the carriage of goods by sea. 1978 Hamburg Rules.* London: Commonwealth Secretariat, 1990. 83 pp.

2077 Waldron, A. J. The Hamburg Rules: A boondoggle for lawyers. *Journal of Business Law,* 1991, 305–19.

Examines Hamburg Rules and likely impact on regulation of liability for carriage of goods by sea. Provisions. General liability. Exceptional liability. Notification of damage or delay. Limitation of actions. Arbitration. Jurisdiction. General average. Overall effect.

2078 Williams, B. K. The consequences of the Hamburg Rules on insurance. In: Samir Mankabady (ed.) *The Hamburg Rules on the carriage of goods by sea.* Leiden and Boston: Sijthoff, 1978, pp. 251–60.

Presents view that the new Convention tends to reduce the overall volume of cargo insurance and would increase liability of carrier. Hamburg Convention – justice to developing countries, injustice to others.

2079 Wilson, J. F. Basic carrier liability and the right of limitation. In: Samir Mankabady (ed.) *The Hamburg Rules on the carriage of goods by sea.* Leiden and Boston: Sijthoff, 1978, pp. 137–50.

Examines Art. 5 – basis of liability; Art. 6 – limits of liability; Art. 8 – loss of right to limit responsibility. Thorough analysis of liability and burden of proof.

2080 Wilson, J. F. *Marine law.* London: Chartered Insurance Institute Tuition Service, 1989. 118 pp.

Coursebook of C.I.I. Tuition Service. Contracts of affreightment. Bills of lading, the Carriage of Goods by Sea Act 1971. COGSA – a historical and comparative survey. Charterparties. Carriage by road, by rail. Combined transport. The ship. Collisions and general average. Salvage, wreck, towage, pollution.

2081 Wilson, John F. *Carriage of goods by sea.* 2nd edn. London: Pitman, 1993. 456 pp.

Comprehensive and critical study of legal principles. Emphasis on English common law approach with appropriate reference to US Commonwealth case material. Legal principles examined against current documentation and contemporary practice. The second edition also deals with the effect of the Hamburg Rules (which came into effect 1 November 1992) and the enactment of the Carriage of Goods by Sea Act 1992.

2082 Zaphiriou, G. A. Amending the Hague Rules. *Journal of Business Law,* 1971, 12–17.

Indicates main amendments informally agreed in Stockholm and finally included in the Protocol, with two exceptions: (i) application of Hague Rules to inward shipments; (ii) amendment concerning unseaworthiness caused by reputable repairers' negligence.

Sale of goods

2083 Hudson, A. H. Sales from bulk. *Lloyd's Maritime and Commercial Law Quarterly,* November 1989, Part 4, 420–30.

Problems presented by sales of unseparated shares in a larger quantity of goods. Outlines law and its defects. Section 16 Sale of Goods Act 1979. Section 1 Bills of Lading Act 1855. The problems. Proposals for reform. The case for no change.

2084 Ivamy, E. R. Hardy. *Casebook on sale of goods.* 5th edn. London: Lloyd's of London Press, 1987. 192 pp.

Sets out the leading and important cases relating to the sale of goods. The Sale of Goods Act 1979. The Sale of Goods Act 1893. The Factors Act 1889. The Hire-Purchase Act 1964. International sales. Bankers' commercial credits. Performance guarantees.

2085 Sassoon, David M. (ed.) *CIF and FOB contracts. British shipping laws.* Vol. 5, 3rd rev. edn. London: Stevens & Sons, 1984. 528 pp.

Law: nature of c.i.f contract; the shipment; the bill of lading; insurance; tender and payment; remedies; conflict of laws. Various f.o.b. contracts; delivery f.o.b.; f.o.b. price term; payment and acceptance; insurance; remedies; conflict of laws.

Charterparties

2086 Astle, W. E. *The safe port or berth reachable on arrival: Charterers' and shipowners' responsibilities and liabilities.* London: Fairplay Publications Ltd, 1986. 109 pp.

Examines questions of whether port or berth 'safe'; 'reachable on arrival'; whether vessel 'arrived ship'. Based on experiences of charterers, shipowners and respective insurers.

2087 Astle, W. E. *Time charter withdrawals: An examination of shipowners' rights to withdraw vessels from the service of time charterers.* London: Fairplay Publications Ltd, 1984. 136 pp.

Examines time-charter clauses providing for owner to withdraw vessel from service of charterers. Problems which can arise, their causes, effect and resolution by reference to disputes before Court of Appeal and House of Lords.

2088 Berg, Jack. Overlap/underlap. In: *Time charters: Why the confusion?* London: Lloyd's of London Press, 1977. 9 pp.

Expressions of duration of charter giving rise to overlap/underlap situations. Essentially treats the American view but in so doing out of necessity reviews major English decisions.

2089 Berlingieri, Francesco. The allocation of risk delay in voyage charter parties. *Journal of Maritime Law and Commerce* 8(4), July 1977, 497–504.

Importance of allocation of risk in maritime law concerning risk of delay. Problem of its allocation between owner and charterer. Economic importance.

2090 Besman, John. 'FONASBATIME' – could it be the solution? In: *Time charters: Why the confusion?* London: Lloyd's of London Press, 1977. 4 pp.

Dry cargo time charters. Necessity for up-to-date and comprehensive charters to avoid arbitration shocks in respect of interpretation of 'commercial intent'. Examines FONASBATIME charter as an example of effort to overcome such problems. Considers areas of major disputes.

2091 BIMCO; CMI; FONASBA and **GCBS.** Charterparty laytime definitions 1980. *Journal of Maritime Law and Commerce* 12(3), April 1981, 421–6.

Joint announcement to diminish confusion and uncertainty of interpretation of words/phrases in charterparties. Set of definitions for voluntary adoption – 'Charterparty Laytime Definitions 1980'.

2092 Bonnick, S. *Gram on chartering documents.* 2nd edn. London: Lloyd's of London Press, 1988. 220 pp.

Familiarization with charterparties and bills of lading. Arranged as a running commentary on specimen standard forms and complemented by list of abbreviations. Contracts for sea transport in detail and reference to important cases. Covers Hague–Visby Rules and Tankervoy 87.

2093 Bonnick, Stan. *Some observations on BPVOY 3.* Oslo: INTERTANKO, 1990. 20 pp.

Commentaries on BPVOY 3. Clause index. Sets out the charterparty. BP Shipping Questionnaire 1990.

2094 Cadwallader, F. J. J. Shipowner and charterer: Third party liabilities. In: *Time charters: Why the confusion?* London: Lloyd's of London Press, 1977. 9 pp.

How relationship between time charterer–shipowner comes into being and what forms it may take. Difficulties which arise between time charterer and shipowner. How shipowner may seek to transfer his liabilities to the charterer.

2095 Cassegrain, Jean. Charterers' responsibilities and liabilities. In: *Time charters: Why the confusion?* London: Lloyd's of London Press, 1977. 10 pp.

Charterer's responsibilities as (i) lessee and (ii) operator of the hired ship. Problems caused by lack of explicit clauses and provisions in different contracts agreed between the parties. Differences between the laws of maritime nations.

2096 *Charter parties: A comparative analysis.* New York: United Nations, 1990. 31 pp.

2097 Cohen, Michael Marks. Off-hire clause. In: *Time charters: Why the confusion?* London: Lloyd's of London Press, 1977. 15 pp.

Examines why principle of off-hire clause, although seemingly straightforward, in practice has proved difficult to apply. Uses structure of off-hire clause in FONASBATIME form in outline for discussing all such clauses.

2098 Cooke, Julian and others. *Voyage charters.* London: Lloyd's of London Press, 1993. 930 pp.

Covers both English and American law. Introduces discussion on some of the unsolved practical and legal problems which emerged from the research. Covers general principles and Gencon charter. The Asbatankvoy charter. The Hague and Hague–Visby Rules.

2099 Curwen, Nicholas. The problems of transferring carriage rights: An equitable solution. *Journal of Business Law,* May 1992, 245–62.

Argues that even without statutory reform, there is already available an important but untested remedy for the buyer in an export sale who finds difficulties with the carrier's contract rights. Suggests regarding the transfer of bill of lading as an equitable assignment of seller's carriage rights.

2100 Davies, Donald. Assessment of damages for inadequate vessel performance under time charters. *Journal of Maritime Law and Commerce* 14(4), October 1983, 595–601.

Considers damages to which a charterer entitled, under English law, when shipowner breaches charterparty terms in relation to vessel's performance.

2101 Davies, Martin. The off-hire clause in the New York Produce Exchange time charterparty. *Lloyd's Maritime and Commercial Law Quarterly,* February 1990, Part 1, 107–13.

Competing judicial views on the meaning of the phrase 'or by any other cause whatsoever' in clause 15 of the NYPE Form. Describes how the two views have evolved. Argues that clause 15 has not achieved desired result and suggests amendment to ensure this.

2102 Diamond, Anthony. Owners' responsibilities and the exceptions clauses relating to them. In: *Time charters: Why the confusion?* London: Lloyd's of London Press, 1977. 10 pp.

Owners' responsibilities under (i) the Baltime 1939 Form; (ii) the NYPE Form; (iii) the Linertime Form. Owners' obligations as to seaworthiness of ship. Owners' and charterers' responsibilities as regards stowage of the cargo. Owners' responsibilities as regards performance of the voyage.

2103 Federation of National Associations of Ship Brokers and Agents. *Time charters: Why the confusion?* London: Lloyd's of London Press, 1977.

2104 Glass, David A. The boxtime charterparty. *Lloyd's Maritime and Commercial Law Quarterly,* November 1992, Part 4, 514–37.

Review of Boxtime charterparty developed by BIMCO in 1990, for charter of complete container ships. Aims and format of Boxtime. Legal significance of the description. Structure of Boxtime. Duties of parties. Liabilities. Employment of the ship. Limitation.

2105 Gorton, Lars. Escalation and currency clauses in shipping contracts. *Journal of World Trade* 12(4), July/August 1978, 319–41.

Examines circumstances which affect the cost structure of a contract. Object of a general escalation clause. Clauses to prevent the effect of currency exchange rate fluctuations. Escalation clauses. Variations in costs.

2106 Holman, Michael. Speed and consumption. In: *Time charters: Why the confusion?* London: Lloyd's of London Press, 1977. 5 pp.

Essential difference between time charterparty and voyage charterparty: vessel's speed in time charterparty of primary importance. Fundamental question of whether reference in time charter to speed and fuel consumption describes capability of vessel or is actually a promise by shipowner that vessel will maintain this throughout performance.

2107 International Association of Independent Tanker Owners. *Observations on tanker time charterparties.* Oslo: INTERTANKO, 1980. 48 pp.

Provides a reference for the many terms peculiar to the subject of time-chartering, often with cross-references to solutions provided in INTERTANKTIME 80.

2108 International Association of Independent Tanker Owners. *Some observations on EXXONVOY 84.* Oslo: INTERTANKO, 1987. 15 pp.

Observations as a guidance to members when negotiating fixtures on EXXONVOY 84 Charter Party.

2109 International Association of Independent Tanker Owners. *Some observations on SHELLTIME 4.* Oslo: INTERTANKO, 1988. 51 pp.

Observations as a guidance to members when negotiating fixtures on SHELLTIME 4.

2110 International Chamber of Commerce. *Bare boat charter registration.* Paris: I.C.C. Publications, 1988. 52 pp.

2111 INTERTANKO. *Comparative observations on tanker voyage charter parties.* Oslo: INTERTANKO, 1979. 47 pp.

Provides a reference booklet on standard tanker voyage forms. The parties and their helpers. The vessel and cargo. The voyage. Laytime, demurrage, freight, dues. Clean seas programme. Cancellations. Ships' agents. Bills of lading and Hague Rules. General average. Law, arbitration.

2112 Koh, Peter S. K. 'Errors of navigation' in NYPE form: reflections on the Emmanuel C. *Lloyd's Maritime and Commercial Law Quarterly*, November 1983, Part 4, 597–602.

Considers the meaning of 'errors of navigation' in clause 16 of the NYPE. The English case. The Emmanuel C.

2113 Kovats, L. J. Substitution clauses in charterparties. *Journal of Maritime Law and Commerce* 8(4), July 1977, 505–10.

Examines the right of a shipowner to substitute a 'similar' vessel for the one chartered in order to keep charterparty alive. Case law and arbitration decisions.

2114 Lloyd's of London Press. *Time charters: Why the confusion?* London: Lloyd's of London Press, 1977. 120 pp.

Papers from international two-day seminar sponsored by FONASBA and organized by Lloyd's, held in London, 24–25 March 1977.

2115 McCune, D. Thomas. *The ASBATANKVOY Charter: A special report.* London: Lloyd's of London Press, 1984. 97 pp.

Study of ASBATANKVOY form itself. Judicial decisions and arbitration awards which have construed various clauses of the ASBATANKVOY form. Considers 17 most disputed clauses of the 26 clauses.

2116 McMahon, John P. The Hague Rules and incorporation of charter party arbitration clauses into bills of lading. *Journal of Maritime Law and Commerce* 2(1), October 1970, 1–16.

Relation of arbitration to the basic 'international legislation' on carriage of goods by sea, Brussels Convention 1924 – the Hague Rules. Comparison of English/American practice.

2117 Morris, Albert. The shipbroker's point of view. In: *Time charters: Why the confusion?* London: Lloyd's of London Press, 1977, pp. 1–8.

Paper from international two-day seminar sponsored by FONASBA, held in London, 24–25 March 1977. Confusion self-induced; confusion charterparty motivated; confusion arbitrarily assisted; confusion judicially imposed.

2118 Packard, William V. *Time chartering.* London: Fairplay Publications, 1980. 99 pp.

Provides specific material on subject of period employment, explaining particular phrases and terms that are encountered. Guidance on practical problems of timechartering ships.

2119 Pritchett, Russell W. The New York Produce Exchange charterparty: The clause paramount and responsibility for cargo operations and seaworthiness. *Journal of Maritime Law and Commerce* 14(1), January 1983, 69–112.

Examines effect of clause paramount provisions of both 1946 and 1981 NYPE forms upon (i) the division of responsibility between shipowner and charterer for cargo operations; (ii) scope of responsibility of shipowner to provide seaworthy vessel.

2120 Richardson, John. *Boxtime charterparty.* London: Lloyd's of London Press, 1990. 92 pp.

A practical guide to the BOXTIME charterparty to explain how and why certain clauses are as they are and why others have been omitted. Relevant case law.

2121 Rokison, Kenneth. Withdrawal and cancellation and let out clauses. In: *Time charters: Why the confusion?* London: Lloyd's of London Press, 1977. 8 pp.

Examines circumstances in which the predetermined period of time that the charterer has for the use of the vessel may be brought to an end prematurely or even before it starts.

2122 Schwampe, D. *Charterers' liability insurance.* London: Lloyd's of London Press, 1988.

2123 Trappe, J. 'The ship': Compliance with charterparty description and condition on re-delivery. In: *Time charters: Why the confusion?* London: Lloyd's of London Press, 1977. 13 pp.

Focuses on two moments in ship's life: (i) her delivery to a time charterer, and (ii) her re-delivery to the owner. Practical difficulties and legal problems connected therewith which may arise on delivery and re-delivery.

2124 Ventris, F. M. *Tanker voyage charterparties.* London: Kluwer Law Publishers, 1986. 413 pp.

Examines the contractual relations established between the parties in the voyage-chartering of oil tankers and the implications and consequences which arise therefrom. Basic document used is the 'STB VOY' form. Detailed examination of the text.

2125 **Waring, Anne.** *Charterparties: A comparative study of South African, English and American law.* Wetton, South Africa: Juta & Company Ltd, 1983. 417 pp.

2126 **Wilford, M., Coghlin, T.** and **Kimball, J. D.** (eds) *Time charters.* 3rd rev. edn. London: Lloyd's of London Press, 1989. 656 pp.

2127 **Williams, Harvey** and **Bonnick, Stan.** *Commentaries on tanker voyage charterparties.* Oslo: INTERTANKO, 1989. 133 pp.

Combines and updates the two 1979 publications of comments and comparative observations on tanker voyage charterparties. Practical issues in modern-day tanker chartering. Clause notes. Charter commentaries. Charter forms.

Bills of lading

2128 **Abol al Shamat, M. F.** The future of the bill of lading as a document of title. Thesis submitted for PhD, University of Southampton, 1986.

2129 **Asser, T. M. C.** Choice of law in bills of lading. *Journal of Maritime Law and Commerce* 5(3), April 1974, 355–405.

Examines purpose of Hague Rules. Impact of reduction of freedom of contract under the Rules. Scope of the Rules. Bills of lading governed by the Rules. Bills of lading not governed; choice of proper law; choice of Rules. Conflict reference to foreign Hague Rules. The 1968 Protocol.

2130 **Astle, W. E.** *International rules of law relating to bills of lading.* London: Fairplay Publications, 1982. 72 pp.

The rules of law relating to bills of lading traced through the years from the introduction of the Harter Act of the United States, 1893, to the events leading to the Hamburg Rules, 1978.

2131 **Bell, Andrew P.** The Bills of Lading Act 1855 to Today. *Journal of Business Law,* 1985, 124–32.

Examines scope of section 1 of the 1855 Act. Why and when the carrier at risk of liability. Considers leading views on debate as to when section 1 will/will not apply.

2132 **Chandler, George F.** The electronic transmission of bills of lading. *Journal of Maritime Law and Commerce* 20(4), October 1989, 571–9.

Development of electronic transmission and registration of negotiable bills of lading in systems that handle all shipping documents. Acronyms of technological advances. Requirements of tying into such systems.

2133 **Chrispeels, Erik** and **Graham, Thomas.** Brussels convention of 1924, the Ocean bills of lading: Further action toward revision. *Journal of World Trade Law* 7(6), November/December 1973, 680–700.

Action by UNCITRAL Working Group on International Legislation on Shipping on four topics in respect of ongoing revision of 1924 Brussels Convention and 1968 Brussels Protocol. Unit limitation of liability; transhipment; deviation; period of limitation.

2134 **Cimokowski, Edwin W.** The U.S. Government bill of lading and related practices and procedures. *Journal of Maritime Law and Commerce* 3(2), January 1972, 213–44.

Explains terms and functions of Government bills of lading. Encompasses some aspects of paying and auditing carrier bills for transportation services furnished in US Government transportation operations. Arrangements with carriers, documentation, other incidents of shipper–carrier relationship.

2135 **Davies, Martin.** What obligations does a shipowner take over from a defaulting time charterer under charterers' bills of lading? *Australian Business Law Review* 21(3), June 1993, 227–30.

2136 **Debattista, Charles.** The bill of lading as the contract of carriage: A reassessment of Leduc v. Ward. *Modern Law Review* 45, November 1982, 652–63.

This lodges the question of whether bills of lading issued under a contract of carriage contain that contract or are merely evidence of its terms.

2137 **Debattista, Charles.** *Sale of goods carried by sea.* Sevenoaks: Butterworth, 1990. 378 pp.

Examines English law principles which govern use of bills of lading in international trade. Emphasis on tender of bills of lading by sellers to buyers and bankers.

2138 Denning, Stephen M. Choice of forum clauses in bills of lading. *Journal of Maritime Law and Commerce* 2(1), October 1970, 17–39.

English experience used as background against which American assaults on choice of forum clauses in bills of lading viewed. Includes Carriage of Goods by Sea Act 1924, Harter Act 1893.

2139 Gandy, Dick. *How to deal with bills of lading: A quick reference book for everyone.* Sydney: Sydney Maritime Private College, 1982. 76 pp.

Practical reference guide on bills of lading law. How goods exchange hands. Forms and status of bills of lading. Functions of bill of lading. Problems with bulk cargo bills. Bill of lading under a charterparty. Transhipment. NVOCC bills. Shipowner's liabilities. Limitation. Stevedores.

2140 ICHCA. *How to deal with bills of lading.* London: ICHCA, 1988.

2141 Kozolchyk, Boris. Evolution and present state of the ocean bill of lading from a banking law perspective. *Journal of Maritime Law and Commerce* 23(2), April 1992, 161–246.

Historical and comparative analysis to identify features of bill of lading and changes with advent of containerization. Certainty and fairness features. Fundamentals of ocean bill of lading law leading to ocean bill of lading as merchantable document. Banking – bill of lading law.

2142 Lennie, J. M. Container bills of lading. *Scottish Bankers Magazine* 61, February 1970, 189–91.

2143 Liao, S.-P. Issues in deviation and in bills of lading under charterparties: A comparative study of Anglo-American and R.O.C. Laws. Thesis submitted for PhD, University of Southampton, 1986.

2144 Lloyd, Anthony. The bill of lading: Do we really need it? *Lloyd's Maritime and Commercial Law Quarterly,* February 1989, Part 1, 47–58.

Paper presented at Sixth Annual Lecture of the Institute of Maritime Law, University of Southampton, 24 November 1988. Describes efforts of and problems encountered by CMI in drafting a new set of Uniform Rules for incorporation in sea waybills. Suggests discouragement of issuing bills of lading when negotiable document not required.

2145 Lloyd's of London Press. *Waybills and short form documents.* London: Lloyd's of London Press, 1979.

2146 Mankabady, Samir. The new Lloyd's policy and cargo clauses. *Journal of Maritime Law and Commerce* 13(4), July–October 1982, 527–35.

Lloyd's S G Policy 1779. Developments. New policy and clauses to be used. The Lloyd's Marine Policy. Five sets of Institute Cargo Clauses. General exclusions.

2147 Mendelsohn, Allan I. Liberalism, choice of forum clauses and the Hague Rules. *Journal of Maritime Law and Commerce* 2(3), April 1971, 661–8.

Takes issue with conclusions and claims of 'liberalism' in Denning's article 'Choice of forum clauses in bills of lading', *Journal of Maritime Law and Commerce* 2(1), October 1970, pp. 17–39.

2148 Mitchellhill, Alan. *Bills of lading: Law and practice.* 2nd rev. edn. London and New York: Chapman & Hall, 1990. 168 pp.

Explains in layman terms application of bill of lading as well as history, development, legal background. Detailed coverage of the rules governing conventional and combined transport bills of lading. UN Convention on multimodal transport. American law.

2149 Mocatta, A. A., Mustill, M. J. and **Boyd, S. C.** *Scrutton on charterparties and bills of lading.* 19th rev. edn. London: Sweet & Maxwell, 1984. 593 pp.

Law and developments in respect of charterparties and bills of lading. Types of contracts; terms and parties; performance; representations; functions of bills of lading; shipowners' liabilities; demurrage; freight; liens; through bills of lading; combined transportation; containers; damages. COGSA 1971.

2150 Naschitz, P. G. Choice of forum clauses in Israel maritime cases. *Journal of Maritime Law and Commerce* 3(4), July 1972, 775–82.

Review of maritime aspect of local jurisdiction and jurisdiction clauses in bills of lading. Case law; case law rules.

2151 Rodger, Alan. Pledge of bills of lading in Scots law. *Juridicial Review,* December 1971, 193–213.

2152 Tetley, William. The Himalaya clause: Heresy or genius? *Journal of Maritime Law and Commerce* 9(1), October 1977, 111–30.

Considers the problem of finding a way to allow third parties who are neither agents nor servants to limit their liability; to find a way to allow stevedores who are not agents or carriers of the cargo or cargo owner to benefit under the law.

2153 Thommen, Jistice T. *Bills of lading in international law and practice.* Eastern Bk. India, 1985.

2154 Todd, Paul. *Bills of lading and bankers' documentary credits.* London: Lloyd's of London Press, 1990. 245 pp.

Central theme is part played in commercial credit agreements by bill of lading. Explains each type of documentary credit in use, how and why developed. Detailed examination of the autonomous contracts in a confirmed irrevocable credit. Relationship between bank and carrier.

2155 Todd, Paul. *Cases and materials on bills of lading.* Oxford, London (*et al*): BSP Professional Books, 1987. 376 pp.

Reference book – major authorities in detail; summaries of facts and decisions. Leading cases; main statutory provisions and international conventions. Views of specialists (e.g. UNCTAD) included.

2156 Todd, Paul. *Modern bills of lading.* 2nd rev. edn. Oxford (*et al.*): BSP Professional Books, 1990. 332 pp.

Second edition extensively rewritten, not just update of 1st edition. Function of bill of lading as negotiable document; contractual relationships revolving around bill of lading. Considers developments of transportation, problems for bills of lading – of new technology.

2157 Treitel, G. H. Bills of lading and implied contracts. *Lloyd's Maritime and Commercial Law Quarterly*, May 1989, Part 2, 162–73.

Examines significance of the *Aramis* (1989) in respect of no contractual relationship to be inferred between shipowners and holders of bill of lading from presentation of bill of lading. Legal background. Facts of the *Aramis.* Bills of Lading Act 1855, section 1. Implied contract. Evaluation.

2158 UNCITRAL. UNCITRAL 1973–4. *Journal of World Trade Law* 9(2), March/April 1975, 209–21.

Report on matters on which Commission engaged. Including uniform rules governing international sale of goods, international legislation on shipping. (Unification of rules relating to bills of lading.)

2159 UNCITRAL. UNCITRAL 1975. *Journal of World Trade Law* 10(2), March/April 1976, 157–61.

Report on Eighth Session of UNCITRAL, April 1975. Conclusion of three texts, one concerning bills of lading and extending weight of information contained therein.

2160 United Nations Conference on Trade and Development. *Bills of lading.* Geneva: UNCTAD, 1970.

2161 Wiwatpattarakul, P. The use of bills of lading in oil trading. Thesis submitted for LL M, University of Dundee, 1988.

Examines some of the legal issues that arise in connection with the use of bills of lading in the oil trading transactions. This is done against a background of the structural changes in the international oil industry, in particular the breakdown of the ownership link between the upstream and downstream sides of the industry.

Freight fixtures

2162 Faghfoori, M. The law of freight. Thesis submitted for PhD, UWIST, Cardiff, 1984.

2163 Rose, F. D. Deductions from freight and hire under English law. *Lloyd's Maritime and Commercial Law Quarterly*, February 1982, Part 1, 33–56.

Considers whether defendant's claim classified as counterclaim or as set-off. Impact on shipping cases: (i) deducting from freight by voyage charterer of sums owned by shipowner; (ii) similar deductions from hire by timecharterer. Case law development.

Laytime

2164 Cohen, Michael Marks. Confusion in the drafting and application of off-hire clauses. *Journal of Maritime Law and Commerce* 9(3), April 1978, 343–61.

Difficulties in application of off-hire clauses. Clause found in FONASBATIME form used as outline for discussing all off-hire clauses.

2165 Davies, Donald. *Commencement of laytime.* 2nd rev. edn. London: Lloyd's of London Press, 1992. 254 pp.

Examines English law in relation to issue of commencement of laytime in laytime/demurrage disputes. Various remedies. Specific consideration of breaches of contract. Cases and arbitration decisions. Incorporation of excerpts from relevant charterparties. ETA section.

2166 Lax, Michael D. Laytime and demurrage: Recent developments. *Marine Policy* 16(2), March 1992, 118–26.

Case development examination on laytime and demurrage.

2167 Packard, William V. *Laytime calculating.* London: Fairplay Publications, 1979. 77 pp.

Provides information on the practical calculation of laytime. Legal references based on English law. The basics. Commencement. Interruptions. Durations and examples. Calculation. Cargo-handling terms. Tanker laytime.

2168 Powles, D. G. Reversing the arrived ship. *Journal of Business Law,* 1974, 282–93.

Examines decision of House of Lords in *E. L. Oldendorff & Co. GmbH* v. *Tradax Export S.A., The Johanna Oldendorff* (1973).

2169 Powles, David G. An anchorage for the arrived ship. *Journal of Business Law,* April 1979, 115–23.

2170 Recent developments in laytime cases. *Lloyd's Maritime and Commercial Law Quarterly,* August 1990, Part 3, 383–400.

Review of selection of interesting cases dealing with laytime awards. Shows how courts and arbitration tribunals in one country may arrive at decisions differing from decisions arrived at in another.

2171 Schofield, John. *Laytime and demurrage.* 2nd rev. edn. London and New York: Lloyd's of London Press, 1990. 402 pp.

Judicial and arbitration decisions which have led to present law relating to laytime and demurrage. Details of known awards from arbitration proceedings.

2172 Summerskill, Michael Brynmor. *Laytime. British shipping laws.* 4th edn. London: Stevens & Sons, 1989. 358 pp.

To provide help with laytime provisions. English law. Effect of laytime clauses; laytime not fixed; reaching the destination; readiness; modification of when laytime begins; bad weather; other suspensions; loading and discharging; demurrage; dispatch.

Demurrage

2173 **Tiberg, Hugo.** *Law of demurrage.* 3rd rev. edn. London: Stevens & Sons, 1979. 676 pp.

Law of demurrage; international case and awards references. General background; laytime; demurrage and damages for detention.

Agency

2174 **Chaplin, Peggy.** The customhouse broker as importer of record. *Journal of Maritime Law and Commerce* 13(3), April 1982, 295–312.

Examines liabilities assumed by customhouse broker and the rights he acquires *vis-à-vis* his principal and the US Customs Service when he becomes importer of record.

2175 **Eadie, J. D.** *Ships' agency.* London: Fairplay Publications, 1985. 10 pp.

A series reprinted from pages of *Fairplay International Shipping Weekly*, 23 September 1982 to 3 February 1983. Definitions and classification of agency. Duties of agent under (i) voyage charterparty; (ii) time charterparty; (iii) liner agency. Shipowners' responsibilities.

2176 **Mankabady, S.** Rights and immunities of the carrier's servant or agents. *Journal of Maritime Law and Commerce* 5(1), October 1973, 111–23.

Considers provision in COGSA 1971 on rights and immunities of the carrier's servant or agents (Art. III). Also, law in some common law countries where problem has received critical attention: English law, American law, Australian law, New Zealand law, Canadian law.

2177 **Schiels, Kenneth T.** *Tramp ship agency practice.* London: Lloyd's of London Press, 1987. 114 pp.

A complete survey of the specialized functions of the tramp ship agent and his relationships with the rest of the shipping community. Function and purpose of the tramp ship agent. Legal aspects of agency relationship and liabilities as between tramp agent, shipowner and charterer.

Liabilities

2178 **Allen, Betty B.** Update maritime liability rules. *Transportation and Distribution* 32(6), June 1991, 50–54.

2179 Astle, W. E. *International cargo carriers' liabilities.* London: Fairplay Publications, 1983. 140 pp.

Determination of variety of contractual liability problems arising out of contracts covering international carriage of goods, marine insurance and other maritime transactions. Considers disputes determined by Court of Appeal or House of Lords, of importance to shipowners, international road carriers, air carriers, cargo interests and respective insurers.

2180 Bauer, R. Glenn. The measure of liability for cargo damage under charter parties: A second look. *Journal of Maritime Law and Commerce* 21(3), July 1990, 397–415.

Distinction between private and common carriers. Burden of proof in private carriage. The 'Clause Paramount'.

2181 Bayer, Nathan J. and **McClain, Ellen Y.** The Anti-Drug Abuse Act of 1986: The ocean carrier's dilemma. *Journal of Maritime Law and Commerce* 20(3), July 1989, 299–325.

Penalties facing carriage companies – possible seizure and forfeiture of vessels in event that unmanifested drugs transported in violation of customs laws. Statutory developments. Impact of 1986 Act.

2182 Bond Smith, J. Recent developments in the law of damages for breach of charter. *Journal of Maritime Law and Commerce* 13(3), April 1982, 313–22.

Case law developments. Recovery of lost profits. The question of mitigation.

2183 Diplock, Kenneth. Conventions and morals: Limitation clauses in international maritime conventions. *Journal of Maritime Law and Commerce* 1(4), July 1970, 525–36.

Sets out necessity of allocation of liability and not morality in international conventions relating to carriage of goods by sea. Economic aim of law.

2184 Gronfors, Kurt. Liability for delay in combined transport. *Journal of Maritime Law and Commerce* 5(3), April 1974, 483–90.

Provisions as to 'delay' in Draft Convention on the combined transport contract favoured at IMCO/ECE meetings. Advises what is meant by delay in combined transport situations and to which segment a delay is to be attributed: Applicable rule of liability, uniform liability.

2185 Henderson, Lewis E. The Wreck Act: Duties to mark and remove. *Journal of Maritime Law and Commerce* 21(3), July 1990, 417–43.

Litigation involving obstructions of navigable waters. Statutory developments. Case law development. Liabilities. The duty to mark. The duty to remove.

2186 Herman, Lewis and **Goldman, Steven E.** The Master's negligence and charterers' warranty of safe port/berth. *Lloyd's Maritime and Commercial Law Quarterly,* November 1983, Part 4, 615–24.

Review of developments and cases in one particular area of safe berth/port disputes – the willingness of courts and arbitrators to look at actions/omissions of vessel's master as intervening cause resulting in owner's liability for casualty.

2187 ICHCA. *The law of loss and damage and carrier liability after the container reaches land or before it leaves it.* London: ICHCA, 1975.

2188 INTERTANKO. *Oil cargo losses and problems with measurement.* Oslo: INTERTANKO, 1986. 15 pp.

A booklet showing that many of the apparent losses of oil cargo either (i) did not, in fact, occur and thus the carriers were the victims of measurement inadequacies, or (ii) that such losses were, in fact, accountable. (A follow up to INTERTANKO's 1978 Cargo Measurement and Cargo Loss and Tanker Cargo Claims.)

2189 Larsen, P. B., Sweeney, J. C. and **Falvey, P. J.** The uniform rules on the liability of operators of transport terminals. *Journal of Maritime Law and Commerce* 20(1), January 1989, 21–54.

Examines issues raised during the discussions of the Working Group on Transport Terminal Operators' Liability and set up by UNCITRAL to propose draft Uniform Rules on liability of such operators.

2190 Mankabady, S. Speed and consumption clauses. *Journal of Maritime Law and Commerce* 6(4), July 1975, 635–42.

Shipowner's liability in respect of terms in contract and reasons for reduction of speed or the excessive bunker consumption. The remedy available for charterer for breach of these clauses. Main criticisms of clauses.

2191 O'Keefe, Patrick and **Colinard, R. J.** The law of the sea and the marine carrier's liability. *Journal of World Trade* 10(4), July/August 1976, 334–54.

Examines relationship between the 1976 Draft Convention on carriage of goods by sea and the 1976 Third United Nations Conference on the law of the sea. To what extent will possible changes in traditionally accepted public international law of the sea interact with recommended modifications in the liability of sea carrier of goods?

2192 **Parrish, Robert B.** Liabilities as between tug and tow. *Federation of Insurance and Corporate Counsel Quarterly* 39(3), Spring 1989, 275–91.

2193 **Rose. F. D.** From bailment to bailment. *Lloyd's Maritime and Commercial Law Quarterly*, May 1982, Part 2, 200–09.

Considers rights and liabilities of carriers, cargo owners and salvors where through no fault of the carrier, salvage services are necessary for ship and cargo. Case law.

2194 **Sassoon, David M.** Liability for the international carriage of goods by sea, land, air: Some comparisons. *Journal of Maritime Law and Commerce* 3(4), July 1972, 759–73.

Highlights some of principal discrepancies between rights and obligations of carriers operating different modes of transport, and the different rules governing their liability for loss/damage to cargo moved in international trade.

2195 **Selvig, Erling.** Towards strict shipowner liability: Recent trends in Norwegian law on maritime torts. *Journal of Maritime Law and Commerce* 2(2), January 1971, 383–98.

Norwegian law on enterprise liability. Case law. A new road to strict liability?

2196 *Shipowners' liability for loss or damage to cargo under the Hague Rules: Gold Clause Agreement.* London: Witherby & Co., 1987.

2197 **Tettenborn, A. M.** The carrier and the non-owning consignee: An inconsequential immunity. *Journal of Business Law*, 1987, 12–17.

Examines the House of Lords' decision in *Leigh & Sullivan* v. *The 'Aliakman'* (1986) which addresses question of who can sue a shipowner in tort for damage done to goods in his care.

2198 **Textor, James M.** Oil shortages caused by the inherent properties of petroleum cargoes. *Journal of Maritime Law and Commerce* 13(3), April 1982, 281–94.

Liabilities concerning oil shortage disputes and the calculation of tanker outages. Case law update.

2199 **Villareal, Dewey R.** Fourteen years with conduct sufficient to preclude indemnity. *Journal of Maritime Law and Commerce* 4(2), January 1973, 309–17.

Examines how US courts have dealt with the concept of 'conduct sufficient to preclude recovery' in respect of shipowners' liability towards injured persons.

Maritime Law

Admiralty law

2200 Albrecht, Hans-Christian. Germany. In: Lennart Hagberg (ed.) *Maritime law. Vol. I. Arrest of vessels.* The Netherlands: Kluwer, 1976, pp. 35–8.

Types of claim. Existence of claim. Security. Court. Form for application. Representation. Appeals. Arrest – enforcement of order; service of notice; maintaining. Foreign vessels, national vessels. Time. 1952 Brussels Convention.

2201 Allbritton, Jack L. Division of damages in admiralty: A rising tide of confusion. *Journal of Maritime Law and Commerce* 2(2), January 1971, 323–47.

Law of division of damages in US admiralty and maritime cases. Commences in time with Halayon decision, 1951. Contribution, indemnity, contributory negligence.

2202 August, Raymond S. Trial by jury in a court of Admiralty: A constitutional right buried under historical ignorance. *Journal of Maritime Law and Commerce* 13(2), January 1982, 149–76.

Examines US holding that there is no constitutional right to a jury trial in admiralty. Shows that historically such a right did exist. Argues that the Constitution did allow defendant the right to a common law jury trial and that developed holding by federal courts erroneous.

2203 Behrendt-Poulsen, E. Denmark. In: Lennart Hagberg (ed.) *Maritime Law. Vol I. Arrest of vessels.* The Netherlands: Kluwer, 1976, pp. 23–6.

Types of claim – pecuniary claims. Existence of claim. Security. Court. Form for application. Representation. Appeals. Arrest – service of notice; maintaining; enforcement of order. Foreign vessels, national vessels. Time. Brussels Convention 1952.

2204 Bentley, Anthony M. Shipowners' responsibility for rescue at sea. *Journal of Maritime Law and Commerce* 3(3), April 1972, 373–86.

Case law development – the duty to rescue; breach of duty is negligence; discharge of the rescue duty; lifesaving equipment.

2205 **Berlingieri, Francesco.** *Arrest of ships.* London: Lloyd's of London Press, 1992. 320 pp.

2206 **Cadwallader, F. J. J.** 'The Court of Admiralty and Equity': Principles and practice of maritime law. *Royal Society of Arts Journal* 138, January 1990, 119–33.

The paper given and following discussion at the Thomas Gray Memorial Lecture, April 1989, to the Society. Development of the Admiralty Court. Its jurisdiction.

2207 **Chorley, Robert Samuel Theodore** and others. *Shipping law.* 9th rev. edn. London: Pitman, 1992.

2208 **Colinvaux, R., Steel, D.** and **Ricks, V. E.** *Forms and precedents. British shipping laws.* Vol. 6. London: Stevens & Sons, 1973. 334 pp.

Precedents for Admiralty practice. Cases, statutes. Model bills of lading. Model passenger ticket. Charterparties. Commercial court pleadings – carriage by sea; marine insurance. Admiralty pleadings. Admiralty forms.

2209 **Cunningham, John R.** Canada. In: Lennart Hagberg (ed.) *Maritime law. Vol I. Arrest of vessels.* The Netherlands: Kluwer, 1976, pp. 17–22.

Types of claim as set out in Federal Court Act of Canada. Existence of claim. Security. Court. Form for application. Representation. Appeals. Enforcement order, maintaining, service of notice *re* arrest. Foreign vessels. National vessels. Time. Brussels International Convention 1952.

2210 **Davies, Martin** and **Dickey, Anthony.** *Shipping law.* Holmes Beach (Florida): William W. Gaunt & Sons, 1990. 431 pp.

2211 **De Koning, John.** Australia. In: Lennart Hagberg (ed.) *Maritime Law. Vol. I. Arrest of vessels.* The Netherlands: Kluwer, 1976, pp. 1–12.

Jurisdiction – courts. Basis and extent of jurisdiction. Types of claims. Evidence of existence of a claim. Security. Competent court of authority. Form for application. Representation by counsel. Power of attorney. Appeals. Enforcement of Order of Arrest. Service of Order. Maintaining arrest. Foreign vessels. National vessels. Time.

2212 De Sieno, Timothy B. Choice of law in maritime torts: An analysis of a recent trend. *Journal of Maritime Law and Commerce* 20(3), July 1989, 375–93.

Examines theory/jurisprudence. Application of law to maritime torts. Critical analysis, policy of choice of law. Reduced role of choice of law in *forum non conveniens* determinations.

2213 Drzal, Michael P. and **Carnilla, Phyllis D.** Documentation of vessels: The fog lifts. *Journal of Maritime Law and Commerce* 13(3), April 1982, 261–79.

Purpose and development of federal documentation. Documentation statutes. Overview of procedural simplification under Vessel Documentation Act 1980. Problems raised by Act.

2214 Fajemirokun, Bola. Jurisdictional aspects of admiralty claims in Nigeria. *Lloyd's Maritime and Commercial Law Quarterly*, August 1992, Part 3, 395–408.

Examines competing arguments about jurisdictional competence for admiralty claims between federal and state high courts. Future direction of admiralty jurisdiction in Nigeria.

2215 Farrugia, R. Malta. In: Lennart Hagberg (ed.) *Maritime law. Vol. I. Arrest of vessels.* The Netherlands: Kluwer, 1976, pp. 55–8.

Types of claims. Existence of claim. Security. Court – in Malta; in Gozo. Form for application. Representation. Appeals. Arrest – enforcement of order; maintaining; service of notice. Foreign vessels, national vessels. Time. Conventions.

2216 Fink, Ze'ev S. Admiralty foreclosure: Sale of a vessel subject to legislation limiting transfer. *Journal of Maritime Law and Commerce* 10(4), July 1979, 555–61.

Considers possible conflict between two policies: (i) encouragement of financing vessels by means of preferred status given to ship mortgages and (ii) administrative limitation on foreclosures of mortgages and change of flag in order to protect national security interests. Israeli High Court's approach in the 'BANACOR', 1972.

2217 Florida Bar. *Maritime law and practice.* 2nd edn. Tallahassee

(Florida): Florida Bar Continuing Legal Education Publications, 1991. 594 pp.

2218 Force, Robert. Determining admiralty tort jurisdiction: An alternative analytical framework. *Journal of Maritime Law and Commerce* 21(1), January 1990, 1–80.

Development of admiralty jurisdiction. Divergent views on proper scope of admiralty jurisdiction. The nexus test/excluding marginal cases from admiralty jurisdiction. An alternative analytical framework. Tort and maritime activity – a matter of perspective. A bright line rule. Easy/difficult cases.

2219 Friedell, Steven F. and **Healy, Nicholas J.** An introduction to in rem jurisdiction and procedure in the United States. *Journal of Maritime Law and Commerce* 20(1), January 1989, 55–73.

Procedure and case law on admiralty practice in federal courts of in rem procedure.

2220 Ganado, Max and others. *Arrest of ships. Vol. 3: Malta, Panama, Sweden, United Arab Emirates.* London: Lloyd's of London Press, 1987. 110 pp.

Third book of series providing practical guidance to maritime claimants on the correct procedure to arrest a ship. Highlights lack of international uniformity in rules for ship arrest. Malta, Panama, Sweden and UAE studied in detail.

2221 Gaskell, N. J. J., Debattista, C. and **Swatton, R. J.** *Chorley and Giles shipping law.* 8th edn. London: Pitman Publishing, 1987. 704 pp.

Established as a standard work – eighth edition restructures the original work to explain what has become a large and complex area of the law. Deals with all aspects of the law with which those engaged in the business of owning and managing ships are likely to meet.

2222 Georghadjis, A. and others. *Arrest of ships.* Vol. 7. London: Lloyd's of London Press, 1988. 98 pp.

Seventh book in series planned to provide maritime claimants with practical guidance on ship arrest. Highlights the lack of international uniformity in rules for ship arrest. In detail: Cyprus, Egypt, Pakistan and Poland.

2223 Gorton, Lars. Joint ventures in shipping law. *Journal of Business Law,* 1975, 238–44.

Principles in relation to international shipping law. International nature of ocean shipping. Diminishing of principle of freedom of the sea.

2224 Griggs, J. G. R. United Kingdom. In: Lennart Hagberg (ed.) *Maritime law. Vol. I. Arrest of vessels.* The Netherlands: Kluwer, 1976, pp. 85–8.

Types of claims. Existence of claim. Security. Court. Form for application. Other formal requirements. Time. Conventions. Enforcement of warrant of arrest.

2225 Grime, Robert. *Shipping law.* 2nd rev. edn. London: Sweet & Maxwell, 1991. 449 pp.

Practical introduction to maritime law. Current law plus historical background to many contemporary issues.

2226 Hagberg, Lennart. *Maritime law. Vol. I. Arrest of vessels.* The Netherlands: Kluwer, 1976. 92 pp.

Handbook on maritime law produced by the Maritime and Transport Law Committee of the Section on Business Law of the International Bar Association. Arrest of vessels. First in series of four. General law issues for different countries.

2227 Hazelwood, Steven J. Gaps in the action in rem. *Lloyd's Maritime and Commercial Law Quarterly,* May 1982, Part 2, 210–17.

Considers 1952 International Convention Relating to the Arrest of Seagoing Ships and problems caused by its only partial implementation into English law by the Administration of Justice Act 1956. Case law interpretation of 1956 Act.

2228 Hazelwood, Steven J. Gaps in the action in rem: Plugged? *Lloyd's Maritime and Commercial Law Quarterly,* August 1982, Part 3, 422–6.

Considers the Court of Appeal decision in *The Span Terza* (1982) and the operation of the 'sister-ship' arrest provisions.

2229 Hill, Christopher. *Maritime law.* 3rd rev. edn. London: Lloyd's of London Press, 1989. 416 pp.

Detailed exposition of both traditional aspects of maritime law and of key international issues which have emerged in recent years. Frequent reference to case law.

2230 Hill, Christopher and others. *Arrest of ships.* London and New York (*et al.*): Lloyds of London Press, 1985. 99 pp.

Considers question of when, how and where a ship can be most advantageously arrested. Law, practice and procedure of arrest. Jurisdictions give comparative examples. England and Wales; Germany; Japan; USA.

2231 Hill, Christopher and others. *Arrest of ships.* Vol. 5. London: Lloyd's of London Press, 1987. 109 pp.

2232 Hill, Christopher and others. *Arrest of ships.* Vol. 6. London: Lloyds of London Press, 1987. 108 pp.

2233 *International shipping law.* Vols. 1–2. London: Euromoney UK, 1990.

2234 Ivamy, E. R. Hardy. *Casebook on shipping law.* 4th rev. edn. London: Lloyd's of London Press, 1987. 212 pp.

Leading and important cases. At head of each case short statement of principle involved. Summary of facts. Decision given. Extracts from judgments given.

2235 Jackson, D. C. Admiralty jurisdiction: The Supreme Court Act 1981. *Lloyd's Maritime and Commercial Law Quarterly,* May 1982, Part 2, 236–48.

Examines impact of 1981 Act. Effect on the in rem procedure. The 1952 Arrest Convention.

2236 Jarvis, Robert M. An introduction to Maritime Attachment Practice under Rule B. *Journal of Maritime Law and Commerce* 20(4), October 1989, 521–39.

Examination of Rule B which enables plaintiff to obtain jurisdiction over defendant by attacking its property. History. Availability. Procedure. Problem areas.

2237 Kan, Vun-Ping. *Arrest of ships.* Vol. 4. London: Lloyd's of London Press, 1987. 107 pp.

Fourth book in a series planned to provide maritime claimants with practical guidance on ship arrest. Highlights lack of international uniformity in rules for ship arrest. In detail: People's Republic of China, Nigeria, Oman, Scotland.

2238 Kegels, Tony. Belgium. In: Lennart Hagberg (ed.) *Maritime law. Vol. I. Arrest of vessels.* The Netherlands: Kluwer, 1976, pp. 13–16.

Belgian Code of Procedure – covers arrest of sea-going vessels. Evidence of claim. Security. Court/authority. Form for application. Appeals. Arrest – enforcement of order for maintaining. Foreign vessels, national vessels. Time. Brussels Convention on arrest of sea-going vessels.

2239 Keriakosk, Mary Jane. The case for recognising the general equitable powers of the admiralty courts. *Journal of Maritime Law and Commerce* 20(3), July 1989, 363–73.

Historical background. Trends – particularly with respect to injunctive relief.

2240 Langenskiold, Henrik. Finland. In: Lennart Hagberg (ed.) *Maritime law. Vol. I. Arrest of vessels.* The Netherlands: Kluwer, 1976, pp. 27–30.

Types of claim. Security. Evidence of claim. Court. Form for application. Representation. Appeals. Arrest – enforcement of order; maintaining; service of notice. Foreign vessels, national vessels. Time. Conventions. Maritime liens, preferred mortgagor securing arrest of vessel ready to sail.

2241 Laughlin, James P. Choice of law in the federal admiralty court. *Journal of Maritime Law and Commerce* 10(2), January 1979, 165–97.

Analysis of choice of law by US federal courts in maritime cases. Tests created by Supreme Court to govern choice of laws. Have tests been applied consistently by the court? How some of inferior federal courts have dealt with problem of choice of determinant law.

2242 Lay, S. H., Churchill, R. and **Nordquist, M.** *New directions in the law of the sea. Documents – Volume II.* New York: Oceana Publications, 1973. 911 pp.

Regulation of marine activities. Relevant documents relating to navigation.

2243 Matthew, Paul. Releasing ships from arrest. *Lloyd's Maritime and Commercial Law Quarterly,* February 1983, Part 1, 99–102.

Examines question of what happens when parties cannot agree as to whether arrested ship should be released.

2244 McArdle, A. D. *International ship arrest: A practical guide.* London: Lloyd's of London Press, 1988. 336 pp.

2245 McGuffie, Kenneth C. and others. *Admiralty practice. British shipping laws.* Vol. 1. London: Stevens and Sons, 1975. 83 pp.

2246 McHose, John C. United States. In: Lennart Hagberg (ed.) *Maritime law. Vol I. Arrest of vessels.* The Netherlands: Kluwer, 1976, pp. 89–92.

Types of claims. Evidence of existence of claim. Security. Court. Form for application. Representation. Appeals. Arrest – enforcement of order; service of notice. Foreign vessels, national vessels. Time. Conventions. Cases.

2247 Meeson, Nigel. *The practice and procedure of the Admiralty Court: Forms and precedents.* London: Lloyd's of London Press, 1986. 270 pp.

First section – succinct guide to procedural matters peculiar to litigation in Admiralty. County Court proceedings dealt with in parallel to High Court. Second section – selection of forms and precedents for claims within Admiralty jurisdiction. Third section – relevant statutes and rules.

2248 Moloney, G. John. Ireland. In: Lennart Hagberg (ed.) *Maritime law. Vol. I. Arrest of vessels.* The Netherlands: Kluwer, 1976, pp. 47–50.

Types of claims. Existence of claim. Security. Court. Form for application. Representation. Appeals. Arrest – enforcement of order; maintaining; service of notice. Foreign vessels, national vessels. Time. Conventions.

2249 Myburgh, Paul. Admiralty jurisdiction in American Samoa: Le Va'a Lagoa. *Journal of Maritime Law and Commerce* 23(1), January 1992, 73–100.

Sets out the background of establishment of the High Court of American Samoa's jurisdiction and the conferring of general admiralty jurisdiction on it in 1975. Difficulties for High Court in justifying its assumption of jurisdiction under federal admiralty statutes. Case law. Options for reform.

2250 Naschitz, Peter Gad. Israel. In: Lennart Hagberg (ed.) *Maritime law. Vol. I. Arrest of vessels.* The Netherlands: Kluwer, 1976, pp. 51–4.

Types of claims. Admiralty Court. Existence of claim. Security. Form for application. Representation. Appeals. Arrest – enforcement of order; service of notice. Foreign vessels, national vessels. Time. Conventions. Legal situation.

2251 Panagopoulos, P. C. Greece. In: Lennart Hagberg (ed.) *Maritime law. Vol. I. Arrest of vessels.* The Netherlands: Kluwer, 1976, pp. 39–46.

Types of claims. Security. Competent Court. Form for application. Provisional order. Representation. Appeals. Lifting of arrest. Arrest – enforcement of order; maintaining; service of notice. Foreign vessels, national vessels. Time. Brussels Convention 1952.

2252 Pardini, Juan. Anatomy of an arrest in Panama. *Lloyd's Maritime and Commercial Law Quarterly*, August 1983, Part 3, 456–63.

Panama Canal Treaties 1978 – Law No. 8 of 30 March 1982, the 'Maritime Law'. Supplementary provision of the Judicial Code of the Republic of Panama. Procedural considerations. Process in rem.

2253 Paulsen, Gordon W. and **Burrick, Robert S.** Forum non conveniens in Admiralty. *Journal of Maritime Law and Commerce* 13(3), April 1982, 343–65.

Discussion of the doctrine and application following US Court of Appeals in *Alcoa Steamship Company Inc.* v. *M/V Nordic Regent* (1980). History of the doctrine. Forum selection clauses and *forum non conveniens.* Conflict among the circuits in application of the doctrine. Application in other countries.

2254 Paulsen, Gordon W. and **Lambert, Leroy.** Admissibility of Government agency reports in maritime cases. *Journal of Maritime Law and Commerce* 21(1), January 1990, 111–28.

Considers effect of *Beech Aircraft Corporation* v. *Rainey* (1989) (evaluative conclusions and opinions in a Navy Judge Advocate General's report admissible evidence – exception to hearsay rule) on admissibility of US Coast Guard and National Transportation Safety Board investigations.

2255 Philip, Allan and others. *Arrest of ships – 6.* London: Lloyd's of London Press, 1987. 89 pp.

2256 Potts, T. H. A. Singapore. In: Lennart Hagberg (ed.) *Maritime law. Vol. I. Arrest of vessels.* The Netherlands: Kluwer, 1976, pp. 67–72.

Types of claims. Existence of claim. Security. Court. Form for application. Representation. Appeals. Arrest – enforcement of order, maintaining, service of notice. Foreign vessels, national vessels. Time. Conventions.

2257 Practicing Law Institute. *Maritime law and practice: Current developments.* New York: Practicing Law Institute, 1970. 224 pp.

2258 Pritchett, Russell W. Ship to shore allision damages: The problem of remedial disuniformity in the federal courts. *Journal of Maritime Law and Commerce* 11(3), April 1980, 333–48.

Admiralty jurisdiction over torts. Evolution of the law regarding measure of damages which owner of a shoreside structure entitled to collect from seaborne tortfeasor. Does choice of forum colour choice of applicable law? Post-1948 cases.

2259 Rein, Alex. Norway. In: Lennart Hagberg (ed.) *Maritime law. Vol. I. Arrest of vessels.* The Netherlands: Kluwer, 1976, pp. 63–6.

Types of claims. Existence of claim. Security. Court. Form of application. Representation. Appeals. Arrest – enforcement of order; maintaining; service of notice. Foreign vessels, national vessels. Conventions. Time.

2260 Robinson, Susan K. Arresting the misconception. *Lloyd's Maritime and Commercial Law Quarterly,* May 1982, Part 2, 261–5.

Examines arrest procedures under Supreme Court Act 1981. Critical consideration of article by Andrew Tettenborn in *LMCLQ* 4, 1981, 507 which examined meaning of 'beneficial ownership' of vessel.

2261 Robol, Richard T. Admiralty's adjudicatory jurisdiction over alien defendants: A functional analysis. *Journal of Maritime Law and Commerce* 11(4), July 1980, 395–451.

Considers problem of admiralty's adjudicatory jurisdiction over alien defendants by constructing a heuristic model of admiralty jurisprudence based upon admiralty's special characteristics.

2262 Romlov, Robert. Sweden. In: Lennart Hagberg (ed.) *Maritime law. Vol. I. Arrest of vessels.* The Netherlands: Kluwer, 1976, pp. 81–4.

Types of claims. Existence of claim. Security. Court. Form for application. Representation. Arrest – enforcement of order; maintaining; service of notice. Foreign vessels, national vessels. Time. Conventions. Forum.

2263 Sampson, D. H. South Africa. In: Lennart Hagberg (ed.) *Maritime law. Vol. I. Arrest of vessels.* The Netherlands: Kluwer, 1976, pp. 73–80.

Types of claims. Existence of claim. Security. Court. Form for application. Representation. Appeals. Arrest – enforcement of order; maintaining; service of notice. Foreign vessels, national vessels. Time. Conventions.

2264 Schlichting, Mathias P. *The arrest of ships in German and South African law.* Frankfurt: Peter Lang Publishing, 1991. 407 pp.

2265 Schoenbaum, Thomas J. *Admiralty and maritime law, 1992: Pocket part.* St Paul (Minnesota): West Publishing Company, 1992. 185 pp.

2266 Sharpe and Spicer. *New directions of maritime law, 1984.* London: Stevens & Sons, 1984.

2267 Spicer, W. Wylie. Some admiralty law issues in Canadian offshore oil and gas development. *Lloyd's Maritime and Commercial Law Quarterly*, February 1982, Part 1, 5–32.

Considers whether offshore rigs are 'ships' within US Admiralty law. Reviews some of the cases on ships and concludes that, with possible exception of jack-ups, offshore oil rigs are ships. Discussion of number of consequences of this – nature of claims; actions for damage by a ship; statutory liabilities and limitation.

2268 Staniland, Hilton. The Halcyon Isle revisited: A South African perspective. *Lloyd's Maritime and Commercial Law Quarterly*, May 1989, Part 2, 174–90.

Considers the decision of Privy Council in the *Halcyon Isle* (1981) and the recognition, enforcement and ranking of foreign maritime liens within South African context.

2269 Stokes, James C. Henry J. Friendly: Opinions in admiralty and maritime law, 1959–1974. *Journal of Maritime Law and Commerce* 7(1), October 1975, 275–92.

Examines decisions of this judge after fifteen years on the bench of the US Court of Appeals, Second Circuit. Cases in respect of choice of law; seaworthiness and seamen; compensation for longshoremen; COGSA. The powers of the federal courts and administrative agencies in maritime matters.

2270 Strasburger, John H. The ABC's of admiralty and bankruptcy in concert or conflict. *Journal of Maritime Law and Commerce* 21(2), April 1990, 273–92.

Jurisdiction of bankruptcy judges to hear and decide admiralty cases. Post-petition maritime liens. A bankruptcy court's in rem jurisdiction of the bankrupt's property. Miscellaneous matters.

2271 Tabbush, S. J. Arrest of ships owned by charterers. *Lloyd's Maritime and Commercial Law Quarterly*, November 1982, Part 4, 585–91.

Examines question of whether injured party can arrest vessel belonging to the relevant person who is not the owner of the vessel in connection with which the action arises. Statutory position and case law development. The impact of the *Span Terza* (1982).

2272 Theunis, Jan and others. *Arrest of ships*. Vol. 2. London: Lloyd's of London Press, 1986. 128 pp.

Second book of a series planned to provide maritime claimants, who have occasion to arrest a ship, with practical guidance on the correct procedure to be followed. Highlights lack of international uniformity in ship arrest laws of Belgium, the Netherlands, India and Yugoslavia.

2273 Van der Feltz, F. The Netherlands. In: Lennart Hagberg (ed.) *Maritime law. Vol. I. Arrest of vessels.* The Netherlands: Kluwer, 1976, pp. 59–62.

Types of claims. Existence of claim. Security. Court. Form for application. Representation. Appeals. Arrest – enforcement of order; maintaining; service of notice. Foreign vessels, national vessels. Conventions.

2274 Villeneau, Jacques. France. In: Lennart Hagberg (ed.) *Maritime law. Vol. I. Arrest of vessels.* The Netherlands: Kluwer, 1976, pp. 31–4.

Types of claims. Existence of claim. Security. Court. Form for application. Representation. Appeals. Arrest – enforcement of order; maintaining; service of order. Foreign vessels, national vessels. Time. Conventions.

2275 Williams, David McI. A comment on the recodification of the United States Ship Mortgage Act. *Journal of Maritime Law and Commerce* 20(2), April 1989, 153–67.

Changes made by the enacting into law of the ship mortgage, maritime lien and vessel identification system. How the 1989 Act marked a watershed for marine financing.

Registration

2276 Akhtar, Saleem. Pakistan. In: Lennart Hagberg (ed.) *Handbook on maritime law. Vol. IIIB. Registration of vessels, mortgages on vessels.* The Netherlands: Kluwer, 1983, pp. 1–16.

Registration of vessels. Registration of mortgages on vessels. Legal effects of a registered mortgage. Enforcement of mortgages. Registration of rights in respect of vessels under construction.

2277 Albrecht, Hans-Christian. Federal Republic of Germany. In: Lennart Hagberg (ed.) *Handbook on maritime law. Vol. IIIA. Registration of vessels, mortgages on vessels.* The Netherlands: Kluwer, 1983, pp. 165–84.

Registration of vessels. Registration of mortgages on vessels. Legal effects of a registered mortgage. Enforcement. Registration of rights in respect of vessels under construction.

2278 Aminoff, Jan. Finland. In: Lennart Hagberg (ed.) *Handbook on maritime law. Vol. IIIA. Registration of vessels, mortgages on vessels.* The Netherlands: Kluwer, 1983, pp. 185–98.

Registration of vessels. Registration of mortgages on vessels. Legal effects of a registered mortgage. Enforcement. Registration of rights in respect of vessels under construction.

Registration of vessels. Registration of mortgages on vessels. Legal effects of a registered mortgage. Enforcement. Registration of rights in respect of vessels under construction.

2279 Baker and **McKenzie.** Venezuela. In: Lennart Hagberg (ed.) *Handbook on maritime law. Vol. IIIB. Registration of vessels, mortgages on vessels.* The Netherlands: Kluwer, 1983, pp. 247–76.

Registration of vessels. Registration of mortgages on vessels. Legal effects of a registered mortgage. Enforcement of mortgages. Registration of rights in respect of vessels under construction.

2280 Ballantyne, W. M. *Register of laws of the Arabian Gulf.* London: Lloyd's of London Press, 1985.

2281 Baran, Metin. Turkey. In: Lennart Hagberg (ed.) *Handbook on maritime law. Vol. IIIB. Registration of vessels, mortgages on vessels.* The Netherlands: Kluwer, 1983, pp. 217–28.

Registration of vessels. Registration of mortgages on vessels. Legal effects of a registered mortgage. Enforcement of mortgages. Registration of rights in respect of vessels under construction.

2282 Berlingieri, Francesco. The new Italian law on temporary registration of bareboat chartered vessels. *Journal of Maritime Law and Commerce* 21(2), April 1990, 199–211.

Background to new rules – amending of provisions which regulate procedure of deregulation of ships from Italian register previously registered in another state. Procedure. Problems of private international law.

2283 Bomchil, M. Argentina. In: Lennart Hagberg (ed.) *Handbook on maritime law. Vol. IIIA. Registration of vessels, mortgages on vessels.* The Netherlands: Kluwer, 1983, pp. 1–16.

Important statutory provisions – Shipping Act, No. 20.094; National Register of Vessels Act, No. 19.170. Registration. Registration of mortgages. Legal effects of registered mortgage. Enforcement. Registration of rights in respect of vessels under construction.

2284 Calder, Douglas. Cayman Islands. In: Lennart Hagberg (ed.) *Handbook on maritime law. Vol. IIIA. Registration of vessels, mortgages on vessels.* The Netherlands: Kluwer, 1983, pp. 85–106.

Registration of vessels. Registration of mortgages on vessels. Legal effects of a registered mortgage. Enforcement. Registration of rights in respect of vessels under construction.

2285 Chesterman, Philip J. Bermuda. In: Lennart Hagberg (ed.) *Handbook on maritime law. Vol. IIIA. Registration of vessels, mortgages on vessels.* The Netherlands: Kluwer, 1983, pp. 65–70.

Similarity with law of UK. Differences – registration of vessels; registration of mortgages on vessels; enforcement; registration of rights in respect of vessels under construction.

2286 Costabel, Attilio M. Italy. In: Lennart Hagberg (ed.) *Handbook on maritime law. Vol. IIIA. Registration of vessels, mortgages on vessels.* The Netherlands: Kluwer, 1983, pp. 287–300.

Registration of vessels. Registration of mortgages on vessels. Legal effects of a registered mortgage. Enforcement. Registration of rights in respect of vessels under construction.

2287 Cunningham, John R. Canada. In: Lennart Hagberg (ed.) *Handbook on maritime law. Vol. IIIA. Registration of vessels, mortgages on vessels.* The Netherlands: Kluwer, 1983, pp. 71–84.

Registration of vessels. Registration of mortgages on vessels. Legal effects of a registered mortgage. Enforcement of mortgages. Registration of rights in respect of vessels under construction.

2288 De Gaetano, George A. Malta. In: Lennart Hagberg (ed.) *Handbook on maritime law. Vol. IIIA. Registration of vessels, mortgages on vessels.* The Netherlands: Kluwer, 1983, pp. 327–44.

Registration of vessels. Registration of mortgages on vessels. Legal effects of registered mortgage. Enforcement. Registration of rights in respect of vessels under construction. Priority between registered mortgages and maritime liens or privileges.

2289 De Koning, J. P. M. Australia. In: Lennart Hagberg (ed.) *Handbook on maritime law. Vol. IIIA. Registration of vessels, mortgages on vessels.* The Netherlands: Kluwer, 1983, pp. 17–30.

Registration of vessels. Registration of mortgages on vessels. Legal effects of registered mortgage. Enforcement. Registration of rights in respect of vessels under construction.

2290 Dias, Mateus Andrade. Portugal. In: Lennart Hagberg (ed.) *Handbook on maritime law. Vol. IIIB. Registration of vessels, mortgages on vessels.* The Netherlands: Kluwer, 1983, pp. 51–64.

Registration of vessels. Registration of mortgages on vessels. Legal effects of a registered mortgage. Enforcement of mortgages. Registration of rights in respect of vessels under construction.

2291 Evangelista, Rafael E. and **Kwan Natividad, B.** Philippines. In: Lennart Hagberg (ed.) *Handbook on maritime law. Vol. IIIB. Registration of vessels, mortgages on vessels.* The Netherlands: Kluwer, 1983, pp. 35–50.

Registration of vessels. Registration of mortgages on vessels. Legal effects of a registered mortgage. Enforcement of mortgages. Registration of rights in respect of vessels under construction.

2292 Frieden, Luc. The new Luxemburg Shipping Register. *Lloyd's Maritime and Commercial Law Quarterly*, May 1991, Part 2, 257–63.

Considers the most important legal provisions of the 1991 law adopted by Luxemburg Parliament providing the state with a shipping register.

2293 Gitsu, M. A. The registration of merchant ships under perestroika: Soviet legislation. *Marine Policy* 16(2), March 1992, 111–14.

Examines laws governing foreign participation in Soviet shipping. The creation and activities of foreign joint ventures. Legal issues relating to the rights of such vessels to sail under the Soviet flag.

2294 Green, Meana Fernando. Spain. In: Lennart Hagberg (ed.) *Handbook on maritime law. Vol. IIIB. Registration of vessels, mortgages on vessels.* The Netherlands: Kluwer, 1983, pp. 121–48.

Registration of vessels. Registration of mortgages on vessels. Legal effects of a registered mortgage. Enforcement of mortgages. Registration of rights in respect of vessels under construction.

2295 Hagberg, Lennart (ed.) *Handbook on maritime law. Vol. IIIA. Registration of vessels, mortgages on vessels. Argentina–Norway.* The Netherlands: Kluwer, 1983. 401 pp.

Papers written by members of the Committee on Maritime and Transport Law, International Bar Association. Overall view of ship registration in different countries, and mortgages.

2296 Hagberg, Lennart (ed.) *Handbook on maritime law. Vol. IIIB.*

Registration of vessels, mortgages on vessels. Pakistan–Yugoslavia. The Netherlands: Kluwer, 1983. 296 pp.

Papers written by members of the Committee on Maritime and Transport Law, International Bar Association. Overall view of ship registration in different countries, and mortgages.

2297 Hamilton, D. O. England and Wales. In: Lennart Hagberg (ed.) *Handbook on maritime law. Vol. IIIA. Registration of vessels, mortgages on vessels.* The Netherlands: Kluwer, 1983, pp. 137–64.

Registration of vessels. Registration of mortgages on vessels. Legal effects of a registered mortgage. Enforcement. Registration of rights in respect of vessels under construction.

2298 Hansen, Victor V. Denmark. In: Lennart Hagberg (ed.) *Handbook on maritime law. Vol. IIIA. Registration of vessels, mortgages on vessels.* The Netherlands: Kluwer, 1983, pp. 127–36.

Registration of vessels. Registration of mortgages on vessels. Legal effects of a registered mortgage. Enforcement. Registration of rights in respect of vessels under construction.

2299 Hjort, Erling C. Norway. In: Lennart Hagberg (ed.) *Handbook on maritime law. Vol. IIIA. Registration of vessels, mortgages on vessels.* The Netherlands: Kluwer, 1983, pp. 385–401.

Registration of vessels. Registration of mortgages on vessels. Legal effects of a registered mortgage. Enforcement. Registration of rights in respect of vessels under construction.

2300 Ivković, Dorde. Yugoslavia. In: Lennart Hagberg (ed.) *Handbook on maritime law. Vol. IIIB. Registration of vessels, mortgages on vessels.* The Netherlands: Kluwer, 1983, pp. 277–96.

The Sea and Inland Navigation Act 1978. Registration of vessels. Registration of hypothecs on vessels. Legal effects of a registered hypothec. Enforcement of hypothec. Registration of rights in respect of vessels under construction.

2301 Jansen, Willem. The Netherlands Antilles. In: Lennart Hagberg (ed.) *Handbook on maritime law. Vol. IIIA. Registration of vessels, mortgages on vessels.* The Netherlands: Kluwer, 1983, pp. 367–74.

Mainly legislation of Holland. Registration of vessels. Registration of mortgages on vessels. Legal effects of registered mortgage. Enforcement. Registration of rights in respect of vessels under construction.

2302 Johnson, Edward L. U.S.A. In: Lennart Hagberg (ed.) *Handbook* *on maritime law. Vol. IIIB. Registration of vessels, mortgages on vessels.* The Netherlands: Kluwer, 1983, pp. 229–46.

Registration of vessels. Registration of mortgages on vessels. Legal effects of a registered mortgage. Enforcement of mortgages. Registration of rights in respect of vessels under construction.

2303 **Johnson, Edward L.** and **Teewia, John.** Liberia. In: Lennart Hagberg (ed.) *Handbook on maritime law. Vol. IIIA. Registration of vessels, mortgages on vessels.* The Netherlands: Kluwer, 1983, pp. 311–26.

Chapters II and III. Liberian maritime law. Equitable principles, natural justice. Section 30 Liberian maritime law. Registration of vessels. Registration of mortgages on vessels. Legal effects of registered mortgage. Enforcement. Registration of rights in respect of vessels under construction.

2304 **Kegels, Tony.** Belgium. In: Lennart Hagberg (ed.) *Handbook on maritime law. Vol. IIIA. Registration of vessels, mortgages on vessels.* The Netherlands: Kluwer, 1983, pp. 55–64.

Registration of vessels. Registration of mortgages on vessels. Legal effects of a registered mortgage. Enforcement of mortgages. Registration of rights in respect of vessels under construction.

2305 **Landau, E.** Israel. In: Lennart Hagberg (ed.) *Handbook on maritime law. Vol. IIIA. Registration of vessels, mortgages on vessels.* The Netherlands: Kluwer, 1983, pp. 273–86.

Registration of vessels. Registration of mortgages on vessels. Legal effects of a registered mortgage. Enforcement. Registration of rights in respect of vessels under construction.

2306 **Lee, John S.** and others. Taiwan. In: Lennart Hagberg (ed.) *Handbook on maritime law. Vol. IIIB. Registration of vessels, mortgages on vessels.* The Netherlands: Kluwer, 1983, pp. 169–94.

Registration of vessels. Registration of mortgages on vessels. Legal effects of a registered mortgage. Enforcement of mortgages. Registration of rights in respect of vessels under construction.

2307 **Lloyd's Register of Shipping.** *Lloyd's Register: Serving world industry.* London: Lloyd's Register of Shipping, 1988. 32 pp.

Booklet describing Lloyd's Register and its services. The constitution. Independent inspection agency. The L. R. surveyor. Shipping information. Classification services. Statutory services. Marine advisory services. Specification services. Yacht and small craft services. Industrial services etc.

2308 Lobosky, Reginald H. Bahamas. In: Lennart Hagberg (ed.) *Handbook on maritime law. Vol. IIIA. Registration of vessels, mortgages on vessels.* The Netherlands: Kluwer, 1983, pp. 31–54.

Registration of vessels. Registration of mortgages on vessels. Legal effects of registered mortgage. Enforcement. Registration of rights in respect of vessels under construction.

2309 Mahassni, Hassan. Saudi Arabia. In: Lennart Hagberg (ed.) *Handbook on maritime law. Vol. IIIB. Registration of vessels, mortgages on vessels.* The Netherlands: Kluwer, 1983, pp. 65–74.

Registration of vessels. Registration of mortgages on vessels. Legal effects of a registered mortgage. Enforcement of mortgages. Registration of rights in respect of vessels under construction.

2310 Maritime Law Association of South Africa. South Africa. In: Lennart Hagberg (ed.) *Handbook on maritime law. Vol. IIIB. Registration of vessels, mortgages on vessels.* The Netherlands: Kluwer, 1983, pp. 105–20.

Republic of South Africa. Merchant Shipping Act (Act No. 57 of 1951). Registration of Ships Regulations 1961. Registration of vessels. Registration of mortgages on vessels. Legal effects of a registered mortgage. Enforcement of mortgages. Registration of rights in respect of vessels under construction.

2311 Marston, Geoffrey. The U.N. Convention on Registration of Ships. *Journal of World Trade Law* 20(5), September/October 1986, 575–80.

Developments leading to 1986 UN Conference on Conditions for Registration of Ships. Considers purpose of Convention – to deal with FOCs and to ensure that there is a genuine link and effective exercise of jurisdiction and control by flag-state.

2312 Moloney, Gerald J. Ireland. In: Lennart Hagberg (ed.) *Handbook on maritime law. Vol. IIIA. Registration of vessels, mortgages on vessels.* The Netherlands: Kluwer, 1983, pp. 251–72.

Registration of vessels. Registration of mortgages on vessels. Legal effects of a registered mortgage. Enforcement. Registration of rights in respect of vessels under construction. Effect of Treaty of Rome on legislation.

2313 Mossack, Jürgen R. D. Panama. In: Lennart Hagberg (ed.) *Handbook on maritime law. Vol. IIIB. Registration of vessels, mortgages on vessels.* The Netherlands: Kluwer, 1983, pp. 17–34.

Registration of vessels. Registration of mortgages on vessels. Legal effects of a registered mortgage. Enforcement of mortgages. Registration of rights in respect of vessels under construction.

2314 **Ninomiya, Seijiro.** Japan. In: Lennart Hagberg (ed.) *Handbook on maritime law. Vol. IIIA. Registration of vessels, mortgages on vessels.* The Netherlands: Kluwer, 1983, pp. 301–10.

Registration of vessels. Registration of mortgages on vessels. Legal effects of a registered mortgage. Enforcement. Registration of rights in respect of vessels under construction.

2315 **Panagopoulos, P. C.** Greece. In: Lennart Hagberg (ed.) *Handbook on maritime law. Vol. IIIA. Registration of vessels, mortgages on vessels.* The Netherlands: Kluwer, 1983, pp. 209–46.

Registration of vessels. Registration of mortgages on vessels. Legal effects of a registered mortgage. Enforcement. Registration of rights in respect of vessels under construction.

2316 **Patrick, Bruce R.** Scotland. In: Lennart Hagberg (ed.) *Handbook on maritime law. Vol. IIIB. Registration of vessels, mortgages on vessels.* The Netherlands: Kluwer, 1983, pp. 75–86.

Registration of vessels. Registration of mortgages on vessels. Legal effects of a registered mortgage. Enforcement of mortgages. Registration of rights in respect of vessels under construction.

2317 **Potts, T. H. A.** and **Selvadurai, P.** Singapore. In: Lennart Hagberg (ed.) *Handbook on maritime law. Vol. IIIB. Registration of vessels, mortgages on vessels.* The Netherlands: Kluwer, 1983, pp. 87–104.

Registration of vessels. Registration of mortgages on vessels. Legal effects of a registered mortgage. Enforcement of mortgages. Registration of rights in respect of vessels under construction.

2318 **Psillaki, Efti.** Cyprus. In: Lennart Hagberg (ed.) *Handbook on maritime law. Vol. IIIA. Registration of vessels, mortgages on vessels.* The Netherlands: Kluwer, 1983, pp. 107–26.

Reliance upon English law. Registration of vessels. Registration of mortgages on vessels. Legal effects of a registered mortgage. Enforcement. Registration of rights in respect of vessels under construction. International Convention.

2319 **Rohart, Jean-Serge.** France. In: Lennart Hagberg (ed.) *Handbook on maritime law. Vol. IIIA. Registration of vessels, mortgages on vessels.* The Netherlands: Kluwer, 1983, pp. 199–208.

Registration of vessels. Registration of mortgages on vessels. Legal effects of a registered mortgage. Enforcement. Registration of rights in respect of vessels under construction.

2320 **Romlov, Robert.** Sweden. In: Lennart Hagberg (ed.) *Handbook on maritime law. Vol. IIIB. Registration of vessels, mortgages on vessels.* The Netherlands: Kluwer, 1983, pp. 149–68.

Registration of vessels. Registration of mortgages on vessels. Legal effects of a registered mortgage. Enforcement of mortgages. Registration of rights in respect of vessels under construction.

2321 **Smeets, G. C. A.** The new Ship Registration Tax Act: The Netherlands Antilles. *Bulletin for International Fiscal Documentation* 41(11), November 1987, 505–8.

2322 **Thompson, Peter.** Hong Kong. In: Lennart Hagberg (ed.) *Handbook on maritime law. Vol. IIIA. Registration of vessels, mortgages on vessels.* The Netherlands: Kluwer, 1983, pp. 247–50.

Registration of vessels. Registration of mortgages on vessels. Legal effects of a registered mortgage. Enforcement of mortgages. Registration of rights in respect of vessels under construction.

2323 **Tiyaphorn, Anurat** and **Vongkiatkachorn, Suparb.** Thailand. In: Lennart Hagberg (ed.) *Handbook on maritime law. Vol. IIIB. Registration of vessels, mortgages on vessels.* The Netherlands: Kluwer, 1983, pp. 195–216.

Registration of vessels. Registration of mortgages on vessels. Legal effects of a registered mortgage. Enforcement of mortgages. Registration of rights in respect of vessels under construction.

2324 **Verhoeven, W.** The Netherlands. In: Lennart Hagberg (ed.) *Handbook on maritime law. Vol. IIIA. Registration of vessels, mortgages on vessels.* The Netherlands: Kluwer, 1983, pp. 345–66.

Commercial Code/Civil Code. Use of Dutch civil law notary. Registration of vessels. Registration of mortgages on vessels. Legal effects of a registered mortgage. Enforcement. Registration of rights in respect of vessels under construction.

2325 **Worth, R. W.** New Zealand. In: Lennart Hagberg (ed.) *Handbook on maritime law. Vol. IIIA. Registration of vessels, mortgages on vessels.* The Netherlands: Kluwer, 1983, pp. 375–84.

Shipping and Seamen Act 1952. Admiralty Act 1973. Registration of vessels.

Registration of mortgages on vessels. Legal effects of a registered mortgage. Enforcement. Registration of rights in respect of vessels under construction.

Flags of convenience

2326 Carlisle, Rodney. *Sovereignty for sale: The origins and evolution of the Panamanian and Liberian flags of convenience.* Annapolis (Maryland): Navez Institute Press, 1981. 278 pp.

Documentation of evolution of FOC system. From days of Prohibition and liquor smuggled through Panamanian registered ships to Second World War; to boom in Middle East oil production, and resultant; growth of supertanker fleets registered FOC. 'Effective control doctrine' and other inherent problems.

Pilotage

2327 Geen, G. K. and **Douglas, R. P. A.** *Law of harbours and pilotage.* 4th rev. edn. London: Lloyd's of London Press, 1993. 344 pp.

2328 Helmick, Jon S. and **Glaskowsky, Nicholas A.** State and federal regulation of pilots, piloting, and pilotage: A study in contrasts and conflict. *Transportation Journal* 32(4), Summer 1992, 47–59.

2329 Hunter, Ormunde. Liability of vessel for negligent act of compulsory pilot. *Journal of Maritime Law and Commerce* 8(1), October 1976, 87–8.

Questions correctness and wisdom of the rule that liability imposed by negligent act of a compulsory pilot is denied when the action is *in personam* against the shipowner, even though the case is laid in admiralty.

2330 Rose, F. D. *The modern law of pilotage.* London: Sweet & Maxwell, 1984. 138 pp.

Commences with general account of law of pilotage. Necessary reference towage, salvage. Continues with the Pilotage Act 1983, with detailed annotations to individual sections and relevant statutory instruments.

2331 Zammit, Joseph P. The validity of the 'pilotage clause' in docking assistance contracts. *Journal of Maritime Law and Commerce* 3(1), October 1971, 71–90.

Examination of law regarding 'pilotage clauses'. Analyses conflicting policies involved. Attempt to formulate theory as to clause's validity for future court guidance. General background on origin and use of pilotage clause.

Salvage

2332 **Alexander, Bruce E.** Treasure salvage beyond the territorial sea: An assessment and recommendations. *Journal of Maritime Law and Commerce* 20(1), January 1989, 1–19.

Examines legal issues that arise from the application of traditional salvage law to abandoned shipwrecks, particularly those located beyond the territorial sea. Focuses on the power of admiralty court to protect priority of first salvor against intervening or competing salvors.

2333 **Arnold, R. A. H.** New policy forms: Development in the British market. In: *Salvage, general average and marine insurance.* London: Lloyd's of London Press, 1982. 15 pp.

Considers the drafts of the Institute Cargo Clauses and the drafts of the proposed Marine Policies and Clauses. Cargo insurance. Hull insurance. Example of sue and labour clause. Example of apportionment of recovery.

2334 **Baptist, C. N. T.** *Salvage operations.* London: Stanford Maritime, 1979. 155 pp.

Accounts of salvage jobs, including *Torrey Canyon* and the rig *Orion*, based on Lloyd's Open Form, 'No Cure, No Pay'. Combination of descriptive text and technical data.

2335 **Bockrath, Joseph.** The American law of life salvage. *Journal of Maritime Law and Commerce* 7(1), October 1975, 207–16.

Considers the reasons for creating incentive to rescuers at sea to save lives. Essence of salvage and the value of life. Life salvage prior to 1912. Salvage Act 1912.

2336 **Bookman, Charles A.** (ed.) *Marine salvage: Proceedings of the Third International Symposium.* Washington (DC): Marine Technology Society, 1985. 188 pp.

2337 **Brice, Geoffrey.** *Maritime law of salvage.* 2nd rev. edn. London: Sweet & Maxwell, 1993. 1141 pp.

The 1989 text of Salvage Convention. Each article considered separately and analysis of impact upon existing salvage law and practice. Focus upon emphasis placed on marine environmental protection, plus other provisions which affect marine casualties threatening damage to environment.

2338 Brice, Geoffrey. The new Salvage Convention: Green seas and grey areas. *Lloyd's Maritime and Commercial Law Quarterly*, February 1990, Part 1, 32–63.

Discusses the more important of the changes made by the 1989 Salvage Convention. Commentary on problems left unresolved. Authority to conclude salvage contracts. New duties under the Convention. Safety nets and special compensation. Misconduct, negligence and damages. (Includes Convention text.)

2339 Coghlin, Terence. Salvage: The attitude of P & I Clubs to present and future changes. In: *Salvage, general average and marine insurance.* London: Lloyd's of London Press, 1982. 11 pp.

Considers the interest of P & I Clubs in LOF and draft salvage convention. Liability cover obtained by salvors. Importance to all concerned of having a Form that can be immediately agreed in all serious casualties. 'Enhancement' of the reward. Problems of concern to Clubs in respect of oil pollution. Convention of 1910.

2340 Collins, Michael G. The salvage of sunken military vessels. Project Jennifer: A dangerous precedent. *Journal of Maritime Law and Commerce* 8(4), July 1977, 433–54.

US undercover operation to recover sunken Soviet submarine. General concepts of salvage. International salvage law. USSR, US concepts of salvage.

2341 Coulthard, Peter. A new cure for salvors?: A comparative analysis of the LOF 1980 and the C.M.I. Draft Salvage Convention. *Journal of Maritime Law and Commerce* 14(1), January 1983, 45–67.

Examines impact of LOF 1980 and the exception to the principle 'No Cure, No Pay'. Historical development of 'No Cure, No Pay doctrine'. Developments leading to LOF 1980. Interface of LOF 1980 and Draft Salvage Convention.

2342 Darling, Gerald. Lloyd's Form of Salvage Agreement: Past, present and future. In: *Salvage, general average and marine insurance.* London: Lloyd's of London Press, 1982. 15 pp.

The past – origins and development of Lloyd's Form. Political aspects, economic aspects. LOF 1980. What future changes?

2343 Darling, Gerald and **Smith, Christopher.** *LOF 90 and the new Salvage Convention.* London: Lloyd's of London Press, 1991. 227 pp.

Discusses both provisions of 1989 London International Convention on Salvage and LOF 90, published one year later. Examination of changes in maritime community that have led to need to revise LOF and the Brussels Convention. Brief discussion of LOF 80. Reception new form received.

2344 Elborne, Robert. The **C.M.I.** Draft Salvage Convention: A comparison with L.O.F. 1980. Salvage and the safety net. In: *Salvage, general average and marine insurance.* London: Lloyd's of London Press, 1982. 22 pp.

Dilemma of encouraging professional salvors to prevent oil pollution – leading to LOF 80. Development of CMI 1981. Conflicts of interests between attending parties – the Third World countries, maritime shipowning communities, countries with littoral coasts. Comparison of LOF 80; CMI Montreal 81 and Brussels 1910.

2345 Gold, Edgar. Marine salvage: Towards a new regime. *Journal of Maritime Law and Commerce* 20(4), October 1989, 487–503.

Examines changes which have evolved from both legal and non-legal basis. Environmental factor. Legal developments. International Convention on Salvage, 1989.

2346 Hudson, Geoffrey N. Salvage: LOF 1980 and the CMI Draft Convention application to general average and policies of insurance. In: *Salvage, general average and marine insurance.* London: Lloyd's of London Press, 1982. 16 pp.

The relationship between salvage and general average. The enhancement of property rewards to take account of the avoidance of environmental damage: (i) avoidance of pollution or environmental damage, from general average contributors' view; (ii) interpretation of Rule VI York–Antwerp Rules 1974. Salvors' expenses falling within the 'safety net'.

2347 Kerr, Donald A. The 1989 Salvage Convention: Expediency or equity? *Journal of Maritime Law and Commerce* 20(4), October 1989, 505–20.

Motivation of the new Convention. Examines Arts 3, 6, 7, 13, 14, 18, 21. Impact of new Convention. Effect on salvors.

2348 Kerr, Donald A. The past and future of 'No cure – no pay'. *Journal of Maritime Law and Commerce* 23(3), July 1992, 411–28.

Examines history of salvage jurisprudence since time of Rhodians and addresses the question of whether traditional philosophy of salvage can, or should, prevail in situations where requirement of a 'cure' is diminished. Reviews quantification of awards in cases of governmental salvage.

2349 Lanier, R. A. Abandon ship? *Journal of Maritime Law and Commerce* 9(1), October 1977, 131–6.

Considers questions of what happens if vessel sinks in a navigable channel. Must something be done about it? What if it is not worth the cost of having it removed and salvaged? Landmark case of *Wyandotte Company* v. *United States* (1967). Responsibilities of owners.

2350 Lloyd's of London Press. *Salvage, general average and marine insurance.* London: Lloyd's of London Press, 1982.

Papers from international congrescentrum held in Amsterdam, 12–13 November 1981. Salvage, general average and marine insurance. Draft convention on salvage.

2351 Miller, Anthony R. Lloyd's standard form of salvage agreement – 'LOF 1980': A commentary. *Journal of Maritime Law and Commerce* 12(2), January 1981, 243–61.

Introductory discussion to salvage jurisprudence. Development of regulatory provisions. Detailed comment on the Salvage Agreement LOF 1980.

2352 Pelaez, Alfred S. Salvage: A new look at an old concept. *Journal of Maritime Law and Commerce* 7(3), April 1976, 505–19.

Examines how far the concept of salvageable property extends and considers the full scope of salvage lien, and possible extension into new areas. Explores peripheries of maritime property.

2353 Redgwell, C. J. The greening of salvage law. *Marine Policy* 14(2), March 1990, 142–50.

Gives a brief consideration of the traditional law of salvage and then the environmental pressures for reform following the *Amoco Cadiz* disaster. The 1989 IMO International Convention on the International Law of Salvage and the 'greening' of salvage law.

2354 Rudolph, James L. Negligent salvage: Reduction of award, forfeiture of award or damages? *Journal of Maritime Law and Commerce* 7(2), January 1976, 419–31.

Case law, statute, jurisprudence on salvage. Considers issue of degree owed by public or private salvor and the effect that negligent salvage should have on any award.

2355 Shah, M. J. The UNCTAD discussions on maritime insurance: Background and principal issues. In: *Salvage, general average and marine insurance.* London: Lloyd's of London Press, 1982. 24 pp.

Text edited from an abstract of the talk given – incorporates questions and replies from the floor: Lloyd's seminar, Amsterdam, 1981. Background and principal issues. Study of marine insurance within UNCTAD. Work accomplished by 1981 by UNCTAD Working Group. Future objectives.

2356 **Shaw, Richard.** The 1989 Salvage Convention and Lloyd's Open Form 1990: Are they working? *Marine policy* 16(2), March 1992, 127–9.

An article which relates the 1989 Salvage Convention to LOF 1980. The changes which have been introduced to LOF 1990 are discussed and criticized and the underlying objectives of the 1989 Salvage Convention are outlined.

2357 **Steel, David W.** and **Rose, Francis D.** *Kennedy's law of salvage. British shipping laws.* 5th edn. London: Stevens & Sons, 1985, 833 pp.

Law of salvage. Necessary elements. Agreements. Duties of parties. Misconduct. Values. Assessments. Contribution. Apportionment of salvage reward. Remedies. Procedure. Public interest.

2358 **Thomas, D. Rhidian.** Life salvage in Anglo-American law. *Journal of Maritime Law and Commerce* 10(1), October 1978, 79–104.

Focuses attention upon salvage of human life and ascertains the extent to which human life and valuable services rendered thereto may be defined and rewarded as salvage.

2359 **Van Beuningen, John.** Is the pen mightier than the tow-rope? In: *Salvage, general average and marine insurance.* London: Lloyd's of London Press, 1982. 10 pp.

Paper given at international conference in Amsterdam, November 1981. Sets out essence of salvage business. Investment in tugs, equipment. Skill and experience of personnel. Problems of 1910 Salvage Convention. Impact of Lloyds Form 1980.

2360 **Vincenzini, Enrico.** *International regulation of salvage at sea.* London: Lloyd's of London Press, 1987. 190 pp.

2361 **Vincenzini, Enrico.** *International salvage law.* 2nd rev. edn. London: Lloyd's of London Press, 1992. 304 pp.

History of development of salvage law. Detailed interpretation of effects of changes, including principles introduced by 1989 Salvage Convention and LOF 90. International in scope. From different legal systems. All recent important cases, including *The Patmos*; *The Choko Star*; *The Goring*; *The Mimosa.* York–Antwerp Rules; TOVALOP; CRISTAL.

2362 Waddams, S. M. Dr. Lushington's contribution to the law of maritime salvage (1838–67). *Lloyd's Maritime and Commercial Law Quarterly*, February 1989, Part 1, 59–80.

Describes influence of Dr Lushington, as judge of Admiralty Court from 1838 to 1867, on development of law of salvage. The judicial process. The incentives. Contracts. Restitution and salvage.

2363 Wildeboer, Ina H. The C.M.I. Draft Salvage Convention: What it omits. In: *Salvage, general average and marine insurance.* London: Lloyd's of London Press, 1982. 12 pp.

Raises question of what Salvage Convention leaves national laws to deal with. Considers (i) who is entitled to salvage reward; (ii) who owes the salvage reward – position of debtor. The 'enhancement' of the reward.

2364 Wooder, James B. The new Salvage Convention: A shipowner's perspective. *Journal of Maritime Law and Commerce* 21(1), January 1990, 81–98.

Examines some of Convention's provisions and omissions from shipowner's perspective.

2365 Wurfbain, J. W. The UNCTAD proposals: Salvage, general average and marine insurance. In: *Salvage, general average and marine insurance.* London: Lloyd's of London Press, 1982. 17 pp.

Examines what UNCTAD is doing to marine insurance. UNCTAD's discussions on ocean hull insurance. 'Off the peg' insurance; 'tailor made' insurance. Standard clauses to be used as non-mandatory model for marine insurance contracts. 'All Risks' approach; 'Named Perils' approach. The language of UNCTAD model clauses.

Limitation of liabilities

2366 Adamezak, Weseiech. Poland and the 1976 Convention. In: Institute of Maritime Law, University of Southampton. *Limitation of shipowners' liability: The new law.* London: Sweet & Maxwell, 1986, pp. 243–6.

Assessment in Polish civil law of compensation for damage. Polish civil code. Problems of particular importance from viewpoint of eventual incorporation of 1976 Convention.

2367 Asser, T. M. C. Golden limitations of liability in international transport conventions and the currency crisis. *Journal of Maritime Law and Commerce* 5(4), July 1974, 645–69.

Poincaré franc used in gold clause limitation of liability provided for in multilateral conventions concerning transportation by sea, air. Impact of gold and currency crisis. Basis of conversion – 'official' value or free market value of gold?

2368 **Astle, W. E.** *Limitation of liability: Merchant Shipping Act, 1894 (section 503 and subsequent amendments).* London: Fairplay Publications, 1985. 113 pp.

Examines the problems that have arisen out of the interpretation of the provisions of s. 503 of the 1894 Act, and amending Acts, concerning the right of the shipowner, wharf owner or others to limit liability arising from negligence of their agents or servants.

2369 **Auchter, Gerard.** France and the 1976 Convention. In: Institute of Maritime Law, University of Southampton. *Limitation of shipowners' liability: The new law.* London: Sweet & Maxwell, 1986, pp. 191–206.

Background and history of limitation from French perspective. Interpretation of 1957 Convention and French 1967 law and decree – particular problems. Implementation of the 1976 Convention in French law.

2370 **Bassoff, Joel M.** Fire losses and statutory fire exceptions. *Journal of Maritime Law and Commerce* 12(4), July 1981, 507–22.

Impact of shipboard fire. Historical context in which statutory fire provisions under the Fire Statute 1851 and COGSA 1936 enacted. Liability of carrier. Case law development. Burden of proof.

2371 **Beddard, Ralph.** The implementation of the Convention. In: Institute of Maritime Law, University of Southampton. *Limitation of shipowners' liability: The new law.* London: Sweet & Maxwell, 1986, pp. 152–60.

Scope of application of 1976 Convention. Amendment. Revision of limitation amounts. Entry into force.

2372 **Brice, Geoffrey.** The scope of the Limitation Action. In: Institute of Maritime Law, University of Southampton. *Limitation of shipowners' liability: The new law.* London: Sweet & Maxwell, 1986, pp. 18–32.

History. Effect of new provisions. Categories of claim. Claims excepted from limitation. Persons entitled to limit liability. The salvor and limitation.

2373 Cadwallader, F. J. J. The shipowner's indemnity. *Journal of Business Law,* 1972, 181–94.

Studies nature and interpretations of the indemnity in time charterparties. Survey of position of shipowner at common law and development of the express indemnity. First of two parts. (Second part, pp. 286–96 in same volume.)

2374 Cadwallader, F. J. J. The shipowner's indemnity. *Journal of Business Law,* 1972, 286–96.

Second part of study of nature and interpretation of the indemnity in time charterparties. Concerned with orders of charterer which relate to general employment of the vessel and application of indemnity to such orders. (First part, pp. 181–94 in same volume.)

2375 Conner, M. Timothy. Maritime tort liability of the United States. *Journal of Maritime Law and Commerce* 11(3), April 1980, 271–92.

Examines unique aspects and developments of specialized law stemming from commission of maritime torts by US Government. Scope of Federal Government's maritime tort liability.

2376 Davies, Martin and **Lawson, Guy.** Limiting shipowners' liability for economic loss. *Australian Business Law Review* 16(4), August 1988, 271–81.

2377 Dykes, Andrew. Limitation and oil pollution. In: Institute of Maritime Law, University of Southampton. *Limitation of shipowners' liability: The new law.* London: Sweet & Maxwell, 1986, pp. 144–51.

Examines reasons for exclusion under Art. 3 of 1976 Convention of certain classes of claim, including claims for oil pollution damage. Practical consequences of this exclusion.

2378 Gaskell, Nicholas. The amount of limitation. In: Institute of Maritime Law, University of Southampton. *Limitation of shipowners' liability: The new law.* London: Sweet & Maxwell, 1986, pp. 33–101.

Considers the size of the limits allowed by the 1976 Limitation Convention; how they are calculated; how much of a difference there will be between 1976 Limitation Convention limits and those under 1957 Limitation Convention.

2379 Gorman, F. J., Alexander, B. E. and **Scheller, D. M.** Exculpatory and limitation of liability provisions in terminal operator tariffs.

Journal of Maritime Law and Commerce 13(4), July–October 1982, 407–54.

Evaluates certain types of common 'liability' provisions which appear in terminal operator tariffs. Exculpatory provisions; limitation clauses; indemnity provisions. Additional clauses which may purport to effect liability to which terminal operator otherwise exposed.

2380 **Griggs, Patrick** and **Williams, Richard.** *Limitation of liability for maritime claims.* 2nd rev. edn. London: Lloyd's of London Press, 1991. 228 pp.

Examines important legislation and case law developments. Reviews working of 1976 Limitation Convention (the London Convention). Hague; Hague–Visby Rules. Hamburg Rules. 1974 (Athens) Convention – Carriage of Passengers and their Luggage by Sea.

2381 **Grime, R. P.** Implementation of the 1976 Limitation Convention. *Marine Policy* 12(3), July 1988, 306–13.

Looks at the effectiveness of implementation of the Convention on Limitation of Liability for Maritime Claims. Outlines changes from 1957 Convention. Discusses some difficulties with the 1976 Convention, such as tonnage measurement and the new system of calculation of the limitation fund.

2382 **Grime, Robert.** The loss of the right to limit. In: Institute of Maritime Law, University of Southampton. *Limitation of shipowners' liability: The new law.* London: Sweet & Maxwell, 1986, pp. 102–12.

The demise of actual fault or privity. Personal act or omission. The disappearance of privity. Introduction of intention and recklessness. Burden of proof. Insurance.

2383 **Hazelwood, Steven J.** The United Kingdom and the 1976 Convention. In: Institute of Maritime Law, University of Southampton. *Limitation of shipowners' liability: The new law.* London: Sweet & Maxwell, 1986, pp. 253–9.

Implementation of 1976 Convention. Entry into force. Repeals and amendments of existing legislation. Reaction of the UK.

2384 **Heller, Paul P.** The value of the gold franc: A different point of view. *Journal of Maritime Law and Commerce* 6(1), October 1974, 73–103.

Should gold franc be converted into national currency according to official par value or according to price of gold on free market? Refers to article in

Journal of Maritime Law and Commerce 5(4), 1974, pp. 645–69 by T. M. C. Asser – same question.

2385 Herber, Rolf. The German Federal Republic and the 1976 Convention. In: Institute of Maritime Law, University of Southampton. *Limitation of shipowners' liability: The new law.* London: Sweet & Maxwell, 1986, pp. 221–30.

History of limitation of liability in Federal Republic of Germany. Existing legal situation. General attitude towards the 1976 Limitation Convention. Draft law to implement 1976 Convention.

2386 Institute of Maritime Law, University of Southampton. *Limitation of shipowners' liability: The new law.* London: Sweet & Maxwell, 1986. 378 pp.

Twenty-one chapters presented by 20 different authors from 12 different countries. Antecedent history leading to 1976 Limitation of Liability Convention. Effect of 1976 Convention. Relationship between 1976 Convention and other International Conventions. Impact on other jurisdictions.

2387 Jackson, David. The 1976 Convention and international uniformity of rules. In: Institute of Maritime Law, University of Southampton. *Limitation of shipowners' liability: The new law.* London: Sweet & Maxwell, 1986, pp. 126–43.

Relationship of 1976 Convention with other conventions and national laws in respect of assessment of liability to which limitation attached. Jurisdiction, choice of law. Rules of 1976 Convention in relation to the national structures for hearing of disputes.

2388 Kačić, Hrvoje. Yugoslavia and the 1976 Convention. In: Institute of Maritime Law, University of Southampton. *Limitation of shipowners' liability: The new law.* London: Sweet & Maxwell, 1986, pp. 279–85.

Historical background to concept of limited liability. Existing system of limitation. The 1976 Convention.

2389 Larsen, Paul B. and others. The 1991 Diplomatic Conference on uniform liability rules for operators of transport terminals. *Journal of Maritime Law and Commerce* 21(4), October 1990, 449–79.

Report on the status of the draft convention, the UNCITRAL negotiations and work of US Delegation in preparation for 1991 Convention. Issues covered. Final clauses.

2390 No entry.

2391 Lukvic, Branimir. Limitation of liability for the raising and removal of ships and wrecks: A comparative study. *Journal of Maritime Law and Commerce* 12(1), October 1980, 47–64.

Draws attention to some of the salient problems relating to limitation of liability of shipowner and salvor for such claims. Persons entitled to limit. The bar to limitation.

2392 Mendelsohn, Allan I. The value of the Poincaré gold franc in limitation of liability conventions. *Journal of Maritime Law and Commerce* 5(1), October 1973, 125–8.

Question of value of Poincaré franc in two-tier gold market. Importance of this in relation to all private international law limitation conventions.

2393 Mendizabal, Marcial J. Argentina and the 1976 Convention. In: Institute of Maritime Law, University of Southampton. *Limitation of shipowners' liability: The new law.* London: Sweet & Maxwell, 1986, pp. 187–90.

System of limitation of liability under Argentine law. Questions relating to a possible adherence to the 1976 Convention.

2394 Parks, Alex L. The United States and the 1976 Convention. In: Institute of Maritime Law, University of Southampton. *Limitation of shipowners' liability: The new law.* London: Sweet & Maxwell, 1986, pp. 260–78.

Judicial attitude to shipowners' limitation of liability. Development of limitation of liability. Synopsis of the US Limitation of Liability Act. Judicial construction of the Act. US response to 1976 Limitation Convention.

2395 Scowcroft, Jerome C. Recent developments concerning the package limitation. *Journal of Maritime Law and Commerce* 20(3), July 1983, 403–19.

Examines statutory valuation of $500 per package under COGSA. Uncertainty of application. Opportunity to declare a higher value. Definition of 'package'. Customary freight unit. The parties' freedom to contract when COGSA does not apply *ex proprio vigore.*

2396 Sedano, Alejandro G. Spain and the 1976 Convention. In: Institute of Maritime Law, University of Southampton. *Limitation of shipowners' liability: The new law.* London: Sweet & Maxwell, 1986, pp. 249–52.

Difference between domestic law and Conventions. Domestic law. Brussels Convention 1957. Limitation Convention 1976.

2397 Selvig, Erling. An introduction to the 1976 Convention. In: Institute of Maritime Law, University of Southampton. *Limitation of shipowners' liability: The new law.* London: Sweet & Maxwell, 1986, pp. 3–17.

(i) The legal background: Principles of limitation; English developments; 1924 and 1957 Conventions; (ii) the need for changes: Risks contemplated; the CLC; (iii) the elaboration of a new convention; (iv) the 1976 Convention – global limitation system for the future?

2398 Selvig, Erling. The Nordic countries and the 1976 Convention. In: Institute of Maritime Law, University of Southampton. *Limitation of shipowners' liability: The new law.* London: Sweet & Maxwell, 1986, pp. 240–42.

Implementation of 1976 Convention – uniform legislation prepared through Nordic co-operation. Substantive provisions. New procedural rules on limitation funds and limitation actions.

2399 Shaw, Richard. Practice and procedure. In: Institute of Maritime Law, University of Southampton. *Limitation of shipowners' liability: The new law.* London: Sweet & Maxwell, 1986, pp. 113–25.

Practical effects of 1976 Convention. Principal changes brought about by 1976 Convention compared with previous regimes. Subordinate changes. Choice of jurisdiction. Administration and distribution of limitation fund. Limitation without constitution of limitation fund – Art. 10. With limitation fund – Art. 11. Distribution of fund – Art. 12. Bar to other actions – Art. 13.

2400 Tetley, William. Shipowners' limitation of liability and conflicts of law: The properly applicable law. *Journal of Maritime Law and Commerce* 23(4), October 1992, 585–606.

Outlines origin and basic characteristics of concept of limitation of shipowners' liability in common law and civil law jurisdictions. Substance/procedure dichotomy. Conflict of law problems. Review of conflict of law directives in international conventions on limitation of shipowners' liability.

2401 Toda, Mitsuhiro. Japan and the 1976 Convention. In: Institute of Maritime Law, University of Southampton. *Limitation of ship-*

owners' liability: The new law. London: Sweet & Maxwell, 1986, pp. 235–9.

Commercial Code 1899 and limitation. Development of the law. The 1976 Convention.

2402 **Trotz, Norbet.** The German Democratic Republic and the 1976 Convention. In: Institute of Maritime Law, University of Southampton. *Limitation of shipowners' liability: The new law.* London: Sweet & Maxwell, 1986, pp. 207–20.

General remarks. Scope of right of limitation. Limitation procedure.

2403 **Vlachos, Gerasimos M.** Greece and the 1976 Convention. In: Institute of Maritime Law, University of Southampton. *Limitation of shipowners' liability: The new law.* London: Sweet & Maxwell, 1986, pp. 231–4.

Background, history of limitation provisions, Greece. Present position. Comments and prospects of implementation of limitation provisions as per 1957 or 1976 Convention.

2404 **Ward, Les.** The SDR in transport liability conventions: Some clarification. *Journal of Maritime Law and Commerce* 13(1), October 1981, 1–20.

Considers previous works on object of the unit of account employed in international transport liability conventions. Examines controversial issues and contentions. Examines the SDR.

2405 **Weil, Prosper.** *Law of maritime delimitation: Reflections.* Cambridge: Grotius Publications, 1989. 320 pp.

2406 **Wilmot, Nicholas.** Who has got his cross liabilities crossed? *Lloyd's Maritime and Commercial Law Quarterly,* November 1989, Part 4, 450–64.

Considers whether it is possible to use the cross liability principle in insurance settlements where the rules as to limitation of liability have been involved in the settlement between the vessels. The basic principles. The problem where there has been a limitation of liability. Settlement under the Norwegian Plan. Other cases.

Maritime liens

2407 Adamczak, Wojciech. Maritime liens in Polish law in the context of the law of the European Socialist States. *Lloyd's Maritime and Commercial Law Quarterly*, August 1982, Part 3, 433–7.

Considers the role of maritime liens in the systems of the Socialist States.

2408 Harley, S. T. *How to secure a maritime lien: A study.* Lampert & Selway Publishing, 1981. 180 pp.

Presents practical guidance on how to secure a maritime lien. Shows the problems that are encountered in various countries throughout the world, citing legal cases for the purposes of illustration. International Conventions.

2409 Hill, D. J. Rights of lien in respect of forwarding and carriage contracts under the civil law. *Journal of Maritime Law and Commerce* 3(1), October 1971, 145–63.

(i) Preferential rights which are analogous to the pledge or charge which may be expressed or implied by statute. (ii) Right of retention which is essentially the same as lien at common law. Consideration of two classes of rights in relation to principal civil law jurisdictions.

2410 Huang, Mao-Ching. Maritime liens in the Republic of China. *Journal of Maritime Law and Commerce* 8(2), January 1977, 227–46.

Examines maritime liens under law of the Republic of China with emphasis on priorities among maritime liens. Considers whether a party has a lien in respect of a claim, type, priority relating to other liens and ship mortgages, how lien enforced/extinguished.

2411 Jackson, D. *Arrest liens and mortgagors' modern needs and modern developments.* Southampton: Institute of Maritime Law, University of Southampton, 1988.

2412 Mattis, Brian. Maritime liens for 'necessaries': A tale of statutory misinterpretation. Part 1. *Journal of Maritime Law and Commerce* 21(2), April 1990, 213–50.

Introduction to granting of maritime liens by US federal courts. Statutory development. History of maritime lien for necessaries. Summary of law prior to Federal Maritime Lien Act 1910.

2413 Mattis, Brian. Maritime liens for 'necessaries': A tale of statutory misinterpretation. Part 2. *Journal of Maritime Law and Commerce* 21(3), July 1990, 331–95.

The first Federal Maritime Lien Act. Background and legislative history. The 1920 Act. Statutory interpretation. Post-enactment developments. Judicial expansion of the legislation. The legitimate scope of section 971.

2414 Mattis, Brian. Maritime liens for 'necessaries': A tale of statutory misinterpretation. Part 3. *Journal of Maritime Law and Commerce* 21(4), October 1990, 481–502.

Liens for unpaid insurance premiums. Case law. Are liens for necessaries still required? The third Federal Maritime Lien Act – the new 1988 version, in the Ship Mortgage Law Codification. Amendment of the new statute.

2415 Naschitz, P. G. Maritime liens and mortgages in Israel: Who has first priority? *Journal of Maritime Law and Commerce* 23(1), January 1992, 123–30.

Priorities of maritime liens and mortgages against background of different legal systems and question of whether classification rests on *lex causae* or *lex fori.* The Nadja S. (*Griffin Corp.* v. *Koor Sachar*), 1990.

2416 Oditah, Fidelis. The judicial nature of a lien on sub-freights. *Lloyd's Maritime and Commercial Law Quarterly,* May 1989, Part 2, 191–7.

Contrasts the extent of the right conferred by a lien over sub-freight clause with the nature of that right. Explores the possible analyses of the nature of the right created by a lien clause in a time charter.

2417 Simmons, Kevin P. Admiralty practice under the Foreign Sovereign Immunities Act: A trap for the unwary. *Journal of Maritime Law and Commerce* 12(1), October 1980, 109–21.

Examines the Act's effect on admiralty practice in two areas: (i) the defence of sovereign immunity; (ii) traditional practice of judicial seizures to establish quasi in rem jurisdiction and to enforce maritime liens.

2418 Simon, Seymour. The law of shipping containers: Part I. A look at the factual and legislative background, and a critique of a new judicial formula. Part II. A reply to 'The container and the package limitation: The search for predictability'. *Journal of Maritime Law and Commerce* 5(3), April 1974, 507–38.

Part I examines US case law consideration of whether container constitutes a 'package' within COGSA 1936. Part II considers article by M. E. De Orchis (see p. 152, entry 1060).

2419 Tetley, W. *Maritime liens and claims.* London: Witherby & Co., 1985.

2420 Tetley, William. Repairmen's liens. *Journal of Maritime Law and Commerce* 13(2), January 1982, 177–221.

Study of repairmen's lien in the US, Great Britain, Canada, France. Looks at Quebec Civil Code. Conflict of law cases. Analyses the solutions of the International Maritime Lien and Ship Mortgage Conventions of 1926 and 1967.

Collisions

2421 Allbritton, Jack L. Contribution among joint tortfeasors in non-collision maritime cases. *Journal of Maritime Law and Commerce* 4(3), April 1973, 425–39.

Examines whether a right of contribution exists among joint tortfeasors in a non-collision maritime case. Case law developments. Controversy from the *Halcyon* (1952) case. US law.

2422 Cahill, Richard A. *Collisions and their causes.* London: Fairplay Publications, 1983. 171 pp.

Analyses a large number of collisions where basic principles and practice flouted or ignored. Separate consideration given to collisions associated with different types of encounter situations in both clear and restricted visibility – in open seas or confined waters.

2423 Cahill, Richard A. *Strandings and their causes.* London: Fairplay Publications, 1985. 239 pp.

Examines a series of strandings, each group of which points up a particular moral. Mistakes of judgement, neglect of fundamentals, bad watchkeeping, poor passage planning and failure to recognize risk. Master/pilot relationship, inadequate manning, doubtful seaworthiness. (Companion and complement to author's study *Collisions and their causes.*)

2424 Chameides, Steven B. Inscrutable fault in collision litigation. *Journal of Maritime Law and Commerce* 9(3), April 1978, 363–75.

Examines doctrine of inscrutable fault in collision cases and how its application may be a powerful trial tactic. Comparison with English rule of letting losses lie where they have fallen. Inscrutable fault cases. Cases which are not properly inscrutable fault cases.

2425 Couper, Frank E. The deep draught vessel and the Proposed International Regulations for Preventing Collisions at Sea, 1972. *Journal of Maritime Law and Commerce* 8(3), April 1977, 295–335.

Considers ambiguities of the Proposed Rules and suggests possible professional and judicial interpretations of the Rules involving deep-draught vessels. Considers how to avoid in some measure uncertainty that accompanies changes in law and tradition.

2426 **Flyntz, Frank J.** Radar plotting: A developing requirement of law and good seamanship. *Journal of Maritime Law and Commerce* 14(4), October 1983, 561–79.

Use of radar where plotting required by 72 COLREGS. Examines the responsibility of the master and crew to construct a plot in order to use the radar properly. (References include 1981 Inland Navigation Rules.)

2427 **Goddard, K. S.** Is there a right to wreck? *Lloyd's Maritime and Commercial Law Quarterly*, November 1983, Part 4, 625–30.

Removal of property by third party in a maritime casualty – whether (i) legal, and if not (ii) such removal constitutes a criminal offence or gives rise to a civil action. The legal classification of shipwrecked property. Statutory provisions.

2428 **Goschka, Dennis A.** Goodbye to all that! The unlamented demise of the divided damages rule. *Journal of Maritime Law and Commerce* 8(1), October 1976, 51–72.

Criticism surrounding divided damages rule in maritime collision cases. Impact of Supreme Court decision in case *United States* v. *Reliable Transfer Co. Inc.* (1975), adopting the proportional fault doctrine.

2429 **Healy, Nicholas J.** and **Koster, John C.** Reliable Transfer Company v. United States: Proportional fault rule. *Journal of Maritime Law and Commerce* 7(1), October 1975, 293–300.

Considers case law development of rule of divided damages in marine collision cases. Impact of decision in *United States* v. *Reliable Transfer Co. Inc.* (1975) where US Supreme Court abandoned rule and adopted proportional fault in its place.

2430 **Healy, Nicholas J.** and **Sweeney, Joseph C.** Establishing fault in collision cases. *Journal of Maritime Law and Commerce* 23(3), July 1992, 337–68.

Burden of proof, presumptions and permissible inferences. Admissions in pleadings. Judicial notice. Evidence admissible in collision litigation.

2431 **Healy, Nicholas J.** and **Sweeney, Joseph C.** The Starboard Hand Rule under the 1972 Collision Regulations. *Journal of Maritime Law and Commerce* 23(2), April 1992, 263–80.

Introduction to the crossing situation in navigation. Applicability of the starboard hand rule. Head-on and crossing distinguished. Crossing and overtaking situations distinguished. The obligations of the port hand vessel. Obligations of the starboard hand vessel.

2432 Holdert, H. M. C. and **Buzek, F. J.** *Collision cases: Judgments and diagrams.* 2nd rev. edn. London: Lloyd's of London Press, 1990. 328 pp.

Using diagrams and descriptions of the principal facts of collisions, this book illustrates the main points of law in more than 130 cases. Divided into type of circumstances under which collisions occur. Part One: Clear weather; Part Two: Restricted visibility.

2433 McGuffie, Kenneth C. *Marsden: The law of collisions at sea. British shipping laws.* 11th edn. plus cumulative supplements. London: Stevens & Sons, 1961. 975 pp.

Considers Notice issued December 1960 by Ministry of Transport on subject of Navigation with Shipborne Radar in Reduced Visibility. Comments on mention made in Notice of proposed 1960 Regulations. Statutory changes in Admiralty law. Measure of damages. British Transport Commission's bye-laws for regulation of canals. St Lawrence Seaway Regulations.

2434 Mankabady, Samir. *Law of collision at sea.* Amsterdam: North Holland Publishing Co., 1987. 608 pp.

2435 Morris, Albert. The port agent: His rights, liabilities and problems. *Lloyd's Maritime and Commercial Law Quarterly*, May 1982, Part 2, 218–31.

Examines the general principles affecting the port agent – common law as amended by statute. What is an agent? Agent or principal? The EEC Directive on commercial agency law. Whose agent? Port agent's right of arrest. Agent's duty to follow instructions. UN Convention on Multimodal Transport 1980.

2436 Mukherjee, Proshanto K. Overtaking or crossing: Judicial interpretation and the mariner's dilemma. *Journal of Maritime Law and Commerce* 23(2), April 1992, 247–62.

Examines Overtaking Rule (Rule 13) and Crossing Rule (Rule 15) of Collision Regulations 1972 from mariner's perspective in light of key judicial decisions. Action by Stand-On Vessel (Rule 17) plus critical study of judicial interpretation of Rule 13.

2437 Murphy, Thomas Francis. The legal implications of marine radar. *Journal of Maritime Law and Commerce* 7(4), July 1976, 573–614.

Examines major aspects of collision law and radar. Judicial trends in the area. Related policies.

2438 **Owen, David R.** Admiralty practice in the 19th century – a little family party: An historical footnote to Professor Tetley's article. *Journal of Maritime Law and Commerce* 13(2), January 1982, 147–8.

Brief commentary on legal representations of parties to the Pennsylvania case, 1874. Comments on relationship between US and English Admiralty Courts.

2439 **Pineus, K.** and **Rohreke, H. G.** *Limited liability in collision cases.* Colchester: Lloyd's of London Press, 1984.

2440 **Singh, Nagendra.** *International maritime law conventions – Vol. 1. Navigation.* 3rd edn. London: Stevens & Sons, 1983. 978 pp.

Collision prevention; navigational aids; maritime communications; telecommunications.

2441 **Sturt, Richard H. B.** *The Collision Regulations: The application and enforcement of the Merchant (Distress Signals and Prevention of Collisions) Regulations 1989.* 3rd rev. edn. London: Lloyd's of London Press, 1991. 255 pp.

Examines how the 1989 Regulations and the Merchant Shipping Act 1988 have affected the Collision Regulations and their enforcement. English and American case decisions. Manslaughter, the Marine Accident Investigation Branch, the I.M.O.'s guidance on Vessel Traffic Systems. The new Codes of Practice for Police Questioning.

2442 **Tetley, William.** The Pennsylvania Rule – an anachronism? The Pennsylvania judgment – an error? *Journal of Maritime Law and Commerce* 13(2), January 1982, 127–46.

Study of the Pennsylvania Rule and the decision itself (1874). Comparison with decision of Supreme Court of United States and also decision of Privy Council, UK. Reasoning of Supreme Court – based on a misapprehension of British statute?

2443 **Thomas, William B.** How narrow is a narrow channel? *Journal of Maritime Law and Commerce* 14(4), October 1983, 537–60.

Collision avoidance in confined waters. The need for predictable vessel movement. The new unified Inland Rules. The narrow channel Rule. What is a narrow channel? The Mississippi River. Problems with available guidance. Need for clarification. The time for action.

2444 Underwood, Eugene. Contribution among joint-tortfeasors in a non collision case. *Journal of Maritime Law and Commerce* 4(2), January 1973, 319–21.

Examines attitude of Supreme Court in respect of issue of contribution between joint tortfeasors in non-collision cases. Case law developments.

2445 Villareal, Dewey R. Halcyon to Ryan to Weyerhause to Cooper: Where do we go from here? *Journal of Maritime Law and Commerce* 6(4), July 1975, 593–602.

The prohibition of contribution in non-collision cases, following *Halcyon Lines* v. *Haenn Ship Ceiling and Refitting Corporation* (1952). Subsequent case law development.

2446 Zapf, Robert J. The growth of the Pennsylvania Rule: Study of causation in maritime law. *Journal of Maritime Law and Commerce* 7(3), April 1976, 521–39.

Discusses the rule of presumption of causation, as handed down by the Supreme Court in the maritime collision case, the Pennsylvania Rule, 1873. Subsequent decisions in which the rule explained. Fate of rule in various circuits in subsequent cases.

General average

2447 Allen, Joseph M. 'Peril' in the Fifth Circuit. *Journal of Maritime Law and Commerce* 7(2), January 1976, 409–17.

Effect on industry of lack of general average litigation in US courts. 'Judicial innocence'. The Fifth Circuit's treatment of the 'peril' issue in the *Orient Transporter, Orient Mid-East Transporter et al.* v. *A Shipment of Rice et al.* (1974).

2448 Bertschuger, P. P. *Introduction to general average.* London: Witherby, 1986.

2449 Buglass, Leslie J. *General average and the York–Antwerp Rules, 1974.* 2nd edn. Centreville (Maryland): Cornell Maritime Press, 1991. 190 pp.

2450 Buglass, Leslie J. *Marine insurance and general average in the*

United States: An average adjuster's viewpoint. 3rd edn. Centreville, (Maryland): Cornell Maritime Press, 1991. 723 pp.

2451 **Clarke, Malcolm.** Non-apparent damage to goods in transit. *Lloyd's Maritime and Commercial Law Quarterly,* November 1982, Part 4, 533–41.

Considers the problems raised by issue of proof of non-apparent damage to goods in transit. Case law. The Hague–Visby Rules. The CIM. The CMR. Apparent and non-apparent damage.

2452 **Denny, Maurice E. V.** *Introduction to general average.* London: Witherby and Co., 1986. 57 pp.

Text directed to students of insurance and shipping. Considers past and present controversial problems. Who or what is general average? General average and contracts of affreightment. Words and phrases in common usage. York–Antwerp Rules 1974. Application to insurance.

2453 **Donaldson, J. F.** and others. *Law of general average and the new York–Antwerp Rules. British Shipping Laws.* Vol. 7. 11th rev. edn. London: Stevens and Sons, 1990. 743 pp.

2454 **Ernst Robert Lindley & Sons, B. V.** *Salvage, general average and marine insurance.* London: Lloyd's of London Press, 1982.

2455 **Goodacre, J. Kenneth.** *York–Antwerp Rules, 1974: Contrasted with York–Antwerp Rules, 1950.* London: Witherby, 1974. 59 pp.

2456 **Hudson, N. G.** *York–Antwerp Rules.* London: Lloyd's of London Press, 1991. 320 pp.

2457 **Hughes, G. S.** and **Corah, R.** *Guide to general average.* 1989.

2458 **Johnson, Patrick.** A comparison of general average law and the status of average adjusters in Sweden and the United States. *Journal of Maritime Law and Commerce* 12(3), April 1981, 363–85.

Development of international regulations. The York–Antwerp Rules. The American and Swedish average adjuster. Effect of fault. Ballast voyages. Periods of limitation. Dangerous goods. Responsibility – liability of cargo owner. Freight pro rata. Criticism of general average.

2459 **Lyons, R. A.** Some thoughts on general average in the container age. *Journal of Maritime Law and Commerce* 2(1), October 1970, 165–72.

Contributing interests and values: The vessel; the container; the cargo; simplification; carriage on deck; jettison; custom of trade; fire on board; the future?

2460 Pineus, Kaj. *General average: The practical problems.* Stockholm: Almqvist, 1985. 58 pp.

2461 Pineus, Kaj. General average deposits. *Journal of Maritime Law and Commerce* 4(4), July 1972, 619–32.

Summary of law and practice in respect of collection of G. A. deposits. Advantages/disadvantages of collecting deposits. Time factor – when deposits are collected. Currency and devaluation risks. Should deposits be 'deepfreezed'? Settlement of excess deposits. Excess deposits when owner not traced.

2462 Pineus, Kaj. Questions as to the carrier's obligation when there is a general average. *Journal of Maritime Law and Commerce* 1(4), July 1970, 615–20.

Common practical questions in practice of an average adjuster. Is carrier bound to insure; average disbursements; cargo remaining on board drydocked vessel; cargo in port of refuge while stored ashore? Must carrier have cargo surveyed after accident? Duties to incur forwarding charges.

2463 Wilson, John and **Cooke, Julian.** *Lowndes and Rudolf: The law of general average and the York–Antwerp Rules.* 11th edn. London: Sweet and Maxwell, 1990.

Ship Sale and Purchase

General

2464 Albrecht, Hans-Christian. Germany. In: Lennart Hagberg (ed.) *Maritime law. Vol. II. Enforced sales of vessels.* The Netherlands: Kluwer, 1977, pp. 31–4.

Grounds for an enforced sale. Authority. Case and maintenance of vessel pending sale. Notice of sale. Form of sale. Price. Payment of price. Distribution of proceeds of sale. Appeals. Conventions.

2465 Arroyo, Ignacio. Spain. In: Lennart Hagberg (ed.) *Maritime law.*

Vol. II. Enforced sales of vessels. The Netherlands: Kluwer, 1977, pp. 81–6.

Grounds for an enforced sale. Authority. Case and maintenance of vessel pending sale. Appraisal. Notice of sale. Form of sale. Price. Payment of price. Distribution of proceeds of sale. Appeals. Conventions.

2466 Costabel, Attilio. Italy. In: Lennart Hagberg (ed.) *Maritime law. Vol. II. Enforced sales of vessels.* The Netherlands: Kluwer, 1977, pp. 49–56.

Grounds for an enforced sale. Care and maintenance of vessel pending sale. Appraisal. Notice of sale. Form of sale. Price. Payment of price. Distribution of proceeds of sale. Appeals. Conventions. Other points of interest.

2467 Cunningham, John R. Canada. In: Lennart Hagberg (ed.) *Maritime law. Vol. II. Enforced sales of vessels.* The Netherlands: Kluwer, 1977, pp. 15–18.

Grounds for an enforced sale. Case law. Authority. Case and maintenance of vessel pending sale. Appraisal. Notice of sale. Authority which performs sale. Price. Payment. Presence of creditors. Distribution of proceeds. Appeals. Conventions.

2468 De Koning, John. Australia. In: Lennart Hagberg (ed.) *Maritime law. Vol. II. Enforced sales of vessels.* The Netherlands: Kluwer, 1977, pp. 1–8.

Grounds for an enforced sale. Authority. Care and maintenance of vessel pending sale. Appraisal – High Court Rules; Supreme Court Admiralty Rules. Notice of sale. Form of sale. Price. Payment. Distribution of proceeds. Appeals. Conventions. Case law.

2469 Erlund, Jan. Denmark. In: Lennart Hagberg (ed.) *Maritime law. Vol. II. Enforced sales of vessels.* The Netherlands: Kluwer, 1977, pp. 19–22.

Grounds for an enforced sale. Authority. Case and maintenance of vessel pending sale. Appraisal. Notice of sale. Form of sale. Price. Distribution of proceeds. Appeals. International Conventions. Other points of interest.

2470 Fairplay Publications. *First sale and purchase review.* London: Fairplay Publications, 1983.

2471 Farrugia, R. Malta. In: Lennart Hagberg (ed.) *Maritime law. Vol. II. Enforced sales of vessels.* The Netherlands: Kluwer, 1977, pp. 63–6.

Grounds for an enforced sale. Authority. Care and maintenance of vessel pending sale. Valuation. Notice of sale. Form of sale. Price. Payment of price. Distribution of proceeds. Appeals. Conventions.

2472 Goldrein, Iain S. (editor-in-chief) *Ship sale and purchase.* 2nd edn. London: Lloyd's of London Press, 1993. 458 pp.

Financing of ship purchase; new building contracts; standard form contracts for sale second-hand tonnage (including NFS 66, NSF 83, London Ship Sale Contract 1985); demolition. Legal decisions from NFS 66 and its contractual terms interpreted.

2473 Griggs, J. G. R. United Kingdom. In: Lennart Hagberg (ed.) *Maritime law. Vol. II. Enforced sales of vessels.* The Netherlands: Kluwer, 1977, pp. 91–4.

Grounds for an enforced sale. Authority. Appraisal. Notice of sale. Form of sale. Price. Payment of price. Distribution of proceeds. Foreign vessels. Appeals. Conventions.

2474 Hagberg, Lennart (ed.) *Maritime law. Vol. II. Enforced sales of vessels.* The Netherlands: Kluwer, 1977. 98 pp.

Handbook on maritime law produced by The Maritime and Transport Law Committee of the Section on Business Law of the International Bar Association. Second in series of four. Enforced sales of vessels. General law issues for different countries.

2475 Kegels, Tony. Belgium. In: Lennart Hagberg (ed.) *Maritime law. Vol. II. Enforced sales of vessels.* The Netherlands: Kluwer, 1977, pp. 9–14.

Grounds for enforced sale. Authority. Appraisal. Notice of sale. Price. Payment of price. Distribution of proceeds. Appeals. Conventions.

2476 Langenskiold, Henrik. Finland. In: Lennart Hagberg (ed.) *Maritime law. Vol. II. Enforced sales of vessels.* The Netherlands: Kluwer, 1977, pp. 23–6.

Grounds for an enforced sale. Authority. Case and maintenance of vessel pending sale. Appraisal. Notice of sale. Form of sale. Price. Payment. Distribution of proceeds. Appeals. Conventions.

2477 Maritime Law Association of the Republic of South Africa, Cape Town. South Africa. In: Lennart Hagberg (ed.) *Maritime law. Vol. II. Enforced sales of vessels.* The Netherlands: Kluwer, 1977, pp. 77–80.

Grounds for an enforced sale. Authority. Care and maintenance of vessel pending sale. Appraisal. Notice of sale. Form of sale. Price. Payment of price. Distribution of proceeds. Appeals. Conventions. Other points of interest.

2478 McHose, John C. United States of America. In: Lennart Hagberg (ed.) *Maritime law. Vol. II. Enforced sales of vessels.* The Netherlands: Kluwer, 1977, pp. 95–8.

Grounds for an enforced sale. Authority. Care and maintenance of vessel pending sale. Appraisal. Notice of sale. Form of sale. Price. Payment of price. Distribution of proceeds. Appeals. Conventions.

2479 Moloney, G. John. Ireland. In: Lennart Hagberg (ed.) *Maritime law. Vol. II. Enforced sales of vessels.* The Netherlands: Kluwer, 1977, pp. 45–8.

Grounds for an enforced sale. Authority. Care and maintenance of vessel pending sale. Appraisal. Notice of sale. Form of sale. Price. Payment of price. Distribution of proceeds. Appeals. Conventions. Other points of interest.

2480 Ogawa, Yoichi. Japan. In: Lennart Hagberg (ed.) *Maritime law. Vol. II. Enforced sales of vessels.* The Netherlands: Kluwer, 1977, pp. 57–62.

Grounds for an enforced sale – foreign judgment; arbitration award. Authority. Care and maintenance of vessel pending sale. Appraisal. Notice of sale. Form of sale. Price. Payment of price. Distribution of proceeds of sale. Appeals. Conventions. Other points of interest.

2481 Packard, William V. *Sale and purchase.* London: Fairplay Publications, 1988. 124 pp.

Examines the course of events involved in buying and selling ships. Each aspect explained and illustrations provided of some of the more prominent documents involved. Compares 'SALEFORM 1966' and 'SALEFORM 1987'; 'SALESCRAP 1987'; 'NIPPONSALE'.

2482 Panagopoulos, P. C. Greece. In: Lennart Hagberg (ed.) *Maritime law. Vol. II. Enforced sales of vessels.* The Netherlands: Kluwer, 1977, pp. 35–44.

Grounds for an enforced sale. Authority. Care and maintenance of vessel pending sale. Appraisal. Notice of sale. Form of sale. Price. Payment of price. Distribution of proceeds of sale. Appeals. Conventions. Other points of interest.

2483 Potts, T. H. A. Singapore. In: Lennart Hagberg (ed.) *Maritime* *law. Vol. II. Enforced sales of vessels.* The Netherlands: Kluwer, 1977, pp. 73–6.

Grounds for an enforced sale. Authority. Care and maintenance of vessel pending sale. Appraisal. Notice of sale. Form of sale. Price. Payment of price. Distribution of proceeds. Appeals. Conventions.

2484 **Rein, Alex.** Norway. In: Lennart Hagberg (ed.) *Maritime law. Vol. II. Enforced sales of vessels.* The Netherlands: Kluwer, 1977, pp. 69–72.

Grounds for an enforced sale. Authority. Care and maintenance of vessel pending sale. Appraisal. Notice of sale. Form of sale. Price. Payment of price. Distribution of proceeds. Appeals. Conventions. Other points of interest.

2485 **Romlov, Robert.** Sweden. In: Lennart Hagberg (ed.) *Maritime law. Vol. II. Enforced sales of vessels.* The Netherlands: Kluwer, 1977, pp. 87–90.

Grounds for an enforced sale. Authority. Care and maintenance of vessel pending sale. Appraisal. Notice of sale. Form of sale. Price. Payment of price. Distribution of proceeds. Appeals. Conventions. Other points of interest.

2486 **Van der Feltz, F.** The Netherlands. In: Lennart Hagberg (ed.) *Maritime law. Vol. II. Enforced sales of vessels.* The Netherlands: Kluwer, 1977, pp. 67–8.

Grounds for an enforced sale. Authority. Care and maintenance of vessel pending sale. Appraisal. Notice of sale. Form of sale. Price. Payment of price. Distribution of proceeds. Appeals.

2487 **Villeneau, Jacques.** France. In: Lennart Hagberg (ed.) *Maritime law. Vol. II. Enforced sales of vessels.* The Netherlands: Kluwer, 1977, pp. 27–30.

Grounds for an enforced sale. Authority. Care and maintenance of vessel pending sale. Appraisal. Notice of sale. Form of sale. Price. Payment of price. Distribution of proceeds of sale. Appeals. Conventions.

Sale of vessel

2488 **Glennie, Angus J.** and **Phelan, Patricia J.** *Ship sales.* Oxford: Butterworth, 1986.

2489 **Spicer, W. Wylie.** Court-ordered sale of vessels. *Journal of Maritime Law and Commerce* 11(2), January 1980, 239–51.

Examines principle that a judicial sale of a vessel gives title free and clear of all liens, claims, etc. Exceptions giving rise to situations in which courts adjudicating the status of foreign orders for sale will not recognize/enforce the principle. Canadian, British, American authority.

Arbitration

2490 **Astle, W. E.** *Right of appeal: A practical digest of charter-party disputes and the right of appeal under the Arbitration Act 1979.* London: Fairplay Publications, 1983. 153 pp.

Provides a practical and concise summary of the cases relative to the question of when the right of appeal against an arbitration award may arise. Provisions of the 1979 Act. Guidelines for commercial interests.

2491 **Berlingieri, Francesco.** International maritime arbitration. *Journal of Maritime Law and Commerce* 10(2), January 1979, 199–247.

Advantages of arbitration over court litigation. National laws. Arbitration clauses in charterparty forms, in salvage, in shipbuilding contracts, in general average clauses, in collision clauses. Institutional arbitration.

2492 **Bernstein, Ronald.** *Handbook of arbitration practice.* 2nd edn. London: Sweet & Maxwell, 1993. 696 pp.

2493 **Butler, W. E.** Soviet maritime arbitration. In: Francis D. Rose (ed.) *International commercial and maritime arbitration.* London: Sweet & Maxwell, 1988, pp. 37–54.

Maritime arbitration under the USSR Merchant Shipping Code. The Soviet Maritime Arbitration Commission (MAK).

2494 **Butler, William E.** (ed.) *International commercial arbitration: Soviet commercial and maritime arbitration.* New York: Oceana Publications, 1980.

2495 **Chandler, George F.** The measure of liability for cargo damage under charter parties. *Journal of Maritime Law and Commerce* 20(3), July 1989, 395–402.

Challenging application of 'due diligence' standards in cargo damage cases where Clause Paramount in charterparty. Examination of arbitration award in *The Marofa*, New York Arbitration 1983.

2496 **Cohen, Michael Marks.** A venue problem with the arbitration clauses found in printed form charters. *Journal of Maritime Law and Commerce* 7(3), April 1976, 541–50.

Considers differences between English and American arbitration laws. Survey of charterparty forms to determine effects caused by substitution of London for New York, or vice versa. Problem of designation of venue for purposes of judicial confirmation of an award.

2497 Collins, Lawrence. Arbitration clauses and forum selecting clauses in the conflict of laws: Some recent developments in England. *Journal of Maritime Law and Commerce* 2(2), January 1971, 363–81.

Choice of law in contract, exercise of jurisdiction. Discusses whether a contractual provision for arbitration in a particular place is invariably regarded as choosing the law of that place to govern the contract, invariably ousting jurisdiction of English courts.

2498 Connell, Raymond. The Federal Arbitration Act: The expanding impact of state law upon rigorous enforcement. *Journal of Maritime Law and Commerce* 20(3), July 1989, 327–61.

Jurisdiction and effect of the Federal Arbitration Act, 1925. Arbitration prior to the Act. Analysis of the Act. Case law. The Act and pre-emption.

2499 Cunningham, John R. A comment on the rules of the Vancouver Maritime Arbitrators Association. *Journal of Maritime Law and Commerce* 20(1), January 1989, 147–51.

The arbitration tribunal. Commencement of arbitration. Arbitration 'time-table'. Qualifications of an arbitrator. Hearings. Close of hearing and award. Fees and expenses.

2500 Darling, Gerald (QC) Salvage arbitrations. In: Francis D. Rose (ed.) *International commercial and maritime arbitration.* London: Sweet & Maxwell, 1988, pp. 95–106.

Procedural and substantive importance of Lloyd's Form system in taking from the courts the role of determining majority of international maritime salvage questions.

2501 Domke, Martin. The enforcement of maritime arbitration agreements with foreign governments. *Journal of Maritime Law and Commerce* 2(3), April 1971, 617–24.

Impact in maritime field of fixing place in US as place of arbitration. Confers *in personam* jurisdiction on federal court district for the purpose of compelling arbitration. Increased participation of foreign governments in international trade transactions.

2502 Donaldson, (Rt. Hon. Sir) John. (PC; Master of the Rolls). Commercial arbitration: 1979 and after. In: Francis D. Rose (ed.) *International commercial and maritime arbitration.* London: Sweet & Maxwell, 1988, pp. 1–13.

Arbitration Act 1979 – essential details. Early judicial responses to the Act.

2503 Healy, Nicholas J. An introduction to the Federal Arbitration Act. *Journal of Maritime Law and Commerce* 13(2), January 1982, 223–34.

Examines the Act and shows how maritime arbitrations conducted under its provisions.

2504 Keane, C. P. Waiver of maritime arbitration. *Journal of Maritime Law and Commerce* 8(2), January 1977, 195–226.

Waiver involving judicial acts by one or both of parties. Considered in the light of ss. 3 and 4 Federal Arbitration Act 1925.

2505 Kimball, John D. Vacating maritime arbitration awards: Is it really possible? *Journal of Maritime Law and Commerce* 13(1), October 1981, 71–88.

Review of case law concerning available grounds for setting aside arbitration awards. Variations of such grounds which have developed. Federal Arbitration Act (1925).

2506 Matthew, Paul. The sleeping and the dead, or when is an arbitration not an arbitration? *Lloyd's Maritime and Commercial Law Quarterly,* August 1982, Part 3, 401–15.

Considers the legal effect of serious delay in the prosecution of an arbitration initiated under an agreement between parties containing an arbitration clause. Case law development. House of Lords, 1981 in *Bremer Vulkan* v. *South India Shipping.* Subsequent cases.

2507 McIntosh, David A. The practice of maritime arbitrations in London: Recent developments in the law. *Lloyd's Maritime and Commercial Law Quarterly,* May 1983, Part 2, 235–47.

Development of arbitration as a means of settling disputes. The reference to arbitration. The law governing arbitration in England. Compatability of third party witness and documents. Injunctive order. Offer of settlement. Developments.

2508 McLaughlin, Susan. Arbitral immunity. In: Francis D. Rose (ed.) *International commercial and maritime arbitration.* London: Sweet & Maxwell, 1988, pp. 55–75.

The extent to which arbitrators are or are not liable for the consequences of their actions.

2509 McMahon, John P. Implementation of the United Nations Convention on Foreign Arbitral Awards in the United States. *Journal of Maritime Law and Commerce* 2(4), July 1971, 735–62.

Analysis of policy of US in response to conventions and uniform rules aimed at facilitating enforcement of arbitration agreements and transnational movement of arbitral awards.

2510 Mustill, (Rt. Hon. Sir) Michael John (PC; Lord Justice of Appeal). Transnational arbitration and English law. In: Francis D. Rose (ed.) *International commercial and maritime arbitration.* London: Sweet & Maxwell, 1988, pp. 15–35.

Theories behind the idea of international commercial arbitration as an autonomous juristic entity independent of all national legal systems. Reference to UNCITRAL Model Law on International Commercial Arbitration.

2511 Ohashi, Mitsuo. *Maritime arbitration in Tokyo.* Tokyo: Japan Shipping Exchange, 1979.

2512 Rose, Francis D. (ed.) *International commercial and maritime arbitration.* London: Sweet & Maxwell, 1988. 118 pp.

Explores in-depth aspects of international arbitration, law and practice.

2513 Shujian, Liu. The new Rules governing maritime arbitration in China. *Journal of Maritime Law and Commerce* 21(1), January 1990, 129–36.

Maritime arbitration as compared with settlement by competent administrative authorities. Compared with lawsuits. The new Rules of Arbitration.

2514 Shum, Clement. Maritime arbitration in the People's Republic of China. *Lloyd's Maritime and Commercial Law Quarterly,* February 1990, Part 1, 114–18.

Arbitration in People's Republic of China. Maritime arbitration, the China Maritime Arbitration Commission. Jurisdiction. Organization. Procedures. Costs. Maritime law. Arbitration fees.

2515 Slabotzky, Albert. *Grain contracts and arbitration: For ship-*

ments from the United States and Canada. London: Lloyd's of London Press, 1984. 109 pp.

Reviews in detail the provision of grain contracts and, in particular, those of the standard forms most important to exporters of grain and soya beans from the US and Canada – GAFTA 27 and 30; FOSFA 24; NAEA 2. A practical exposition with insights into the intent and formulation of contractual terms.

2516 Smedresman, Peter S. The Arbitration Act, 1979. *Journal of Maritime Law and Commerce* 11(3), April 1980, 319–31.

Analysis of the Act. Considers whether it reaches aim of drafters to attract arbitrations to London. Are arbitrations under the 1979 Act free of some of court supervision imposed by 1950 Act?

2517 Society of Maritime Arbitrators (New York). A comment on the rules of the Society of Maritime Arbitrators, Inc. (New York). *Journal of Maritime Law and Commerce* 20(2), April 1989, 199–203.

History of the SMA Rules. Code of Ethics. Roster of Arbitrators. SMA educational programmes. Publications. Initiating on arbitration. Conduct of the arbitration. The award.

2518 Textor, James M. Petroleum shortage disputes: The difference between the legal and arbitration approach. *Lloyd's Maritime and Commercial Law Quarterly,* August 1983, Part 3, 392–422.

Considers the various procedural and substantive variations associated with the litigation and arbitration forums for petroleum cargo disputes. How these distinctive features, respectively, can affect the outcome of such a dispute.

2519 Thomajan, Robert. Tanker problems in arbitration: The 0.5% allowance. *Journal of Maritime Law and Commerce* 14(2), April 1983, 225–41.

Questions whether there is a 0.5 per cent trade allowance of the bill of lading quantity. Case law development. Legal significance of allowance. Effect of COGSA. Cargo retention clauses. Author's conclusions – no allowance.

2520 Thomas, D. Rhidian. Admiralty security and the arbitral process. *Lloyd's Maritime and Commercial Law Quarterly,* August 1983, Part 3, 495–512.

Section 26, Civil Jurisdiction and Judgments Act 1982. Significance in commercial and maritime arbitration, in respect of potential association between security obtained in an Admiralty proceeding in rem and an arbitral award.

2521 Thomas, D. Rhidian. The Antaios: The Nema guidelines reconsidered. *Journal of Business Law,* 1985, 200–8.

Operation of Arbitration Act 1979. Guidelines laid down in *The Nema,* 1982. Policy underlying the 1979 Act. Factors prevailing to determine relationship between arbitration and questions of law. Developed practice of Commercial Court.

2522 Young, James. Arbitration and the European Communities' Judgments Convention. In: Francis D. Rose (ed.) *International commercial and maritime arbitration.* London: Sweet & Maxwell, 1988, pp. 77–93.

Relationship between failure to deal with the recognition and enforcement of arbitration awards and arbitration awards within the European Communities.

Seaworthiness

2523 Baer, Herbert R. Admiralty law and the Burger U.S. Supreme Court. *Journal of Maritime Law and Commerce* 3(1), October 1971, 91–105.

The work of the Burger Court in Admiralty in respect of development of cases regarding concept of unseaworthiness. Problems of anticipation of future cases when past decisions not unanimous, but rather of 5–4 variety.

2524 Clarke, A. P. The resolution of disputes. In: *The mariner and the maritime law. Seminar 3: Seaworthiness.* London: Nautical Institute, 1992, pp. 49–65.

Focuses upon role of master and officers of a vessel in providing evidence for the resolution of disputes where seaworthiness of their vessel in dispute. From viewpoint of tribunal. Under application of Hague–Visby Rules. Evidence – as to seaworthiness of vessel; as to cause of damage; as to due diligence.

2525 Ellman, Mark. Instant unseaworthiness: Mascuilli revisited. *Journal of Maritime Law and Commerce* 1(4), July 1970, 573–601.

US liability imposed for unseaworthiness. Operational negligence introduced 1944 *Mahnich* v. *Southern Steamship Company.* Impact on distinction between operational negligence and strict liability of *Mascuilli* v. *U.S.* (1967).

2526 George, James A. Ship's liability to longshoremen based on unseaworthiness: Sieracki through Usner. *Journal of Maritime Law and Commerce* 3(1), October 1971, 45–70.

Historical development of basic principles of doctrine of liability based on unseaworthiness. Decisions through to 1971 *Usner* v. *Luckenback Overseas Corporation* which eliminated 'instantaneous unseaworthiness'.

2527 Hakiki, M. A. The carrier's undertaking of seaworthiness: A comparative study. Thesis submitted for MA, University of Glasgow, 1990.

In this work the carrier's undertaking to provide a seaworthy ship is studied comparatively under the different regimes of Common and Conventional Laws.

2528 Hilton, Chris. A legal perspective. In: *The mariner and the maritime law. Seminar 3: Seaworthiness.* London: Nautical Institute, 1992, pp. 3–28.

Legal issues arising out of concept of seaworthiness. What is seaworthiness? Seaworthiness and cargo claims. Case law. Exceptions from liability. Hague–Visby Rules. Burden of proof, examples of operation of burden of proof. Hamburg Rules.

2529 Knott, John. A technical perspective. In: *The mariner and the maritime law. Seminar 3: Seaworthiness.* London: Nautical Institute, 1992, pp. 29–48.

Considers concept of seaworthiness from technological sense. Keeping the water out. Hatch covers – a case study. Other typical aspects of seaworthiness – charts, stowage, bilge systems.

2530 *The mariner and the maritime law. Seminar 3: Seaworthiness.* London: Nautical Institute, 1992. 65 pp.

Contains three papers given at a seminar on concept of seaworthiness. A legal perspective; a technical perspective; the resolution of disputes.

2531 Pritchett, Russell W. The implied warranty of seaworthiness in time policies: The American view. *Lloyd's Maritime and Commercial Law Quarterly*, May 1983, Part 2, 195–208.

Comparison of English law under Marine Insurance Act 1906 and US position in relation to whether warranty of seaworthiness implied into time policy of marine insurance. The 'American Rule'. History of British approach. Evolution of American law.

Criminal Activity

Fraud

2532 Conway, Barbara. *Maritime fraud.* London: Lloyd's of London Press, 1990. 170 pp.

Detailed investigation into maritime fraud industry. Describes (with specific examples) the nature of maritime fraud, its major forms and the growth in this kind of crime since the mid-1970s.

2533 Ellen, Eric and **Campbell, Donald.** *International maritime fraud.* London: Sweet and Maxwell, 1981. 174 pp.

Nature and extent of maritime fraud against background of international trade – laws, regulations, procedures. Practical advice on how fraud can be identified and prevented. Describes typical frauds for future recognition. Reviews organizations and institutions whose job it is to facilitate international trade.

2534 Horton Rogers, W. V. The late ship and the non existent cargo. *Modern Law Review* 34, March 1971, 190–94.

Under Notes on Cases – this piece goes into the details which the title indicates.

2535 International Chamber of Commerce. *Guide to the prevention of international trade fraud.* Paris: I.C.C. Publications, 1985. 20 pp.

A guide designed to help the maritime industry guard against fraud. Especially intended for buyers, traders, sellers, shipowners, insurers and bankers doing business in the Middle East, West Africa and South-East Asia. History and types of international trade fraud. Precautionary measures.

2536 International Chamber of Commerce. *Scuttling: A special study.* Paris: I.C.C. Publications, 1982. 15 pp.

Examines developing sophistication of scuttling schemes. The first three post-war decades. The 1970s. Subsequent scuttling. The new dimension – a syndicate of fraudsters.

2537 Kapoon, Peter. Definition and classification of maritime fraud. *Lloyd's Maritime and Commercial Law Quarterly,* February 1983, Part 1, 29–44.

The increasing incidence of maritime fraud. Classification of fraud. Definition of maritime fraud. Documentary maritime fraud.

2538 Kapoon, Peter and **Gray, Richard.** *Maritime fraud: An overview.*

Plymouth: Plymouth Polytechnic, Department of Shipping and Transport, 1985. 57 pp.

Examines the historical background to determine the causes which have contributed to the resurgence of maritime fraud. Types of maritime fraud, problems faced by investigators, recommendations for curbing such fraud.

2539 **Klinghoffer, Arthur Jay.** *Fraud of the century: Case of the mysterious supertanker 'Salem'.* London and New York: Routledge, 1988. 224 pp.

2540 **Powles, D. G.** and **Hazelwood, S. J.** Maritime fraud – I. *Journal of Business Law*, 1984, 31–41.

First of four papers in same volume (part II, pp. 137–44; part III, pp. 234–47; part IV, pp. 403–9) examining legal consequences of fraudulent dealings in connection with maritime activities. Considers bills of lading and documentary credit frauds.

2541 **Powles, D. G.** and **Hazelwood, S. J.** Maritime fraud – II. *Journal of Business Law*, 1984, 137–44.

Second of four papers in same volume (part I, pp. 31–41; part III, pp. 234–47; part IV, pp. 403–9) examining legal consequences of fraudulent dealings in connection with maritime activities. Considers letters of indemnity, charterparty frauds, marine mortgage frauds.

2542 **Powles, D. G.** and **Hazelwood, S. J.** Maritime fraud – III. *Journal of Business Law*, 1984, 234–47.

Third of four papers in same volume (part I, pp. 31–41; part II, pp. 137–44; part IV, pp. 403–9) examining legal consequences of fraudulent dealings in connection with maritime activities. Marine insurance frauds.

2543 **Powles, D. G.** and **Hazelwood, S. J.** Maritime fraud – IV. *Journal of Business Law*, 1984, 403–9.

Fourth and concluding paper (parts I–III in same volume, respectively pp. 31–41; 137–44; 234–47) examining legal consequences of fraudulent dealings in connection with maritime activities. Enforcement of remedies; investigation, prevention, detection of maritime fraud.

2544 **Powles, David G.** Ship sales and frauds by intermediaries. *Journal of Business Law*, 1987, 337–43.

Frauds of intermediaries in relation to sales of vessel – whether such intermediaries be acting in employment of one of parties to the sale or as an independent broker.

Piracy

2545 **Birnie, P. W.** Piracy past, present and future. In: Eric Ellen (ed.) *Piracy at sea.* Paris: I.C.C. Publications, 1989, pp. 131–58.

An examination of the present state of the law relating to piracy and its relevance to violent acts performed at sea. International and municipal law perspectives of the offence. A study of the kinds of violence occurring.

2546 **Blaney, Harry C.** Anti-piracy in South East Asia: US and international efforts and programmes. In: Eric Ellen (ed.) *Piracy at sea.* Paris: I.C.C. Publications, 1989, pp. 101–6.

Examines pirate attacks on Vietnamese boat refugees in the Gulf of Thailand and the US attempts to eliminate such piracy.

2547 **Brittin, Burdick H.** The law of piracy: Does it meet the present and potential challenges? In: Eric Ellen (ed.) *Piracy at sea.* Paris: I.C.C. Publications, 1989, pp. 159–67.

Raises some issues that are suggestive of desirable changes to the law of piracy.

2548 **Brown, E. D.** Maritime commercial malpractices and piracy under international law. *Maritime Policy and Management* 8(2), 1981, 99–107.

Examines concepts of piracy, hijacking and mutiny under international law. Shows that various acts of violence or lawlessness at sea are not acts of piracy in law.

2549 **Clingan, Thomas A.** The law of piracy. In: Eric Ellen (ed.) *Piracy at sea.* Paris: I.C.C. Publications, 1989, pp. 168–72.

Examination of the law of piracy.

2550 **Conway, Barbara.** *The piracy business.* Feltham: Hamlyn Paperbacks, 1981. 160 pp.

General description of types of maritime fraud and conditions which have helped them to flourish, as well as examination of specific incidents.

2551 **Dubner, B. H.** *Law of international sea piracy.* The Hague: Nijhoff, 1979. 194 pp.

2552 **Ellen, Eric** (ed.) *Piracy at sea.* Paris: I.C.C. Publications, 1989. 330 pp.

Collection of papers which address the modern problem of piracy at sea. Emphasis placed on the economics of ocean space, such as costs and inhibitions still imposed by the recurring problem of piracy.

2553 **Harlow, Bruce.** Rethinking piracy control in a modern maritime context. In: Eric Ellen (ed.) *Piracy at sea.* Paris: I.C.C. Publications, 1989, pp. 201–5.

Considers maritime crime as being in four categories – cargo fraud, internal waters crime, territorial seas, high seas or EEZ problem. Impact on maritime trade.

2554 **Hazelwood, Steven J.** The peril of 'pirates': All 'at sea'. *Lloyd's Maritime and Commercial Law Quarterly,* May 1983, Part 2, 283–8.

Modern indicia of 'pirate' for the purposes of policy of marine insurance. Case law.

2555 **Hyslop, I. R.** Contemporary piracy. In: Eric Ellen (ed.) *Piracy at sea.* Paris: I.C.C. Publications, 1989, pp. 3–40.

An introductory paper to Part 1 of this book – 'The reality of piracy'. Consolidates the information which the IMB collated for its three updating reports to the I.M.O., with insights required of maritime problems.

2556 **Ignarski, Jonathan.** Piracy, law and marine insurance. In: Eric Ellen (ed.) *Piracy at sea.* Paris: I.C.C. Publications, 1989, pp. 181–7.

Examines the definitional difficulties which arise in respect of piratical acts and the impact in respect of marine insurance.

2557 **Menefee, Samuel P.** The 'Achille Lauro' and similar incidents as piracy: Two arguments. In: Eric Ellen (ed.) *Piracy at sea.* Paris: I.C.C. Publications, 1989, pp. 179–80.

Questions whether passenger ship takeovers are piracy or terrorism or both. Impact on penalties.

2558 **Menefee, Samuel P.** The United States and post-war piracy. In: Eric Ellen (ed.) *Piracy at sea.* Paris: I.C.C. Publications, 1989, pp. 61–79.

Examines several patterns of piratical behaviour which have affected post-war American interests.

2559 **Office of the Defence Attaché, Permanent Mission of Nigeria to the United Nations.** Piracy control in Nigeria's territorial seas. In:

Eric Ellen (ed.) *Piracy at sea.* Paris: I.C.C. Publications, 1989, pp. 219–23.

Examines the problem faced by Nigeria in particular, and on the west coast of Africa in general, in respect of piracy and armed violence in creeks, quaysides.

2560 Parritt, Brian A. H. *Security at sea: Terrorism, piracy and drugs.* London: Nautical Institute, 1991. 104 pp.

Highlights dangers of piracy and armed robbery against ships. Impact on shipping industry and profitability. Strategies to prevent acts of terrorism. Contingency plans. Action when ship taken over by force. Anti-drug smuggling protection measures.

2561 Pentti, Frank. The MARAD view on maritime policy. In: Eric Ellen (ed.) *Piracy at sea.* Paris: I.C.C. Publications, 1989, pp. 206–12.

Outlines the nature of the interest of the Maritime Administration in the problem of piracy.

2562 Peterson, M. J. An historical perspective on the incidence of piracy. In: Eric Ellen (ed.) *Piracy at sea.* Paris: I.C.C. Publications, 1989, pp. 41–60.

A paper which pays attention to all forms of robbery at sea – attacks on one ship by persons aboard another; raids on ships lying in harbours; capture of ships by persons on board; raids on coastal towns by ship-borne marauders.

2563 Terazaki, Naomichi H. The UNCTAD perspective on piracy. In: Eric Ellen (ed.) *Piracy at sea.* Paris: I.C.C. Publications, 1989, pp. 173–8.

Outlines the activities and concerns of the United Nations Conference on Trade and Development in respect of piracy at sea/maritime fraud including piracy.

2564 Villar, Roger. *Piracy today: Robbery and violence at sea since 1980.* London: Conway Maritime Press, 1985. 160 pp.

Comprehensive report of armed robbery at sea. Detailed analysis and practical suggestions of precautions. Piracy world-wide. Atlas of piracy. Log of attacks.

2565 Walkate, Jaap A. Aspects of international action to combat piracy at sea. In: Eric Ellen (ed.) *Piracy at sea.* Paris: I.C.C. Publications, 1989, pp. 213–18.

A working paper examining action against piracy. Action taken by international organizations up to 1985. Action taken in the Framework of the International Maritime Organisation since April 1985.

Terrorism

2566 Ellen, Eric F. Piracy. In: Eric F. Ellen (ed.) *Violence at sea: A review of terrorism, acts of war and piracy, and countermeasures to prevent terrorism.* Paris: I.C.C. Publishing, 1987, pp. 225–34.

A paper based upon reports published by the United Nations agencies on the problems of piracy and armed robbery at sea.

2567 Ellen, Eric F. (ed.) *Violence at sea: A review of terrorism, acts of war and piracy, and countermeasures to prevent terrorism.* 2nd edn. Paris: I.C.C. Publishing, 1987. 267 pp.

Collection of papers for this book, prepared by the IMB. Examines the implications of terrorism at sea and suggests how risks can be reduced.

2568 Haines, Stephen W. Criminal violence at sea: Observations on the threat and appropriate responses. In: Eric F. Ellen (ed.) *Violence at sea: A review of terrorism, acts of war and piracy, and countermeasures to prevent terrorism.* Paris: I.C.C. Publishing, 1987, pp. 93–107.

Aim of paper is to place the problem of maritime terrorism in overall context. Identifies the scope of low intensity sub-state criminal violence at sea. Identifies the industry's role in the provision of an overall response to the threat.

2569 Harlow, Bruce. The role of military force to protect ships and ports against terrorists attacks. In: Eric F. Ellen (ed.) *Violence at sea: A review of terrorism, acts of war and piracy, and countermeasures to prevent terrorism.* Paris: I.C.C. Publishing, 1987, pp. 143–8.

Examines the difficulties in the involvement of military forces in civil affairs. Distinction between acts of terrorism and acts of piracy, mutiny or insurrection. Issues which arise.

2570 Havens, George R. Proactive measures to protect a seaport. In: Eric F. Ellen (ed.) *Violence at sea: A review of terrorism, acts of war and piracy, and countermeasures to prevent terrorism.* Paris: I.C.C. Publishing, 1987, pp. 175–8.

Examines measures and procedures to prevent attacks against ships and seaports.

2571 Hewer, Christopher. Precautions that shipowners could take to protect their vessels. In: Eric F. Ellen (ed.) *Violence at sea: A review of terrorism, acts of war and piracy, and countermeasures to prevent terrorism.* Paris: I.C.C. Publishing, 1987, pp. 169–74.

What safeguards are available to shipowners and why these are rarely adopted.

2572 Ignarski, Jonathan. Terrorism in a maritime context: Law, insurance and the legal implications of armed merchant ships. In: Eric F. Ellen (ed.) *Violence at sea: A review of terrorism, acts of war and piracy, and countermeasures to prevent terrorism.* Paris: I.C.C. Publishing, 1987, pp. 181–90.

Examines terrorism in a maritime context with reference to public international law and the English law of marine insurance. Considers the legality of armed merchant ships.

2573 Jenkins, Brian M. Future trends in international terrorism. In: Eric F. Ellen (ed.) *Violence at sea: A review of terrorism, acts of war and piracy, and countermeasures to prevent terrorism.* Paris: I.C.C. Publishing, 1987, pp. 41–61.

A paper which questions when terrorism will end.

2574 Jenkins, Brian Michael and others. A chronology of terrorist attacks and other criminal actions against maritime targets. In: Eric F. Ellen (ed.) *Violence at sea: A review of terrorism, acts of war and piracy, and countermeasures to prevent terrorism.* Paris: I.C.C. Publishing, 1987, pp. 63–85.

Reviews growing terrorism on a world-wide basis and the increasing concern about potential attacks on targets in the maritime environment.

2575 Mazzone, F. Prevention of terrorist attack of a ship while at port or sea. In: Eric F. Ellen (ed.) *Violence at sea: A review of terrorism, acts of war and piracy, and countermeasures to prevent terrorism.* Paris: I.C.C. Publishing, 1987, pp. 149–67.

Examines effective security measures to help keep the threat of maritime terrorism in a proper perspective.

2576 Menefee, Samuel P. Terrorism at sea: The historical development of an international legal response. In: Eric F. Ellen (ed.) *Violence at sea: A review of terrorism, acts of war and piracy, and countermea-*

sures to prevent terrorism. Paris: I.C.C. Publishing, 1987, pp. 191–220.

Examines the question of how the international legal community should respond to terrorism at sea. Review of political piracy.

2577 Ronzitt, Natalino (ed.) *Maritime terrorism and international law.* The Hague: Nijhoff, 1990. 200 pp.

2578 Schiller, Thomas S. The Gulf War and shipping: Recent developments. In: Eric F. Ellen (ed.) *Violence at sea: A review of terrorism, acts of war and piracy, and countermeasures to prevent terrorism.* Paris: I.C.C. Publishing, 1987, pp. 109–27.

Examines what the paper calls the economic warfare against shipping in the Persian Gulf by both Iran and Iraq.

2579 Schiller, Thomas S. Maritime terrorism: The threat. In: Eric F. Ellen (ed.) *Violence at sea: A review of terrorism, acts of war and piracy, and countermeasures to prevent terrorism.* Paris: I.C.C. Publishing, 1987, pp. 87–92.

Considers the problem of maritime terrorism both in isolation and in the overall context of violence against shipping, including conflict in the Persian Gulf and piracy.

2580 Simon, Jeffrey D. The implications of the 'Achille Lauro' hijacking for the maritime community. In: Eric F. Ellen (ed.) *Violence at sea: A review of terrorism, acts of war and piracy, and countermeasures to prevent terrorism.* Paris: I.C.C. Publishing, 1987, pp. 17–25.

Paper addresses the prospects that terrorist attacks at sea will become a regular form of international terrorism. Need for improved security. Feasibility of a shared intelligence network for the maritime community.

2581 Stav, Chaim. Practical measures to be taken by ports and ships' crews to prevent an attack and to minimise the risk when an attack occurs. In: Eric F. Ellen (ed.) *Violence at sea: A review of terrorism, acts of war and piracy, and countermeasures to prevent terrorism.* Paris: I.C.C. Publishing, 1987, pp. 131–41.

Examines ways of planning security measures for ports and for ships at sea or docked in harbour.

2582 Wilkinson, Paul. Terrorism and the maritime environment. In: Eric F. Ellen (ed.) *Violence at sea: A review of terrorism, acts of war* *and piracy, and countermeasures to prevent terrorism.* Paris: I.C.C. Publishing, 1987, pp. 27–40.

A paper which examines the key trends in international terrorism. How serious is the security threat? How vulnerable are maritime targets to terrorist attack? Role of navies in combating terrorism.

Criminal law

2583 Bentil, J. Kodwo. Crimes committed on board foreign ships on the high seas and the jurisdiction of the English courts. *Lloyd's Maritime and Commercial Law Quarterly,* February 1983, Part 1, 60–70.

Jurisdiction of national courts to hear cases of crimes committed on board ships on the high seas. Public international law. Relevant English statutory provisions. Basis and nature of jurisdiction of English courts.

2584 Cadwallader, F. J. J. The art of the scuttler: Inference and proof. *Maritime Policy and Management* 8(3), 1981, 163–77.

Examines the legal problems involved in scuttling of ships and how the question of liability depends on the correct inference of the proof available. Several cases examined.

2585 Clark, Peter D. Criminal jurisdiction over merchant vessels engaged in international trade. *Journal of Maritime Law and Commerce* 11(2), January 1980, 219–37.

Overview of development of criminal jurisdiction as it pertains to merchant vessels engaged in international trade. International criminal law. Maritime law. Legislative attempts to deal with jurisdictional problems.

2586 Jarvis, Robert M. Common ethical problems in maritime criminal cases. *Journal of Maritime Law and Commerce* 23(3), July 1992, 387–410.

Examines the most common ethical issues in criminal practice in light of recent convergence of maritime law and criminal law. Deals with American Bar Associations' Rules of Professional Conduct. Client counselling. Conflicts of interest. Attorneys' fees. Trial publicity. Client perjury. Plea bargains. Confidential information.

2587 Potter, Douglas. Safety afloat: Criminal penalties and investigations into marine casualties. *Journal of Maritime Law and Commerce* 23(4), October 1992, 607–20.

Criminal penalties imposed upon shipowners and crew. Historical survey. Public interest in loss of life at sea. Criminal liability under English law. *Herald of Free Enterprise* 1987. Sections 30–32 Merchant Shipping Act 1988. Inquiries into casualties. Rule against self-incrimination. Maritime inquiries.

13 Marine Insurance

General

General

2588 Altahuri, A. M. Impact of the carrier's liability under the Hamburg Rules on marine insurance. Thesis submitted for MSc, UWIST, Cardiff, 1987.

2589 American Bar Association, Tort and Insurance Practice Section. *Damages recoverable in maritime matters.* Chicago: American Bar Association, 1984. 153 pp.

2590 Arnould, Sir Joseph. *Arnould's law of marine insurance and average.* 16th edn. London: Stevens & Sons, 1981.

Vol. 1: Nature, formation and subject-matter of contract of marine insurance. Vol. 2: What renders contract of insurance void or unavailable. Losses and ensuing relations of assured and underwriter. Procedure and evidence. Overall is a treatise on principles of law of marine insurance.

2591 Badger, Dennis and **Whitehead, Geoffrey.** *Elements of cargo insurance.* Cambridge: Woodhead-Faulkner, 1983. 154 pp.

Fundamental principles of insurance in general and cargo insurance. Summary of ways in which insurance effected. Traditional marine insurance law and SG policy and 1.1.63 Institute Cargo Clauses. Assessment of new 1.1.82 Institute Clauses. General average; containerization and insurance impact.

2592 Bashford, A. S. *Marine liabilities: Guidelines to exposures and insurances of port authorities and other port related industries or activities.* London: Witherby and Co., 1989. 41 pp.

2593 Brock, Denis. Title to sue and the story of the third musketeer.

In: *Leading developments in international marine insurance: An industry report.* London: Lloyd's of London Press, 1991. 10 pp.

Discussion of who may/may not have title to sue. Case law examination.

2594 Brown, R. H. *The insurance of merchant ships and shipowners' interests.* 3 vols. London: Witherby & Co., 1990. 200 pp.

2595 Brown, Robert H. *Marine insurance. Vol. I: Principles and basic practice.* 5th rev. edn. London: Witherby & Co., 1986. 343 pp.

Combines description of principles of marine insurance with principles on which the practice is based. Structure of Lloyd's market. Insurance conditions following withdrawal of SG form of policy. MAR form of policy.

2596 Brown, Robert H. *Marine insurance. Vol. 3: Hull practice.* 2nd edn. London: Witherby & Co., 1993. 1187 pp.

This second edition is a complete rewrite of the first. A practical reference work which is suitable for insurance practitioners world-wide. Introduces the new Institute hull insurance conditions whilst retaining such material from the first edition as is of value to markets where SG form still in use.

2597 Brown, Robert H. *Marine reinsurance terms and abbreviations.* London: Witherby & Co., 1981. 32 pp.

2598 Chalmers, Sir Mackenzie D. and **Ivamy, E. R. Hardy.** *Chalmers' Marine Insurance Act 1906.* 9th rev. edn. London: Butterworths, 1983. 228 pp.

Consideration of Marine Insurance Act 1906, supported by reference to leading cases. Illustrations drawn from decided cases to show application of difficult sections. Original text rewritten and set in modern form.

2599 Chartered Insurance Institute. *Introduction to marine and aviation insurance.* London: Chartered Insurance Institute, 1992.

2600 Chartered Insurance Institute. *Marine and aviation insurance.* London: C.I.I. Tuition Service, 1988. 205 pp.

Coursebook for the Chartered Insurance Institute examinations. Geography. Trades and commodities. International trade. Marine insurance. Marine Insurance Act 1906.

2601 Chartered Insurance Institute. *Marine insurance and world trade.* London: Chartered Insurance Institute, 1991. 265 pp.

Based on material supplied by Institute of Chartered Shipbrokers. Coursebook for C.I.I. Maritime geography. World trade. International trade. Balance of payments, rates of exchange. Economies of scale. Liner conferences. Freight market. Layup and break up. Transportation. Ports. Principal shipping organizations. Sale of goods. Payments, finance. Ship sale and purchase. Marine insurance.

2602 Chartered Insurance Institute. *Marine risks and their assessment.* London: Chartered Insurance Institute, 1991. 273 pp.

Study coursebook for the Distance Learning Division of the C.I.I. Ships. Ship construction, classification. Commodities and their characteristics. World trades. Tanker trade cargoes. Liner operations. Ship management and operations. Ship agency. Offshore oil/gas exploration and production units. Underwriting marine risks. Insurance rate. Interests insured as cargo. Loss prevention.

2603 Chartered Insurance Institute Tuition Services. *Marine underwriting. Study course 190M/073. The C.I.I. Tuition Service.* London: Chartered Insurance Institute, 1985. 299 pp.

Coursebook for the Chartered Insurance Institute examinations. Marine insurance. The markets. The Marine Insurance Act 1906. Cargo underwriting. Principal cargo clauses. Institute clauses. Vessels. Hull underwriting. Offshore oil/gas exploration and production units. Reinsurance.

2604 Clarke, Malcolm. Insurance law: Recent causes. *Lloyd's Maritime and Commercial Law Quarterly,* November 1983, Part 4, 576–86.

Causal connection between breach and loss – direct causes; indirect causes; inherent vice; inducement.

2605 *Convention on limitation for maritime claims, 1976.* London: HMSO, 1990.

2606 Dover, Victor and **Brown, Robert Henry.** *A handbook to marine insurance.* 8th rev. edn. London: Witherby & Co., 1975. 870 pp.

Textbook of the history, law and practice of marine insurance. Considerations affecting underwriters when estimating risks. Marine insurance policy forms. Marine Insurance Acts 1906–9. Reinsurance. Protection and Indemnity. Collision liabilities. Claims. General average. Maritime salvage.

2607 Down, Paul M. Marine market on the rocks? *Canadian Insurance* 96(2), February 1991, 12–14.

2608 Fernandes, Rui M. *Marine insurance law in Canada.* Canada: Butterworth, 1987. 304 pp.

2609 **Flitner, Arthur L.** and **Brunck, Arthur E.** *Ocean Marine Insurance.* 2 vols. 2nd edn. Insurance Institute of America, 1992. 466 pp.

2610 **Ford, Michael.** *Marine insurance fraud in international trade.* London: Witherby & Co., 1993. 200 pp.

Examines the legal structures to be found in the law of marine insurance which have a bearing upon fraud. The extent to which fraudulent activity is covered by standard marine policies. The effect of frauds upon validity of both contracts of insurance and claims made on them.

2611 **Goodacre, J. Kenneth.** *Collected papers on marine claims.* London: Witherby & Co., 1980. 243 pp.

Collection of papers representing areas of controversy and discussion in law and practice of marine insurance. Fundamental principles; maritime perils; hull claims; general average, salvage, sue and labour; case summaries.

2612 **Goodacre, J. Kenneth.** *Collected papers on marine claims.* Vol. 2. London: Witherby & Co., 1992. 108 pp.

A collection of papers presented by the author. Examines events which took place in the 1980s.

2613 **Goodacre, J. Kenneth.** *Goodbye to the memorandum: An in-depth study of the standard cargo, war and strikes clauses.* London: Witherby & Co., 1988. 271 pp.

Cargo insurance. New clauses following 1978 UNCTAD report. Recent and past incidents examined in context of the 1982 Clauses. Differences brought about by 1982 Clauses. References to Marine Insurance Act. Cases. Associated war and strikes clauses.

2614 **Goodacre, J. Kenneth.** *Marine insurance claims.* London: Witherby & Co., 1981. 1013 pp.

Law and practice of marine insurance claims. Cases; legislation; international conventions. Disputes and settlements reached without litigation. Non-disclosure, coverage for war, associated perils. Comparison between Institute Time Clauses and American Hull Form.

2615 **Gough, Neville.** *Institute cargo and related trade clauses: A detailed analysis and commentary.* London: Insurance and Reinsurance Research Group, 1988. 352 pp.

Collection of articles *re* development of clauses following UNCTAD 1978 proposals. Specific new clauses – analysis of cover afforded. Specific exclusions in new clauses. General topics. Clauses set out.

2616 **Hall, Geoffrey Michael.** *Reinsurance law casebook.* Chesham, Bucks: Business and Maritime Publications, 1989. 291 pp.

2617 **Hewer, Chris** and **Grey, Michael.** *On the rocks: Tales of shipping and insurance.* London: Fairplay Publications, 1982. 129 pp.

A collection of articles which have appeared in *Fairplay.* Politics and law. Lloyd's. The environment. Insurance and money. Marine technology. Marine fraud. Shipbuilding and design. Job satisfaction. Ship operations and management.

2618 **Huang, C. T.** A programme of international shipping and insurance laws for our future world. Thesis submitted for PhD, University College, London, 1988.

2619 **Hudson, N. G.** *Insurance of average disbursements.* London: Lloyd's of London Press, 1987.

2620 **Hudson, N. G.** and **Allen, J. C.** *The Institute Clauses handbook.* London: Lloyd's of London Press, 1986. 222 pp.

Practical approach. Examines and explains the Institute Marine Clauses in context of everyday work – background and development. Development of the Clauses following UNCTAD document 1978. Scrutinizes Marine Cargo Forms. Examines occasional clauses. Marine Hull clauses. Additional insurances. War and strike risks.

2621 **Hudson, N. G.** and **Allen, J. C.** *Marine claims handbook.* 4th edn. London: Lloyd's of London Press, 1984. 102 pp.

Provides background information and advice for liner operators and their agents. Claims on policies of insurance on ship. Steps from casualty to collation of claim. General average, salvage. English, American Hull clauses compared. Particular average. Shipowners' liabilities. Claims presentation, evidence. Procedure. Relevant texts and forms.

2622 **Institute of London Underwriters.** *Reference book of marine insurance clauses.* 64th rev. edn. London: Witherby & Co., 1992. 266 pp.

Reference book containing all Institute Clauses, some UK and American Clauses and forms, including a comprehensive index of American Clauses. Published annually in October.

2623 Ivamy, E. R. Hardy. *Chalmers' Marine Insurance Act 1906.* 9th edn. London: Butterworth, 1983. 228 pp.

Examination of the Marine Insurance Act 1906. Reference to leading cases. Illustrations from cases to show application of difficult rules of law. Relevant provisions of Sale of Goods Act 1979. Relevant Institute Clauses. Rules of Practice – Association of Average Adjusters.

2624 Ivamy, E. R. Hardy. *Marine insurance.* 4th rev. edn. London: Butterworth, 1985. 613 pp.

Law relating to marine insurance. Case law references. Emphasis on effect of the Institute Clauses on the Marine Insurance Act 1906 and the common law.

2625 Kouladis, Nicholas (principal author) and **Grime, Robert P.** (consultant and contributor). *Marine insurance: The legal and documentary framework.* London: Chartered Insurance Institute, 1991. 408 pp.

Study course 710. Distance Learning Division of Chartered Insurance Institute. Marine and transport law; carriage of goods by road, rail and air; contracts of affreightment and charterparties. Marine Insurance Act 1906; related issues.

2626 Lambeth, R. J. (ed.) *Templeman on marine insurance: Its principles and practice.* 6th edn. London: Pitman, 1986. 624 pp.

Comprehensive treatment of marine insurance. Principles and practice. Law and legislation, cases. Changes following replacement of SG policy form. Limitation of shipowners' liability. Limitation Convention 1976.

2627 *Leading developments in international marine insurance: An industry report.* London: Lloyd's of London Press, 1991. 86 pp.

A selection of papers presented on 28 October 1991 to the Asia–Pacific Insurance Conference in Bangkok.

2628 Leloir, Nicole. The Lloyd's marine policy and the Institute cargo clauses. *Journal of Business Law* 1985, 288–93.

Lloyd's Marine Policy 1982. Rules of Marine Insurance Act 1906. The A, B and C Clauses.

2629 Lloyd's of London. *An introduction to Lloyd's market procedures and practices: Textbook for the Lloyd's introductory test.* 2nd edn. London: Witherby & Co., 1987. 121 pp.

Prepared for candidates for the Lloyd's Introductory Test. Provides a general description of the operations and procedures at Lloyd's covered by the areas of study listed in the syllabus.

2630 **Lux, Jonathan** (ed.) *Classification societies.* London: Lloyd's of London Press, 1993. 132 pp.

An examination of the obligations and liabilities of classification societies. Based on papers from a conference held by the Maritime and Transport Committee of the International Bar Association, October 1991. Examines the mutual obligations of societies and shipowners and their obligations to third parties.

2631 **Mallon, Paul.** Loss prevention in a multi-modal context. In: *Leading developments in international marine insurance: An industry report.* London: Lloyd's of London Press, 1991. 4 pp.

Why is loss prevention necessary? Why now? What form should loss prevention take? 'Stoploss': The results.

2632 **Mavrakis, J. A.** Marine insurance rates and claims. Thesis submitted for MSc, University of Strathclyde, 1976.

2633 **Mead, Ray.** Oil shortage claims: Recent legal developments. In: *Leading developments in international marine insurance: An industry report.* London: Lloyd's of London Press, 1991. 8 pp.

Title to sue; burden of proof/documentary evidence; related topics.

2634 **Mesrati, S. al-A.** The principle of indemnity in marine insurance: A comparative study. Thesis submitted for PhD, University of London, 1983.

2635 **Mohammed, F. J.** Developing countries and the international maritime transport regime. Thesis submitted for M.Phil, University of Salford, 1989.

A study of the influence of various conventions on the carriage of goods by sea on the developing countries and on cargo and liability insurance practices.

2636 **Mustill, Sir Michael J.** and **Gilman, Jonathan C. B.** *Arnould's law of marine insurance and average.* 16th edn. London: Stevens & Sons, 1981. 1242 pp.

Vol. 1: Nature, formation, subject-matter of contract of marine insurance. Vol. 2: What renders contract of insurance void or unavailable. Losses and ensuring relations of assured and underwriter. Procedure and evidence. Overall is a treatise on principles of law of marine insurance.

2637 **Mustill, Sir Michael J.** and **Gilman, Jonathan C. B.** *Law of marine insurance. British shipping laws.* 1st Supplement to 16th rev. edn. London: Stevens & Sons, 1991. 100 pp.

2638 O'May, Donald. *Marine insurance law.* London: Chartered Insurance Institute, 1992.

2639 O'Shea, James A. *Marine claims.* London: Chartered Insurance Institute, 1991. 216 pp.

Coursebook for C.I.I. Claims on policies of marine insurance. Particular average – partial loss. Total loss. Abandonment. Sue and labour. Subrogation. Reinsurance. Collisions. Cargo claims. Commodity clauses. Freight. Salvage. General average. Collection of security. Application of general average to policies of insurance.

2640 O'Sullivan, Patrick. The scope of the Sue and Labor clause. *Journal of Maritime Law and Commerce* 21(4), October 1990, 545–68.

Preview of cases to outline the parameters of 'sue and labor' clause in marine insurance. Origin and history of the clause. Survey of specific clause.

2641 Parks, A. *Marine insurance and average.* London: Sweet & Maxwell, 1987.

2642 Parks, Alex L. *Law and practice of marine insurance and average.* 2 vols. London: Stevens and Sons, 1988.

2643 Pinckernelle, Erdewin (tr.) *German general rules of marine insurance.* 5th rev. edn. New York: De Gruyter, 1991. 77 pp.

2644 Rodda, William H. *Marine insurance: Ocean and inland.* 3rd edn. New Jersey: Prentice-Hall, Inc., 1970. 522 pp.

Covers common law and statute of marine insurance and maritime law in relation to liabilities of land and sea carriers and their insurance structure. Includes a comprehensive presentation of inland marine insurance. Bailee liability. Risks faced by insurance companies. Includes risks in covering diamonds, furs and fine arts.

2645 Rogers, Nigel C. and **Ahern, John R.** *A guide to marine hull insurance claims.* Richard Hoggs Ltd, 1989. 101 pp.

Explains various types of marine hull insurance claims which might arise under standard insurance conditions. Steps which shipowner should consider taking in anticipation of and in response to casualties giving rise to such claims. Case study in practical format.

2646 Selvig, Erling. The Hamburg Rules, the Hague Rules and marine insurance practice. *Journal of Maritime Law and Commerce* 12(3), April 1981, 299–326.

Analysis to establish factual basis for appraisal of Hamburg Rules in respect of marine insurance claims as compared with the law based on Hague and Hague-Visby Rules.

2647 Seward, R. C. The insurance viewpoint. In: Institute of Maritime Law, University of Southampton. *Limitation of shipowners' liability: The new law.* London: Sweet & Maxwell, 1986, pp. 161–86.

Principle of limitation of liability. Sub-standard ships. Importance of limitation in P & I insurance. Expanding liabilities for shipowners. Limitation for shipowners. Alternatives. Problems and attempts to resolve by different legal systems. Direct action against insurers. Insurer's right to deny cover. Advantages of 1976 Convention.

2648 Sturley, Michael F. Changing liability rules and marine insurance: Conflicting empirical arguments about Hague, Visby, and Hamburg in a vacuum of empirical evidence. *Journal of Maritime Law and Commerce* 24(1), January 1993, 119–51.

Focuses on insurance issues which arise in respect of modification of liability regime. Impact of Hamburg Rules on overall insurance costs. Theoretical overview. Incentives for loss avoidance. Ambiguity and 'friction' costs. Shifting the burden of insurance. The empirical problem.

2649 Terry, J. P. D. *Marine account management and reinsurance.* London: Chartered Insurance Institute, 1991. 155 pp.

Coursebook for C.I.I. Functions of reinsurance, why reinsure? Use and underwriting of marine insurance. Reinsurance contracts. Non-proportional treaty wordings. Excess of loss. Constructing the programme. Accounting. Reinsurance claims. Practical examples.

2650 Turner, Harold A. *The principles of marine insurance: A primer.* 7th edn. London: Stone and Cox, 1986. 236 pp.

2651 United Nations Conference on Trade and Development Secretariat. *Legal and documentary aspects of the French Latin American marine insurance legal regimes: Report.* New York: United Nations, 1983. 39 pp.

2652 Winter, W. D. *Marine insurance.* 3rd edn. McGraw, 1952.

Premiums

2653 Beenstock, Michael and **Khajuria, Sajay.** The determination of marine insurance premiums: An international analysis, 1963–1981. *Maritime Policy and Management* 13(1), January-March 1986, 45–52.

Paper examines determinants of marine insurance premiums. Proposes theory of supply and demand for marine insurance from which theoretical determinants of demand for premiums made.

2654 Great Britain. Department of Transport. *Reinsurance of British and other ships against war risks.* London: HMSO, 1988.

2655 Proctor, N. C. Marine insurance. *Maritime Policy and Management* 12(1), 1985, 61–6.

History, definition, mode of operation, benefits, costings, of marine insurance.

Frauds

2656 Asokan, Govindarajalu. The insurance of maritime fraud: The law in Singapore. In: *Leading developments in international marine insurance: An industry report.* London: Lloyd's of London Press, 1991. 15 pp.

Disputes concerning marine policies. Uncertainties pertaining to insurance of maritime fraud.

2657 Kubota, Takeo. Comparison of a subrogation claim between Japanese and English law. In: *Leading developments in international marine insurance: An industry report.* London: Lloyd's of London Press, 1991. 4 pp.

To prevent assured from obtaining duplicated benefit in subject matter insured.

Protection and indemnity (P & I)

2658 Goghlin, T. G. Protection and indemnity insurance: The P & I clubs. *Journal of World Trade Law* 5(6), November/December 1971, 591–608.

Risks carried by P & I Clubs. Concept of mutuality. Liability of members. Scale of operations. Extent of claims (with special reference to oil pollution). Supplementary services.

2659 Goldie, C. W. H. Effect of the Hamburg Rules on shipowners' liability insurance. *Journal of Maritime Law and Commerce* 24(1), January 1993, 111–18.

Hamburg Rules and shipowners' liability insurance. Analysis of claims paid by a P & I Club.

2660 Hazelwood, S. J. *P. & I. Clubs: The law and practice.* London: Lloyd's of London Press, 1989. 366 pp.

2661 Hill, C., Robertson, B. and **Hazelwood, S. J.** *An introduction to P & I.* London: Lloyd's of London Press, 1988. 176 pp.

Comprehensive and practical handbook on specialized world of P & I insurance. Concise outline of origin and history of the clubs. Review of modern day clubs – constitution; membership; management. Offshore associations. Underwriting and call; costs; renewals; loss ratios; club funds; reserves. Specific risks; exceptions; exclusion. Claim handling.

2662 Houdlett, Ronald R. Direct action statutes and marine P & I insurance. *Journal of Maritime Law and Commerce* 3(3), April 1972, 559–72.

Applicability of state 'direct action statutes' to marine indemnity policies. Whether the direct action statute can reach a marine indemnity insurer. *Maryland Casualty Co.* v. *Cushing* (1954); *Wilburn Boat Co.* v. *Fireman's Fund Insurance Co.* (1955).

2663 Kieding, Nicolay. *Introduction to P & I Insurance.* Oslo: Assuranceforeningen Skuld, 1975. 186 pp.

An introduction to P & I insurance, dealing with terms of cover with comments and practical examples. Based on lectures and discourses, chiefly on Norwegian statutes and legal wage.

2664 Kingsley, Jeremy. *Braekhus and Rein handbook on P & I insurance.* 3rd edn. Oslo: Assuranceforeningen Gard, 1988. 496 pp.

Commentary to the Statutes and Rules of Assuranceforeningen Gard. A practical guide to the maritime laws which form the background to P & I insurance.

2665 Leader, P. G. F. Protection and indemnity insurance. *Maritime Policy and Management* 12(1), 1985, 71–89.

History and present position regarding insuring ships and contents for loss of or damage to cargo, and for third party risks required by legislation.

2666 Ledwith, Frank. *Ships afloat in the city.* London: Robert Hale, 1977. 203 pp.

A second book (follows *Ships that go bump in the night*) of personal reminiscences of the author's work in a P & I Club. Considers a variety of issues including attitudes of trade unionists – asks what is the real menace of sea pollution, considers solutions to the energy problem.

2667 Ledwith, Frank. *Ships that go bump in the night.* London: Robert Hale, 1974. 223 pp.

Personal reminiscences of the author's work in a P & I Club. Assesses the value of financial institutions' functions. Management training. Industrial relations. Some historic disasters at sea and the legal battles which followed.

2668 Lemon, Robert T. The Lanasse Rule and the additional assured's dilemma under American-Form P & I policies. *Journal of Maritime Law and Commerce* 21(4), October 1990, 503–43.

Explores the mechanics of the rule from *Lanasse* v. *Travelers Insurance Co.* (1971) where the courts adopted a constrictive scope of the P & I policy's insuring clause, refusing to afford the additional assured broader policy protection than that provided to named assured.

2669 Ueland, Svein. P & I insurance issues and claims handling: Control in the mass casualty cruise ship setting. In: *Leading developments in international marine insurance: An industry report.* London: Lloyd's of London Press, 1991. 7 pp.

Issues arising from the fire on board the *Scandinavian Star*, 7 April 1990.

Brokers' education

2670 *Training notes for brokers: Introduction to marine insurance.* London: Witherby & Co., 1987. 166 pp.

Hull

General

2671 Brown, Robert H. *Analysis of marine insurance clauses. Book 2: The Institute Time Clauses: Hulls, 1983.* London: Witherby & Co., 1984. 50 pp.

Analysis of hull time clauses covering hull and machinery on full conditions. Clauses C1 280. Institute Time Clauses – Hulls 1.10.83. The purpose of a particular clause and its effect on the cover provided.

2672 Goodacre, J. Kenneth. *Institute time clauses. Hulls: A comparison between the 1983 clauses and the 1970 clauses with subsequent amendments.* London: Witherby & Co., 1983. 67 pp.

Explains differences in hull insurance brought about by the revised wording and changes in format of the new cargo clauses introduced in January 1983 for use with the new Marine Policy Form.

2673 Hicks, R. M. The American Institute Hull Clauses. *Journal of Maritime Law and Commerce* 2(4), July 1971, 787–813.

Examination of revision of the 'American Institute Time (Hulls), January 1, 1964 form'.

2674 Massenburg, J. W. *Massenburg's handbook on marine hull insurance for operating personnel.* London: Institute of Marine Engineering (Marine Management), 1983. 94 pp.

Terms and definitions which are necessary for the proper preparation and evaluation of potential hull claims. General guide to approaches that must be taken.

War risk

2675 McDowell, Carl E. Containerisation: Comments on insurance and liability. *Journal of Maritime Law and Commerce* 3(3), April 1972, 503–12.

Examination of insurance purchased by shippers and consignees. Variations in exposure of cargo owner relying solely on carrier liability to protect goods during multimodal through shipment. Legal problems; major legal liability limits.

2676 Miller, Michael D. *Marine war risks.* London: Lloyd's of London Press, 1990. 369 pp.

Explores war risk insurance of hull and freight which protects shipowner from consequences of what other people do to his ship. History; perils; war risk cover; judicial process; legislation; held covered; hostilities; war-like operations; insurance in time of war.

2677 Reynardson, W. Birch. The reaction of Protection and Indemnity

Associations to through transit risks. *Journal of Maritime Law and Commerce* 2(4), July 1971, 837–41.

Formation of the Through Transit Association, June 1979. Manner in which Rules of P & I Clubs extended.

Cargo

General

2678 Brown, Robert H. *Analysis of marine insurance clauses. Book 1: The Institute Cargo Clauses, 1982.* 2nd rev. edn. London: Witherby & Co., 1983. 59 pp.

Analysis of cargo clauses, published by Institute of London Underwriters for use of cargo insurers with effect from 1 January 1982. Embraces all clauses issued under that date and published at time of writing.

2679 Brown, Robert H. *Cargo insurance: The Institute cargo clauses.* London: J. H. Minet & Co., 1977. 16 pp.

Cargo insurance as it relates to the Institute Cargo Clauses. The basic cover and difference between (i) Institute Cargo Clauses – All Risks; (ii) Institute Cargo Clauses – FPA; (iii) Institute Cargo Clauses – WA.

2680 Brown, Robert H. *Marine insurance. Vol. 2: Cargo practice.* 4th rev. edn. London: Witherby & Co., 1985. 479 pp.

Cargo insurance and cargo practice. Contracts of carriage. London market conditions. Practical claims and advice. Trade clauses; comparison charts. Comparison of SG form of policy (1963 clauses) with cover under MAR policy (1982 clauses).

2681 Brown, Robert H. and **Reed, Peter B.** *Marine reinsurance.* London: Witherby & Co., 1981. 343 pp.

Published for marine practitioner. Direct insurance that underlies reinsurance protection. Relationship between direct risk and reinsurance protection. How and why action necessary. Entire field of non-proportional and proportional marine insurance. Treaty accounting.

2682 Kotsovilis, V. The revised law of cargo marine insurance: The Institute Cargo Clauses (A), (B) and (C) – 1/1/82. Thesis submitted for PhD, University of Essex, 1988.

2683 Rossmere, Andrew E. Cargo insurance and carriers' liability: A new approach. *Journal of Maritime Law and Commerce* 6(3), April 1975, 425–34.

Questions whether interrelated system of cargo insurance and carrier liability in fact best for parties involved. Discusses whether introduction of strict and full carrier liability would be better.

Containers

2684 Bannister, J. E. Containerisation and marine insurance. *Journal of Maritime Law and Commerce* 5(3), April 1974, 463–81.

Examines problems in respect of insurance and liabilities for loss or damage with containerized shipments. Insurance implications. Technical changes. Stranding. Heavy weather. Collision. Fire. Insurance of containers. General average. Salvage. Types of loss/damage.

2685 Insurance Institute of London. *Containerisation and marine insurance.* London: Insurance Institute of London, 1972. 80 pp.

A document which looks at insurance of containerization transport in broad terms. The container business. Overseas trade. Liabilities. Casualties – hulls and containers. Cargo casualties. Loss prevention.

Freight

2686 Denny, Maurice E. V. *Freight insurance: A commentary. (The insurance of shipowners' interest in freight for the carriage of goods.)* London: Witherby & Co., 1986. 237 pp.

Summary of current conditions in difficult area of freight insurance. Practical examples of calculations involved. Important judgments. Critical appraisal of modern procedures, with consideration of revised standard freight clauses.

14 Flag and Fleet Development

General

General policy and theory

2687 Canna, Elizabeth. Shipping developments. *American shipper*, December 1992, 43–50, 55, 61–2.

2688 Drewry Shipping Consultants. *The rise of national fleets.* London: Drewry Shipping Consultants, 1973. 37 pp.

Report – part of series covering world's bulk shipping. The fleets and international seaborne trade of the developing countries. Government measures to encourage national fleets. Rise of national fleets of specific countries. Perceived growth of national flags.

2689 Nilsson, Bertil. *Fleet development and new building requirements.* Finspong, Sweden: Stal-Laval Turbin A. B., 1976. 42 pp.

A marine market analysis. Part of series of annual forecasts. Quantifies the future markets for medium to high power marine propulsion machinery, the plus 10 MW range. Considers: Container ships of/more than 5,000 d.w.t.; bulk carriers and tankers of/more than 50,000 d.w.t.; gas carriers of/more than 20,000 d.w.t.

2690 Rosenstein, E. and **Goldstein, A. M.** (eds) *Organizational changes in shipping in Europe and Israel. Symposium Proceedings, November 17, 1981.* Haifa: Israel Shipping Research Institute, 1983. 114 pp.

Papers and transcripts of discussions from the symposium at Haifa University, November 1981. Explores questions in respect of the technological changes in the shipping industry and impact on shipping industry.

2691 Serafetinidis, M. and others. The development of Greek shipping capital and its implications for the political economy of Greece. *Cambridge Journal of Economics* 5(3), September 1981, 289–310.

This paper attempts to demonstrate the serious misconceptions as to the nature and development of Greek shipping capitalism; the significance of US assistance to the Greek maritime industry is also assessed.

2692 Sletmo, G. K. Shipping's fourth wave: Ship management and Vernon's trade cycles. *Maritime Policy and Management* 16(4), 1989, 293–303.

If shipping recovers in 1990s will it revert to pre-crisis hegemony of OECD shipping countries, or has it changed irrevocably, with new players replacing the old?

2693 *The U.N. Ship Registration Convention: Twelve months on.* Oslo: INTERTANKO, 1987. 36 pp.

2694 *United Nations Conference on conditions of registration of ships: A free market perspective.* Oslo: INTERTANKO, 1984. 12 pp.

2695 United Nations Conference on Trade and Development. *National shipping lines.* New York: United Nations, 1975.

United Kingdom

2696 Anon. Discussion review: Venture capital in British shipping: A discussion organized by the UK Centre for Maritime Policy Studies. *Maritime Policy and Management* 18(1), 1991, 11–14.

Asks should apparent shortage of venture capital and private direct investment in UK shipping be researched under auspices of UK Centre for Maritime Policy Studies.

2697 Bonwick, George J. *British shipping: An independent study.* Henley-on-Thames: Fennbond. 48 pp.

Examines ascent of British shipping and its long relative descent towards end of the nineteenth century. The post-war years. UK shipping today.

2698 British Maritime Charitable Foundation. *Merchant fleet and Britain's economy.* London: British Maritime Charitable Foundation, 1988. 100 pp.

Examines the economic effects of possessing a merchant fleet under the headings: value added; importance of maritime skills to the wider economy; threats, opportunities and influence; linkages with other industries; effect on employment.

2699 British Maritime Charitable Foundation. *Why the ships went: Fully quantified analysis of the causes of decline of the United Kingdom registered merchant fleet.* London: Lloyd's of London Press, 1986. 99 pp.

Examines the decline of UK registered fleet – route by route, trade by trade and ship by ship to evaluate and quantify the many factors at work in each sector of the industry. Restricted to commercial shipping. Covers period 1978–83 but includes post-1983 developments and owners' intentions where such were available.

2700 British Maritime Charitable Foundation. *Why the ships went: Fully quantified analysis of the causes of decline of the United Kingdom registered merchant fleet. Suppt: 1983–87. Report including comparison with 1978–83.* London: British Maritime Charitable Foundation, 1988. 41 pp.

2701 The changing shape of British shipping. *Midland Bank Review,* November 1970, 15–22.

This argues that compared with other activities shipping is not particularly profitable, and that despite the introduction of special arrangements the industry will probably have to continue to look mainly to internal funds for financial development. The 1960s is seen as a decade of rapid change in overseas transport technology.

2702 Davies, Michael. *Belief in the sea: State encouragement of British merchant shipping and shipbuilding.* London: Lloyd's of London Press, 1992. 371 pp.

Brings together maritime policy over the last five centuries. Detailed account of support given to shipping and shipbuilding industries by successive British governments – from the protectionist policies through the era of free trade, to the *laissez-faire* position at time of writing.

2703 Galbraith's Shipping Advisory Services. *British flag: An outline of its scope and advantages.* London: GSAS, 1977. 4 pp.

Outlines the workings and relative advantages of British flag operation.

2704 Garratt, M. G. and **Wingfield, M.** *An analysis of shipping to the British Isles.* Liverpool: Marine Transport Centre, University of Liverpool, 1981. 128 pp.

Examines data drawn from that collected by Lloyd's List on ship arrivals and departures at ports, by Lloyd's Register on the quantitative description of each ship and by a variety of commercial registers of unitized ships. Supply and productivity of ships, relationship between ship characteristics, demands upon British ports.

2705 General Council of British Shipping. *The future of the British merchant fleet: An analysis of policy options.* London: General Council of British Shipping, 1987. 31 pp.

A paper setting out the factors which bear on national decisions affecting Britain's merchant fleet, and analysing the options open to Government to influence the size, composition and commercial viability of that fleet in the future.

2706 Goss, R. The decline of British shipping: A case for action? A comment on 'The decline of the UK merchant fleet: An assessment of government policies in recent years'. *Maritime Policy and Management* 20(2), April–June 1993, 93–100.

Comment on article by Ledger and Roe, in the same journal, 19(3), 1992, pp. 239–51. Aim is to clarify arguments rather than to support or oppose any particular case. Comparative advantages of tradable services. Balance of payments. Employment, training. Fostering trade. Taxation. Collective utility. Defence.

2707 Great Britain. Department of Trade and Industry. *Report on the U.K. merchant fleet, world shipping on seaborne trade.* London: HMSO, 1979.

2708 Great Britain. Department of Transport and **General Council of British Shipping.** *British shipping: Challenges and opportunities.* London: HMSO, 1990. 71 pp.

The world markets and the British interest. Obstacles and options. The competitive environment. Maritime standards. New initiatives on training. Flag, ownership and management – the benefits to the UK.

2709 Greenwich Forum Conference on Britain and the sea. Britain and the sea: The challenges for shipping in the 1990s [papers presented at the Greenwich Forum Conference, Royal Naval College, Greenwich, 23, 24 and 25 April 1980]. Unpublished. 106 pp.

Papers from the international conference, London 1980. Considers trade and shipping patterns, potential threats to established shipping, maritime and technological developments, economics. Background papers on the US and USSR and CMEA. Reports of discussion groups.

2710 Moreby, David Henry and **Springett, Peter.** *United Kingdom shipping industry: Critical levels study.* London: British Maritime Charitable Foundation, 1990. 100 pp.

2711 Ranken, M. B. F. *The Greenwich Forum IX. Britain and the sea: Future dependence – future opportunities.* Edinburgh: Scottish Academic Press, 1984. 364 pp.

Conference papers delivered 14–18 September 1983, at the Greenwich Forum's ninth conference at the Royal Naval College, London. Reviews the state of Britain's commitment to marine activities – their supporting services, examining those which are in difficulty or decline.

2712 Shore, P. Sunset over the Red Ensign: The decline of British deepsea shipping 1945–1989. Thesis submitted for PhD, University of Kent at Canterbury, 1990.

It argues that in analysing the industry's poor performance it is necessary to divide the period into three distinct phases – post-war reconstruction without rapid innovation (1945–65); rapid technological and market developments (1966–73); and severe, prolonged depression (1974–89). Methodologically, therefore, explanations valid in one phase need not apply throughout the whole period.

2713 Sturmey, S. G. 'British shipping and world competition' revisited. *Maritime Policy and Management* 18(4), 1991, 263–82.

Author's book, published in 1962, tried to account for decline in tonnage of UK registered shipping. Reviews chapters on applied economics and sees decline in terms of internal constraints.

Western Europe and Scandinavia

2714 Gardiner, Paul. Greece: Shipping enters uncertain waters. *Euromoney,* September 1990, 50–57.

With the move from the Greek national flag to cheaper registers, the Greek shipping industry is under threat. The pressure for change is coming from the banks, although shipowners are reluctant to change from the traditional ways.

2715 No entry.

2716 Honko, Jaakko. What role for Finland and her shipping industry in the world economy? *Liiketaloudellinen-Aikakauskirja* 32(4), 1983, 305–16.

2717 Line, G. The use of futures and options in the Norwegian shipping industry. Thesis submitted for MSc, Heriot-Watt University, Edinburgh, 1991.

This looks at futures and options and how these instruments can be used in the shipping industry. It attempts to remove misunderstandings and myths about these financial derivatives. It concludes that shipowners should be more active in hedging their portfolios as freight rates and fuel oil costs are often exposed to risk.

2718 Los, Matheos D. *Post-war Greek shipping: The road to the top.* Piraeus, 1991. 48 pp.

Lecture presented in November 1990 at the Onassis Center of Hellenic Studies of New York University. Developments in the Greek maritime industry since the Second World War.

2719 Mikelis, N. E. and **Dean, C. J.** *The fleet controlled by the European Community: Present status and some historic trends.* (A paper for the 5th International Congress of the International Maritime Association of East Mediterranean – Athens 28 May–1 June 1990.) London: Lloyd's Register of Shipping, 1990.

Examines the past and present status of the fleets of the EC nations. Historic development of (i) fleets flying national flags of Member States; (ii) the composite EC-flagged fleet. Comparison with fleet development of other maritime nations and also FOC fleets. Fleets owned by EC parent companies. Ship numbers, gross and d.w.t.

2720 Netherlands Maritime Institute, Institute of Shipping Economics, Institute for Shipping Research, Maritime Institute. *Maritime research and European shipping and shipbuilding, 1978. International symposium – papers and discussions.* Rotterdam: 1978.

2721 Norwegian Trade Council. *Norway: Your maritime partner.* London: Industry and Investor Relations. 56 pp.

Magazine to provide independent view of Norwegian maritime industry. Relationship with the I.M.O. Integration with European Community. Safety. The environment. The move to integrated systems. Coping with recession.

Central and Eastern Europe and Middle East

2722 Antoun, Y. Y. Shipping and the development of the Gulf Arab region. Thesis submitted for PhD, UWIST, Cardiff, 1980.

2723 **Armstrong, Terence.** Soviet ideas for pioneering shipping in the Central Arctic Basin. In: Johannes F. Tismer, John Ambler and Leslie Symons (eds) *Transport and economic development: Soviet Union and Eastern Europe.* Berlin: Duncker and Humblot, 1987, pp. 175–84.

2724 **Azzee, B. B. H.** Shipping and the development of Saudi Arabia. Thesis submitted for PhD, UWIST, Cardiff, 1981.

2725 **Azzee, Baha Bin Hussein.** Shipping and development in Saudi Arabia. Thesis submitted for M.Phil, UWIST, Cardiff, 1980. 359 pp.

Discusses process of economic development in Saudi Arabia and the factors which require to be added to the development plans. Argues that a comprehensive shipping industry is required. Identifies the main constraint as manpower for maritime occupations and discusses how to overcome this in a practical way.

2726 **Bergstrand, S. J.** Soviet merchant shipping: Some European attitudes explored. *Marine Policy* 14(2), March 1990, 112–19.

Traces and comments upon some of the Western European views of Soviet shipping expressed in recent years and examines the issues behind them. Focuses on liner shipping, the cruise market and the Soviet bilateral trades.

2727 **Bergstrand, Simon** and **Doganis, Rigas.** *The impact of Soviet shipping.* London: Allen & Unwin, 1987. 184 pp.

Detailed analysis of the operations and impact of Soviet merchant shipping. Draws on discussions with Soviet shipping managers. Examines economic and commercial basis of Soviet shipping operations.

2728 **Blake, Gerald H.** and **Davies, Marian** (eds) *Maritime aspects of Arabian geopolitics.* London: Arab Research Centre, 1982. 36 pp.

2729 **Bohme, Hans.** The organisation, management system, and market behaviour of the Soviet merchant marine. In: Curtis Cate (ed.) *The challenge of Soviet shipping.* London: Aims of Industry (in association with the National Strategy Information Center, New York), 1985, pp. 17–24.

Examines shipping as part of the Soviet economy. The organization of the Soviet maritime complex. International co-operation of Soviet shipping. Management procedures in the Morflot organization. A supplier, *sui generis,* of shipping services.

2730 **Buxton, I. L.** and **Akgul, B. U.** The comparison of general cargo ship economic performance by simulation. *Maritime Policy and Management* 16(1), 1989, 27–44.

Compares by simulation a number of multi-purpose general cargo vessels suitable for the trade of a Turkish shipowner. Includes variables such as ship size, speed, cost, fuel use, etc.

2731 **Cate, Curtis** (ed.) *The challenge of Soviet shipping.* London: Aims of Industry (in association with the National Strategy Information Center, New York), 1985. 46 pp.

Book of papers covering Soviet shipping. The nature of the challenge. The threat of the Soviet merchant fleet. How Soviet seamen live. Some Morflot subsidiaries. The US Longshoremen's boycott. Threat of Trans-Siberian Railway. Soviet oceanographic efforts. The Soviet maritime offensive. Morflot and the KGB.

2732 **Curtis, Jeffery** and **Greenway, Ambrose.** *Soviet merchant ships.* 6th rev. edn. Emsworth, Hants: K. Mason, 1990. 288 pp.

2733 **Dale, M.** The Soviet merchant marine: An historical, commercial and military strategic study. Thesis submitted for MSc, UWIST, Cardiff, 1978.

2734 **Dedial, Jurg.** The threat of the Soviet merchant fleet. In: Curtis Cate (ed.) *The challenge of Soviet shipping.* London: Aims of Industry (in association with the National Strategy Information Center, New York), 1985, pp. 6–9.

Considers the expansion of Soviet naval forces and of Soviet merchant marine, fishing and research fleets. Aims of Soviet planners. Impact on the West.

2735 **Din, M. A. E.** The transport importance of the Arabian (Persian) Gulf. *Transport Reviews* 10(2), April–June 1990, 127–48.

Development of the Gulf region and the importance of transport, especially sea transport, to the region and to the rest of the world.

2736 **Greenway, Ambrose.** *Comecon merchant ships.* 4th rev. edn. Emsworth, Hants: K. Mason, 1990. 190 pp.

2737 **Guzhenko, T.** Soviet merchant marine and world shipping. *Marine policy* 1, 1977, 102–9.

2738 Haellquist, Karl Reinhold (ed.) *Asian trade routes: Continental and maritime.* London: Curzon Press, 1991. 298 pp.

2739 Hanson, Philip. The Soviet Union and world shipping. *Soviet Studies* 22, July 1970, 44–60.

2740 Hayman, Christopher (ed.) *Arab shipping: Poised for take off. A seatrade study.* Colchester: Seatrade Publications, 1975. 36 pp.

Examines developing horizon of Arab involvement in shipping. Sizing up the potential tanker fleet. Influence on shipping demand of money from oil market – emergence of new industrial markets/dry cargo shipping. North African States. Congestion and development of ports. Repair projects/labour supply.

2741 Koide, Shuzo. *Behaviors of Soviet shipping and its business management.* Tokyo: Japan Maritime Research Institute, 1989. 89 pp.

Examines organization and management of Soviet shipping. Soviet external trade. Composition of the Soviet merchant fleet and status of its vessel employment. Economic foundation of Soviet shipping enterprises.

2742 Lambert Brothers Shipping. *Russian shipping: Background data and analysis.* London: Lambert Brothers Shipping, 1976. 56 pp.

A study which analyses background data and information on the expansion of the Russian fleet of general cargo vessels. Employment of fleet in cross trades. Absence of some or all of the usual commercial criteria in operation of Russian merchant ships. Threat to Western shipping.

2743 Levikov, G. Financing of Soviet shipping companies under perestroika. *Maritime Policy and Management* 19(1), 1992, 63–8.

On sources of finance for USSR shipping companies, including their foreign currency funds, loans and centralized sources. Activities of MORBANK for commercial shipping described.

2744 Long, D. M. *Soviet merchant fleet: Its growth, strategy, strength and weaknesses, 1920–99. A special report.* London: Lloyd's of London Press, 1986. 163 pp.

Report of period of socialist maritime growth. Review of Soviet tonnage, route growth, etc. Covers economic and political strategies leading Soviets to invest heavily in maritime transport. Operating cost factors. Soviet perceptions of profit and loss.

2745 Long, D. M. A study of the development and impact of the

Soviet Baltic Steamship Company. Thesis submitted for M.Phil, CNAA (Plymouth Polytechnic), 1985.

2746 Lydolph, Paul E. Soviet maritime transport in an overcrowded market. In: Johannes F. Tismer, John Ambler and Leslie Symons (eds) *Transport and economic development: Soviet Union and Eastern Europe.* Berlin: Duncker and Humblot, 1989, pp. 140–74.

2747 Maslov, G. A. Soviet liner shipping in the 80's. In: Hans Ludwig Beth (ed.) *Liner shipping in the eighties: Report on the International Symposium, Bremen, October 24th–26th 1979.* Bremen: Institute of Shipping Economics, 1980, pp. 83–97.

Looks at developments in international liner shipping in the 1970s. Development of Soviet foreign trade and merchant marine in the 1970s. Carriage of general cargoes between USSR and non-socialist countries.

2748 Nagatsuka, Seiji. *Impacts caused by the collapse of the former Soviet Union on the world shipping and shipbuilding: Changes in CIS' shipping and shipbuilding and their future.* Tokyo: Japan Maritime Research Institute, 1992. 48 pp.

The roles played by former Soviet Union in the world economy and ship-demanding industries. Likely changes. Changes in the shipping of the former Soviet Union. Present state of shipbuilding and its future. Other problems.

2749 Peugniez, Dominique. The Soviet maritime threat: What to do. In: Curtis Cate (ed.) *The challenge of Soviet shipping.* London: Aims of Industry (in association with the National Strategy Information Center, New York), 1985, p. 46.

Questions how expansion of Soviet merchant marine has affected Europe. What threat to the West?

2750 Sawiczewska, Z. Reconstructing Polish ports and shipping. *Maritime Policy and Management* 19(1), 1992, 69–76.

Aspects include: Changing ownership and structure of companies; modernization of tonnage; improvement of infrastructures; proper manning policies in fleets; better management of ports and ships.

2751 Seatrade Publications. *Soviet shipping.* Colchester: Seatrade Publications, 1976. 32 pp.

Examines development of Soviet shipping and impact upon world shipping market. Considers signs that future confrontation can be avoided.

2752 Swayne, Ronald. Introduction: The nature of the challenge. In: Curtis Cate (ed.) *The challenge of Soviet shipping.* London: Aims of Industry (in association with the National Strategy Information Center, New York), 1985, pp. 3–5.

Introductory paper to a set dealing with Soviet shipping. Expansion of Soviet shipping. Price structure. Subsidies.

2753 Wojewodka, C. Fleet statistics in socialist countries: A review. (Commentary.) *Maritime Policy and Management* 10(1), 1983, 67–72.

Considers characteristics and scope of the various published fleet statistics in socialist countries.

North America

2754 American Bureau of Shipping. *The history of American Bureau of Shipping on its one hundred and twenty-fifth anniversary.* New York: American Bureau of Shipping, 1987. 64 pp.

Sets out the story of how the American Shipmasters' Association, a small organization founded in the clipper ship era, participated in and stayed at the forefront of changing times to emerge as the American Bureau of Shipping.

2755 Blackwell, Robert J. Implementation of the Merchant Marine Act of 1970. *Journal of Maritime Law and Commerce* 5(2), January 1974, 165–81.

Examines implementation of programme contained in the Merchant Marine Act 1970 – introduced to revitalize US maritime industry.

2756 Donn, Clifford B. Foreign competition, technological change, and the decline in U.S. maritime employment. *Transportation Journal* 27(4), Summer 1988, 31–41.

2757 Heaver, Trevor D. National flag shipping: An appraisal of policy options from a Canadian perspective. *Maritime Policy and Management* 10(3), July–September 1983, 199–206.

Review of developments in Canadian government policy to promote investment in Canadian shipping. Fiscal incentives. Effect on use of capital and other resources. Arguments for expanding national shipping.

2758 Kilgour, John G. *The U.S. merchant marine: National maritime policy and industrial relations.* New York: Praeger, 1975. 231 pp.

A study which explains the view that the US merchant marine has been rendered non-competitive by a misdirected national maritime policy. Concentrates on the US-flag commercial ocean-shipping industry. Focus is on 1965–70.

2759 Marti, B. E. Patterns of United States–Canada maritime container flows. *Maritime Policy and Management* 8(4), 1981, 253–9.

Identifies origins and destinations of containerized movements between USA and Canada during 1977. Also examines nationality or flag of carrier involved.

2760 Whitehurst, Clinton H. *United States merchant marine: In search of an enduring maritime policy.* Annapolis (Maryland): Naval Institute Press, 1984.

Japan

2761 Yoshida, Shigeru. *Outlook of Japan's steam coal trade and her merchant shipping.* Tokyo: Japan Maritime Research Institute, 1984. 28 pp.

Energy situation, coal situation in the world. Long-term outlook of steam coal trade in the world. Japan's steam coal import, its present and future. Development of steam coal utilizing techniques. Marine transport of coal. Needs for steam coal and demand.

Socialist Asia

2762 Chinese Institute of Navigation. *Ships of China.* Beijing: The Chinese Institute of Navigation, 1988. 261 pp.

A comprehensive collection of the types of Chinese ships known in history. About 1,000 pictures showing the history of navigation and shipbuilding in China.

2763 Drewry Shipping Consultants. *The role of China in international seaborne trade and shipping.* London: Drewry Shipping Consultants, 1979.

2764 Liang, G. Y. The development of Chinese shipping policy. Thesis submitted for PhD, UWIST, Cardiff, 1988.

This explores the distinct development of Chinese shipping policy from remote antiquity to the present day. It further suggests developments in the existing shipping policy which can be used in forecasting.

2765 Oiyu, Wei. The development of water transport in China. *Transport Reviews* 7(1), January–March 1987, 1–16.

General picture of present situation and development of water transport in China. Size of problems, organizations concerned and achievements since 1949. Considers introduction of ocean-going and coastal shipping, ports, inland water transport, other equipment.

2766 Sokoloff, Valentin A. *Ships of China.* Sokoloff, 1982. 54 pp.

2767 Song, Yann-Huei Billy. Shipping and shipbuilding policies in P.R. China. *Marine Policy* 14(1), January 1990, 53–70.

Examines the development of China's policies in the shipping and shipbuilding industries.

2768 Tampoe, Moira. *Maritime trade between China and the West.* British Archaeological Reports, 1989.

2769 Wan, H. Organisation and management of Chinese deep-sea shipping companies: With special reference to the impact of China's reform. *Transport Reviews* 10(1), January–March 1990, 1–28.

A review of changes in the business environment of Chinese deep-sea shipping companies in light of economic reform. Organization and management of these companies. Impact of reform.

2770 Watson, Andrew (tr.) *Transport in transition: The evolution of traditional shipping in China.* Center for Chinese Studies, University of Michigan, 1972. 93 pp.

Africa

2771 Drewry Shipping Consultants. *West African liner shipping.* London: Drewry Shipping Consultants, 1979. 67 pp.

Report – part of series covering world's bulk shipping. Port and transport development. Traffic. Shipping services. Prospects.

2772 Essien, E. E. Competition between air and sea transport in the overseas trade of West Africa. Thesis submitted for PhD, UWIST, Cardiff, 1985.

This argues that during the last fifteen years, air freight has carried an increasing volume of high-bulk and low-value cargoes and is now competing effectively with liner shipping in the context of Nigeria and West Africa.

2773 Hilling, D. African development and the maritime sector. *Dock and Harbour Authority* 70, 1990, 330–35.

2774 Imoyin-Omene, O. W. Shipping and the developing countries: Cost/benefits analysis: Special reference to Nigeria. Thesis submitted for PhD, UWIST, Cardiff, 1986.

This explores the structure and growth strategies adopted by developing countries in the establishment and expansion of the merchant marine. It focuses primarily on Nigerian shipping and its background.

2775 Kokolo, T. Shipping, trade and development: Zaire's experience. In: G. C. Obiozor and others (eds) *Shipping, trade, and development in West and Central African sub-region.* Lagos: Nigerian Shippers Council, 1989.

2776 Mobereola, A. E. The establishment of a national shipping line – an asset or a liability? The case of Nigeria. Thesis submitted for PhD, UWIST, Cardiff, 1988.

This examines the performance of developing countries' national shipping lines to assess whether investment in shipping by developing countries is an asset or a liability. It reveals that for Nigeria, shipping has achieved little towards economic growth.

Central and South America

2777 Emerson, E. C. A systems approach to Caribbean transport planning. Thesis submitted for MSc, Aston University, Birmingham, 1975.

Asia, Australasia and Oceania

2778 Australian National Maritime Association. *Australian shipping: Structure, history and future.* Melbourne: Australian National Maritime Association, 1989.

Review of Australian shipping industry. Identifies the historical and structural factors which must be understood in order to implement effectively the objectives established by Government and industry initiatives of late 1980s. Benefits of reversing decline in shipping industry.

2779 **Babbage, Ross** and **Bateman, Sam** (eds) *Maritime change: Issues for Asia.* Australia: Allen & Co., 1993. 192 pp.

2780 **Brooks, Mary R.** *Fleet development and the control of shipping in Southeast Asia.* Singapore: Institute of Southeast Asian Studies, 1985. 114 pp.

2781 **Desai, H. B.** *Current issues in Indian shipping.* Columbia (Missouri): South Asia Books, 1989.

2782 **Dick, H. W.** *The Indonesian interisland shipping industry: An analysis of competition and regulation.* Singapore: Institute of Southeast Asian Studies, 1987. 217 pp.

2783 **Dick, H. W.** The rise and fall of dualism: The Indonesian interisland shipping industry. In: R. G. Garnaut and P. T. McCawley (eds) *Dualism, growth and poverty, in Australian perspectives.* Canberra: Research School of Pacific Studies, The Australian National University, 1982, pp. 349–71.

2784 **Drewry Shipping Consultants.** *Pacific Rim trade and shipping: The powerhouse of world shipping in the 21st century.* London: Drewry Shipping Consultants.

2785 **Economic and Social Commission for Western Asia.** *The possibilities of expanding the bulk fleets in Western Asia.* New York: United Nations, 1986. 161 pp.

Study prepared by Division of UN Economic and Social Commission for Western Asia in accordance with its 1984–5 programme of work. Development of world bulk seaborne trade. Development of ESCWA countries' bulk seaborne trade and bulk fleets. Future prospects.

2786 **Gino, Paul.** Domestic shipping in the South Pacific. In: Christopher C. Kissling (ed.) *Transport and communications for Pacific microstates.* Suva: The Institute of Pacific Studies of the University of the South Pacific, 1984, pp. 135–42.

Considers why problems arise in the development and maintenance of satisfactory shipping services. Examines need for efficient and reliable inter-island shipping in programme to develop export markets.

2787 **Hong Kong Register of Shipping.** *A consultative paper on the establishment of the Hong Kong Register of Shipping.* Hong Kong: Economic Services Branch, Government Secretariat, 1985. 14 pp.

Suggests the establishment of a separate and autonomous shipping register in a manner which will meet Hong Kong's specific requirements and which will be acceptable to the shipping industry as a whole, for post-1997.

2788 **Hong Kong Shipowners' Association.** *Response: To the consultative document on the establishment of the Hong Kong Shipping Register.* Hong Kong: Hong Kong Shipowners' Association, 1985. 35 pp.

Reviews principal points of submission by Hong Kong Shipowners' Association to Secretary for Economic Services' preliminary proposals for establishment of autonomous Shipping Registry in Hong Kong for post-1997.

2789 *Issues in Australian shipping. Occasional papers in Maritime Affairs: 2.* Canberra: Australian Centre for Maritime Studies, 1983.

2790 **Jon, J. S.** The development of shipping in Korea and the containerisation era. Thesis submitted for PhD, UWIST, Cardiff, 1987.

This argues that the Korean shipping industry suffers from two burdens: Heavy capital costs and high operating costs due to its uneconomic vessels. These factors are severely damaging the prospects for developing a profitable containership fleet even though the technical problems of operating container vessels have been successfully overcome through various forms of co-operation with foreign lines.

2791 **Kissling, Christopher C.** *Transport and communications for Pacific microstates.* Suva: The Institute of Pacific Studies of the University of the South Pacific, 1984. 192 pp.

A collection of essays which range over the broad spectrum of telecommunications, aviation and shipping concerns of Pacific Islanders.

2792 **Lee, Chung H.** and **Naya, Seiji** (eds) *Trade and investment in services in the Asia–Pacific region.* Boulder (Colorado): Westview Press, 1988. 216 pp.

2793 **Marsh, James Barney** (ed.) *Resources and environment in Asia's marine sector.* US: Taylor and Francis, 1992. 608 pp.

2794 Middleton, Bruce W. Development of a regional shipping line in the South Pacific. In: Christopher C. Kissling (ed.) *Transport and communications for Pacific microstates.* Suva: The Institute of Pacific Studies of the University of the South Pacific, 1984, pp. 127–34.

Examines development of new shipping line, 'The Pacific Forum Line'. Considers aims of PFL – reliable service, keeping freight rates down. Developmental needs of the small island countries. Institutional barriers to development of the island economies.

2795 Stubbs, P. *Australia and the maritime industries.* Melbourne: AIDA, 1983.

2796 Wan, H. The effects of the Chinese fiscal regime on shipping: A comparison with the UK regime. *Maritime Policy and Management* 15(4), 1988, 299–308.

Effects of fiscal systems on shipping investment using the discounted cash flow technique much used recently. Argues that Chinese situation can now be examined in same way.

Open Registry

General

2797 Bergstrand, S. J. *Buy the flag: Developments in the open registry debate.* London: Polytechnic of Central London Transport Studies Group, 1983. 77 pp.

Analyses changes in the open registry debate. Reassesses casualty statistics allowing for varying fleet composition. Brings together some crew cost data from various sources enabling possible cost savings under FOC's to be assessed.

2798 Boczek, Boleslaw Adam. *Flag of convenience: An international legal study.* Cambridge (Massachusetts): Harvard University, 1962. 323 pp.

A comprehensive discussion of problem of flags of convenience – the flags of Panama, Liberia, Honduras, as well as others. Outlines origin and growth of the practice, giving size and composition of the various fleets (at time of writing.) Considers whether FOC nationality should be recognized by international law. Labour jurisdiction.

2799 **Carlisle, Rodney.** The 'American Century' implemented: Stettinius and the Liberian flag of convenience. *Business History Review* 54(2), Summer 1980, 175–91.

2800 **Doganis, R. S.** and **Metaxas, B. N.** *The impact of flags of convenience.* London: Polytechnic of Central London, 1976. 137 pp.

Aims to provide an objective insight into the flags of convenience phenomenon in order to provide a sounder basis for the formulation of policies towards this phenomenon by governments and other bodies. Examines the impact of FOCs on pattern and structure of world shipping, and the relationship between FOCs and operating costs.

2801 **Drewry Shipping Consultants.** *Performance of 'open registry' bulk fleets.* London: H. P. Drewry, 1980. 57 pp.

Report – part of series covering world's bulk shipping. Evolution of flag of convenience concept. Growth of world bulk fleet 1970–79. Bulk fleet losses during the 1970s. Control and elimination of sub-standard tonnage.

2802 **Drewry Shipping Consultants.** *World shipping under flags of convenience.* London: H. P. Drewry Shipping Consultants, 1975.

2803 **Dunbar-Nobes, Anne C.** Vanuatu: Domestic shipping in a newly independent state. In: Christopher C. Kissling (ed.) *Transport and communications for Pacific microstates.* Suva: The Institute of Pacific Studies of the University of the South Pacific, 1984, pp. 143–56.

Examines some of the issues of restructuring of domestic shipping sector and eventual restructuring of polarized economy.

2804 **Economic and Social Committee of the European Communities.** *E.E.C. shipping policy: Flags of convenience.* Brussels: Economic and Social Committee – Press, Information and Publications Division, 1979. 166 pp.

Committee's own-initiative opinion on problems facing shipping policy, especially maritime safety. Growing importance of new shipping nations. Development of flags of convenience. Discrimination against certain flags.

2805 **Economist Intelligence Unit.** *Open registry shipping: Some economic considerations.* London: Economist Intelligence Unit, 1979. 69 pp.

2806 Egiyan, G. S. 'Flag of Convenience' or 'Open Registration' of ships. *Marine Policy* 14(2), March 1990, 106–11.

Provides a comparative examination of flag of convenience, open registration, and tax haven countries, with a view to establishing the criteria for distinguishing effectively amongst the three categories.

2807 Farrell, S. The use of flags of convenience by Latin American shipping. *Maritime Policy and Management* 11(1), 1984, 15–20.

Why some Latin American shipowners register their vessels under flags of convenience. Highlights contrasting attitudes of these shipowners and those of their respective governments.

2808 Grundy, Ken. *Flags of convenience in 1978.* London: Polytechnic of Central London, 1978. 58 pp.

Explores changes and developments in respect of impact of FOCs since 1976. Evaluates implications of such changes and developments. Progress of convenience registries. Safety and the convenience fleets. Attitudes towards FOCs.

2809 Hubbard, Vincent K. Registration of a vessel under Vanuatu law. *Journal of Maritime Law and Commerce* 13(2), January 1982, 235–44.

Development of Vanuatu flag and registry. Consideration of open registry shipping. Requirements for registration under Vanuatu flag.

2810 Kilgour, John G. Effective United States control? *Journal of Maritime Law and Commerce* 8(3), April 1977, 337–47.

Examines Effective United States Control (EUSC) programme relating to other terms that have been used, such as 'flags of convenience', 'flags of necessity', 'Pan.Lib.Hon.' fleet. Controversy surrounding EUSC programme, and importance of control of sizeable fleet of foreign-flag vessels.

2811 Lazaratos, G. Greek shipping policy and flags of convenience. Thesis submitted for M.Phil, University of London, 1972.

2812 Lloyd's of London. *Forty years of the Liberian ship register.* London: Lloyd's of London Press, 1989. 16 pp.

A *Lloyd's Ship Manager and Shipping News International* supplement. Reviews Liberian shipping register. History, services, economy, fees, examinations, profile, safety, investigations, documentation, owners' council, contacts.

2813 Lones, Trevor. Flags of convenience: Increasing threat from a rapidly growing fleet. In: *EEC shipping: A guide to the sea transport industries of the European Economic Community.* Colchester: Seatrade Publications, 1979, pp. 27–9.

Reviews problem for EEC in respect of the increasing use of FOC registries and the growing respectability of such flags. Impact on Western European shipowners and operating costs.

2814 McCaul, J. R., Phillips, S. W. and **Rechell, R. E.** *Economic impact of open registry shipping.* 1979.

2815 Metaxas, B. N. *Flags of convenience: A study of internationalisation.* Aldershot: Gower Pub. Co., 1985. 107 pp.

Discusses the hypothesis that flags of convenience lead to the formation of a regime of immunity in the shipping world and in the long run to higher real costs and freight rates.

2816 Metaxas, B. N. Some thoughts on flags of convenience. *Maritime Studies and Management* 1(3), January 1974, 162–77.

Formulates problem of FOCs; assesses its magnitude; can practical policy recommendations be made?

2817 New, Christopher. *Change of flag.* London: Bantam Press, 1990. 409 pp.

2818 OECD. OECD study on flags of convenience. *Journal of Maritime Law and Commerce* 4(2), January 1973, 231–54.

Study published as Part V 'Maritime Transport 1971' by OECD. Growth of FOC fleets. Characteristics. Advantages to shipowners of operating under FOC. Disadvantages. Consequences to world shipping of large FOC fleets. International measures to counter registration under FOC.

2819 Packer, Bernard. *Flags of convenience.* New York: Ticknor and Fields, 1992.

2820 Ready, N. P. *Ship registration.* London: Lloyd's of London Press, 1991. 216 pp.

Provides a concise overview of ship registration, its fundamental concepts, factors governing a shipowner's choice of flag and the impact of registration formalities on ship sale and purchase transactions. Emergence of flags of convenience, offshore national registries, dual registration. Major flag registration procedures.

2821 Schmidt, Folke. *Ships flying flags of convenience.* 1972.

2822 Sturmey, S. G. *The open registry controversy and the development issue.* Bremen: Institute of Shipping Economics, 1983.

2823 Thompson, Dennis. UNCTAD: Flags of convenience. *Journal of World Trade* 15(5), September/October 1981, 466–8.

Committee on shipping resolution on flags of convenience. Reasons for non-observance identified.

2824 Tolofari, S. R. Open registry costs and freight rates: Are they related? *International Journal of Transport Economics* 14(1), 1987, 85–103.

2825 Tolofari, S. R. *Open registry shipping: A comparative study of costs and freight rates.* Switzerland: Gordon B., 1989. 400 pp.

2826 Tolofari, S. R. Open registry shipping: An econometric study of the cost structure of open registry shipping and its impact on freight rates. Thesis submitted for PhD, Loughborough University, 1986.

This provides evidence of the possible costs to international trade of phasing out open registers. Such costs of themselves do not provide an argument for retaining the system.

2827 Tolofari, S. R., Button, K. J. and **Pitfield, D. E.** Shipping costs and the controversy over open registry. *Journal of Industrial Economics* 34(4), June 1986, 409–27.

This article looks at safety from the point of view of phasing out open registry vessels. It compares the cost function of open registry ships with those of ships registered under traditional maritime nations in order to offer some insight into the possible impact of hung shipping markets or withdrawing the opportunities for shippers to elect for open registration.

2828 United Nations Conference on Trade and Development Secretariat. *Economic consequences of the existence or lack of a genuine link between vessel and flag of registry: Report.* New York: United Nations, 1977. 82 pp.

Examines consequences of existence or lack of genuine link between vessels and flag of registry. Legal aspects of the genuine link. Development of world merchant fleet – fleets registered in countries lacking genuine link. General economic considerations. Economic impact on international shipping of existence or lack of genuine link.

2829 *Which register? Which flag? Speakers' papers from a two-day conference at the Gloucester Hotel, London on 2/3 September, 1987.* London: Lloyd's of London Press, 1988.

Flagging out

2830 Egiyan, G. S. The principle of genuine link and the 1986 U.N. Convention on the registration of ships. *Marine Policy* 12(3), July 1988, 314–21.

Looks at the details concerning the principle of genuine link between a ship and the flag state. Steps taken towards codification of ship registration and the 1986 Convention.

2831 Eyre, J. L. A ship's flag – who cares? *Maritime Policy and Management* 16(3), 1989, 179–87.

Over 60 per cent of ocean trade moves under FOC. Shipowners and clients need structure assuring safety, payment, etc. All maritime nations should observe International Conventions.

2832 United Nations Conference on Trade and Development. *Beneficial ownership of open-registry fleets.* Geneva: UNCTAD, 1980.

2833 Yannopoulos, G. N. The economics of 'flagging out'. *Journal of Transport Economics and Policy* XXII(2), May 1988, 197–208.

Equilibrium in maritime transport market between established flag sector and FOC sector explained by reference to comparative levels of efficiency and real cost of labour. Established flag sector would increase if labour costs reduced.

Argument for and against

2834 Argiroffo, Enrico. Flags of convenience and substandard vessels: A review of the ILO's approach to the problem. *International Labour Review* 110(5), November 1974, 437–53.

2835 Aspinwall, M. Kuwaiti tanker reflagging: A confluence of foreign and maritime policies. *Maritime Policy and Management* 16(3), 1989, 257–62.

On 1987 reflagging of 11 Kuwaiti tankers to US, for gaining protection from Iranian attacks. Waivers allowed vessels to operate without inspection etc. Congress acted to end waivers.

2836 Liberian Shipowners' Council. *Open registries and the merchant fleets of developing countries.* New York: Liberian Shipowners' Council, 1981. 20 pp.

Paper which aims to refute some of the arguments put forward by the UNCTAD Secretariat for phasing out open registries. Suggests areas of study relevant to the development of the maritime resources of developing countries.

2837 Naess, Erling D. *The great PanLibHon controversy: The fight over the flags of shipping.* Epping, Essex: Gower Press, 1972. 175 pp.

Full-scale account of formative years of the Pan.Lib.Hon. shipping fleets. True story by author with power politics, courtroom confrontations and clash of personalities. Shows conflict of interests; financial interest on all sides.

2838 National Union of Seamen. *Flags of convenience.* London: National Union of Seamen, 1981. 16 pp.

The FOC scandal. What are FOCs? The state of the world fleet. Shipping, the UK economy and FOCs. Ship losses and casualties. International regulation. The way ahead.

2839 Rowlinson, Mervyn. Flags of convenience: The UNCTAD case. *Maritime Policy and Management* 12(3), July–September 1985, 241–4.

Paper considers opposition to FOC operation in contrast to support in Developed Market Economy Countries. Argues that polemical nature of FOC debate partly explained by DMECs' failure to appreciate reservations of maritime lesser-developed countries.

2840 Sinan, I. M. UNCTAD and flags of convenience. *Journal of World Trade Law* 18(2), March–April 1984, 95–109.

Firstly argues the use of 'flags of convenience' has in the long term been detrimental to the shipping industry, and, secondly, it considers the current work of UNCTAD and its attempts to ameliorate the situation.

2841 Thuong, Le T. From flags of convenience to captive ship registries. *Transportation Journal* 27(2), Winter 1987, 22–34.

2842 Toh, Rex S. and **Susilowidjojo, Henry.** Flags of convenience shipping in the 1980's: The American perspective. *Transportation Journal* 26(4), Summer 1987, 34–42.

2843 Trade Union Research Unit. *The position of the sea transport industry in the EEC, with special reference to flags of convenience.* Oxford: Trade Union Research Unit, Ruskin College, 1983. 74 pp.

International Transport Workers' Federation background paper submitted to the Special Hearing convened by the European Parliament Socialist Group Transport Sub-Committee, London, 22 and 23 November 1983. Reviews position of the Sea Transport Sector in Europe with particular reference to flags of convenience.

2844 United Nations Conference on Trade and Development. *Repercussions of phasing out open registry.* Geneva: UNCTAD, 1979.

2845 Vitek, J. Great flags of convenience controversy. *Contemporary Review* 227, December 1975, 293–5.

Trade union activity

2846 Donn, C. B. *Flag of convenience registry and industrial relations: The internationalization of national flag fleets.* Syracuse (New York): Le Moyne College, Institute of Industrial Relations, 1988.

2847 Donn, C. B. and **Phelan, G.** Australian maritime unions and flag-of-convenience vessels. *Journal of Industrial Relations* 33(3), 1991, 329–40.

2848 Ewing, Keith. Union action against flags of convenience: The legal position in Great Britain. *Journal of Maritime Law and Commerce* 11(4), July 1980, 503–8.

Case law development and legislative protection for trade union action against FOCs. The four torts which are protected by statutory immunity.

2849 Heinemann, Guy L. and **Moss, Donald C.** Federal labor law and the foreign-flag vessel – an inversion of the doctrine of pre-emptive jurisdiction. *Journal of Maritime Law and Commerce* 1(3), April 1970, 415–42.

Examines jurisdictional problems facing owners of foreign-flag vessels picketed by American unions.

2850 International Transport Workers' Federation. *Flags of convenience: The ITF's campaign.* London: International Transport Workers' Federation, 1993. 36 pp.

Booklet to provide seafarers of all nationalities with basic information about ITF's campaign against FOCs.

2851 Laffer, K. Australian maritime unions and the International Transport Workers' Federation. *Journal of Industrial Relations* 19, 1977, 113–32.

2852 Northrup, H. R. and **Rowan, Richard L.** *The International Transport Workers' Federation and flags of convenience.* Philadelphia: University of Pennsylvania, The Wharton School International Research Unit, 1984. 251 pp.

15 The State and Policy

General

General policy and theory

2853 **Abrahamsson, B. J.** Economics of regulation of shipping. *Maritime Policy and Management* 9(3), 1982, 219–27.

Explores nature and scope of social regulation in shipping, comparing its nature with that of economic regulation *per se.* Concept of social regulation has a valid, humanitarian base but is expensive.

2854 **Amorouayeche, F.** State intervention in shipping. Thesis submitted for MSc, University of Birmingham, 1975.

2855 **Behnam, A.** Political factors and the evolution of national fleets in developing countries. *Maritime Studies and Management* 3(3), January 1976, 131–40.

Examination of the behaviour of developing countries in relation to investment in shipping. Desire for nationalism evidenced. Theory of economic nationalism and conclusions.

2856 **Bohme, Hans.** The changing framework of shipping: Trends in trade, technology and policies. *Marine Policy* 8(3), July 1984, 229–38.

Analyses prospects and policy options for shipping in the context of three major influences: Changes in seaborne trade; technical and organizational innovation; changes in the political environment of shipping. The need to remove tonnage surplus.

2857 **Bouayad, A.** Co-operation among developing countries in world shipping. (Editorial.) *Maritime Policy and Management* 14(4), 1987, 267–9.

Co-operation not limited to resources or factor endowments and can extend to security of investments, access to markets, creation of linkages and efficient vessel use.

2858 Bouzoubaa, A. F. Tonnage acquisition and shipping operations and management in developing countries. *Maritime Policy and Management* 8(4), 1981, 229–33.

Development of merchant navies in newly independent countries has been fraught with political and economic problems.

2859 Brown, E. D. Clean seas versus freedom of navigation? British policy at Caracas. (Editorial.) *Maritime Studies and Management* 2(2), October 1974, 69–70.

Considers question of where United Kingdom stands on issue of respective roles of Flag State, Coastal State and Port State in respect of marine pollution.

2860 Cafruny, Alan W. The political economy of international shipping: Europe versus America. *International Organization* 39(1), Winter 1985, 79–119.

2861 Cafruny, Alan W. *Ruling the waves: The political economy of international shipping.* Berkeley and Los Angeles: University of California Press, 1987. 323 pp.

Explores the structure and history of international shipping to illuminate important aspects of the international political economy. International shipping regimes from sixteenth century to date of writing. How political and economic developments in late nineteenth century rendered classical shipping liberalism obsolete. Decline of international power of United States in 1960.

2862 Drewry Shipping Consultants. *Governments and dry bulk shipping.* London: Drewry Shipping Consultants, 1981.

2863 Drewry Shipping Consultants. *Shipping in the third world.* London: Drewry Shipping Consultants, 1976.

2864 Frankel, E. G. Hierarchical logic in shipping policy and decision-making. *Maritime Policy and Management* 19(3), 1992, 211–21.

Proposes hierarchical network approach to determining most effective strategy for showing effect of shipping policy alternatives on desired outcomes and impact on policy objectives.

2865 Frankel, Ernst G. Shipping and its role in economic development. *Marine Policy* 13(1), January 1989, 22–42.

Discusses the issues and options concerned with shipping and its contribution to advancement in developing countries. Considers the various aspects involved, including marketing, technology changes and finance.

2866 Gajawira, B. P. Shipping policy: Concepts and case studies of shipping policies of national governments and international governmental organisations. Thesis submitted for MSc, UWIST, Cardiff, 1986.

This attempts to develop a systematic organizing framework valid not only conceptually but also for the conduct of empirical research to explain some of the central dimensions of shipping policy, conflict and consensus, and the spatial and temporal differences.

2867 Gardner, B. and **Marlow, P.** An international comparison of the fiscal treatment of shipping. *Journal of Industrial Economics* XXXI(4), June 1983, 397–415.

2868 Gardner, B. M. *Ship finance and fiscal policy.* Cardiff: Department of Maritime Studies, University of Wales, 1982. 58 pp.

Considers whether and to what extent various nations' shipping industries are benefiting or suffering financially from the fiscal systems applied to them.

2869 Gautrat, Patrick. Eastern bloc shipping policies and the European shipping industries. In: George N. Yannopoulos (ed.) *Shipping policies for an open world economy.* London and New York: Routledge, 1989, pp. 145–75.

Maritime relations with state trading countries. Availability of countervailing measures in the 1986 EEC common maritime policy. Regulations to ensure, through negotiations, normalization of EEC's maritime relations with USSR and elimination of disruptive effects of their policies.

2870 Gold, E. The international law and policy of ocean transportation: A historical examination of marine policy and shipping law, with particular reference to the Third World. Thesis submitted for PhD, UWIST, Cardiff, 1980.

2871 Goss, R. St. Ursula sails again. *Maritime Policy and Management* 18(2), 1991, 139–43.

Uses legend of St Ursula to suggest that morals may be drawn and applied to

Lord Rochdale's Committee of Inquiry into Shipping, 1970, which complained about absence of statistics in major areas.

2872 Goss, R. O. Politics and paradigms, or: Self-interest and the sea. *Maritime Policy and Management* 9(3), 1982, 165–73.

Criticizes existing foundations of international law of the sea, preferring it should be on the uses of the sea, and seeks to substitute economic principles.

2873 Goss, R. O. and **Marlow, P. B.** Internationalism, protectionism and interventionism in shipping. In: K. M. Gwilliam (ed.) *Current issues in maritime economics.* Dordrecht: Kluwer Academic Publishers, 1993, pp. 45–67.

A paper which considers whether shipping policies should be developed on essentially international basis or whether shipping industries should be protected from competition in the interests of particular nations or groups of countries like the European Community.

2874 Goss, Richard. Sense and shipping policies. In: George N. Yannopoulos (ed.) *Shipping policies for an open world economy.* London and New York: Routledge, 1989, pp. 61–87.

Examines some ideas put forward in field of shipping policy by General Council of British Shipping and by the European Economic Community. Especially efficiency implications of piecemeal liberalization; relevance of common policy in improving competitiveness of European shipping.

2875 Grey, Michael. *Shipping and the environment. (A Lloyd's shipping economist supplement.)* Colchester: Lloyd's of London Press, 1991. 32 pp.

Examines number of issues of importance to the shipping industry in light of concern over its environmental impact. Regulation. US Oil Pollution Act. Emissions. Cruiseships. Regional compacts. Gulf crisis. Offshore operators. Ageing fleet/salvage. Ports. Energy. Financial implications.

2876 Hansen, H. *The developing countries and international shipping.* Washington (DC): World Bank, 1981.

2877 Heaver, Trevor D. The effects of fiscal policies on the development of international shipping. *Logistics and Transportation Review* 21(1), March 1985, 77–91.

2878 Hughes, D. E. The Indonesian cargo sailing vessels and the problem of technology choice for sea transport in a developing country: A study of the consequences of Perahu motorisation policy in the context of the economic regulation of inter-island shipping. Thesis submitted for PhD, UWIST, Cardiff, 1984.

2879 International Maritime Seminar, 1988, Seoul. *Changes in the world shipping environments and counter-strategies toward the year 2000: International Maritime Seminar, 25–26 July, 1988, Seoul, Korea.* Seoul: Korea Maritime Institute, 1988. 366 pp.

Collection of papers giving an in-depth analysis of world shipping. Aims to provide shipping firms and government agencies with the means to develop long-term shipping policy and business strategies.

2880 Knudsen, O. *The politics of international shipping.* New York: Lexington, 1973.

2881 Krommenacker, Raymond J. Maritime transport services in the Uruguay Round. In: George N. Yannopoulos (ed.) *Shipping policies for an open world economy.* London and New York: Routledge, 1989, pp. 188–203.

Explores issues likely to be raised during Uruguay Round on basis that shipping included as priority sector in negotiations for liberalization of trade in services.

2882 Marlow, P. B. Shipping and investment incentives: A trilogy. Part 1: Investment incentives for industry. *Maritime Policy and Management* 18(2), 1991, 123–38.

Discusses form which governmental incentives to invest in shipping take and looks at how such incentives are perceived by businessmen whose actions they are intended to influence.

2883 Marlow, P. B. Shipping and investment incentives: A trilogy. Part 2: Investment incentives for shipping. *Maritime Policy and Management* 18(3), 1991, 201–16.

Considers forms of fiscal and financial incentives used by governments to encourage investment in shipping industry. Makes a common measurement and draws international comparisons.

2884 Marlow, P. B. Shipping and investment incentives: A trilogy. Part 3: The effectiveness of investment incentives for shipping: The UK experience, 1950–1987. *Maritime Policy and Management* 18(4), 1991, 283–311.

Looks at determinants of investment and examines investment function. Unable to find real link between value of incentives for UK shipping industry and fleet size for years 1950–87.

2885 **Moodie, M.** and **Cottrell, Alvin J.** *Geopolitics and maritime power. Washington papers.* Bethesda (Maryland): University Publications of America, 1987. 80 pp.

2886 **Moreby, D. H.** Capitalism, communism and confusionism [*sic*]: The three ideologies affecting modern shipping. (Editorial.) *Maritime Policy and Management* 16(1), 1989, 1–2.

Shipping dominated by 'confusionism', many participants wanting to operate free-market forces while at the same time wanting some degree of help and support from governments.

2887 **Moreby, D. H.** *Global behaviour in shipping.* London: Lloyd's of London Press, 1990.

2888 **Morris, Michael A.** (ed.) *North/South perspectives on marine policy.* Boulder (Colorado): Westview Press, 1988. 200 pp.

2889 **Oceans Institute of Canada.** *Maritime affairs: A world handbook. A reference guide for modern ocean policy and management.* 2nd edn. Harlow, Essex: Longman Group UK, 1991. 479 pp.

A guide to the present state of international maritime law; sea transport and communications; the exploitation and conservation of maritime natural resources; scientific research in marine environment; maritime boundary disputes and military tension areas at sea. International organizations, with addresses. Other textual data.

2890 **Prescott, John Robert Victor.** *Maritime political boundaries of the world.* London: Methuen, 1985. 550 pp.

2891 **Rajwar, L. M. S.** and others. *Shipping and developing countries.* New York: International Conciliation, 1971. 86 pp.

2892 **Ranken, M. B. F.** (ed.) *Greenwich Forum VI: World shipping in the 1990's. Records of a conference at the Royal Naval College, Greenwich, 23–25 April 1980.* Guildford: IPC Science and Technology Press, 1981. 234 pp.

Papers presented at the sixth annual conference of the Greenwich Forum.

Theme of conference was challenges for shipping, the revolutionary changes affecting world shipping. Sorts of changes and those likely to happen.

2893 Ryden, Inger and **Schirach-Szmigiel, Christopher Von** (eds) *Shipping and ships for the 1990's: Proceedings of the international conference 'Supply and Demand of Water Transport' June 18–19, 1979.* Stockholm: Stockholm School of Economics, 1980. 340 pp.

Papers from an international conference in Stockholm, June 1979. Considered the main trends in respect of international shipping policy in the major maritime blocks; demand for sea transport; technology changes in shipping and market impact; changes in the economics and organization of shipping markets.

2894 Schrier, E., Nadel, E. and **Rifas, B. E.** Forces shaping international maritime transport. *World Economy* 7(1), March 1984, 87–102.

This discusses the growing role of governments which has resulted in increased friction between governments as policies in one country conflict with those in another. Thus the current direction of change in the structure of international maritime transport services is towards a regime dominated by governments, where disputes are settled by governments rather than by market forces.

2895 Smith, Brian D. *State responsibility and the marine environment: The rules of decision.* Oxford: Oxford University Press, 1988. 304 pp.

2896 Swayne, R. Shipping: Some political and economic consequences of changes in technology and ownership. *Transport Reviews* 3(2), 1983, 139–56.

2897 Tae-Woo, L. The Gerschenkron model and its applicability to the shipping industry: The case of Korea. *Journal of Korea Association of Shipping Studies* 10, May 1990, 325–55.

2898 United Nations Conference on Trade and Development. *Economic co-operation among developing countries in marine affairs.* New York: United Nations, 1987. 68 pp.

2899 United States Department of Justice. *The regulated ocean shipping industry.* Washington (DC): Government Printing Office, 1977.

2900 VanderZwaag, David L. and **Lamson, Cynthia** (eds) *The challenge of Arctic shipping: Science, environmental assessment and human values.* Montreal and Kingston: McGill–Queen's University Press, 1990. 288 pp.

A collection of essays on the future of the Arctic. The development of the Arctic is addressed in scientific, environmental and human terms.

2901 Waters, C. D. J. and **Soman, S.** Sea transport and industrialising countries. *International Journal of Physical Distribution and Materials Management* 19(8), 1989, 18–23.

2902 Yannopoulos, George N. (ed.) *Shipping policies for an open world economy.* London and New York: Routledge, 1989. 206 pp.

Gathers together eight especially commissioned papers whose aim is to appraise international maritime transport policies and developments in the European Community, the US, non-EEC Scandinavian countries; state trading countries in light of GATT initiative to liberalize world trade services.

2903 Yeats, Alexander J. *Trade barriers facing developing countries: Commercial policy measures and shipping.* Basingstoke: Macmillan Press, 1979. 262 pp.

Study of various barriers to the trade of developing countries. Influence of industrial country tariffs; non-tariff barriers; related policy measure on trade flows of less-developed countries. Factors which retard developing country export growth.

2904 Yeshou, A. B. A study of the value of national fleets and shipping policies in developing countries. Thesis submitted for MSc, UWIST, Cardiff, 1974.

United Kingdom

2905 Aldcroft, Derek H. Reflections on the Rochdale Inquiry into shipping: A review article. *Maritime History* 1, September 1971, 199–217.

2906 British Maritime League. *British owned merchant shipping: The outline of a proposed national policy.* London: British Maritime League, 1983. 16 pp.

Outlines trends affecting British-owned merchant shipping in order to determine whether the national interest has been adversely affected by national and international developments – such as increases in world oil prices; emergence of Soviet maritime power; Falklands campaign; 'open registration', etc.

2907 Colvin, Michael and **Marks, Jonathan.** *British shipping: The right course.* London: Centre for Policy Studies, 1984. 41 pp.

Considers the problems of the British shipping industry and suggests practical proposals to solve them. The decline of the British fleet. Competition. Taxation. Registration. Regulations and safety. Manning. Defence. Government's role.

2908 Denholm, I. The decline of British shipping: A personal view. (Editorial.) *Maritime Policy and Management* 17(1), 1990, 1–2.

Greater maturity of UK shipping industry means it takes fewer risks; foreign competitors' companies are younger and more willing to take risks in the market.

2909 Du Cann, Edward. *Time to hoist the red ensign: The decline in Britain's merchant fleet and how to restore it.* London: Bow Group Publications, 1986. 11 pp.

Paper highlighting dangers resulting from decline in British merchant fleet. Proposed remedies and proposals for adoption of comprehensive and clear maritime policy.

2910 Goss, R. O. Some challenges for British shipping. (Editorial.) *Maritime Policy and Management* 18(1), 1991, 1–9.

Comments on *British Shipping: Challenges and opportunities. Report of working party chaired by Secretary of State for Transport and President of General Council of British Shipping*, London, HMSO, September 1990.

2911 Great Britain. Department of Transport. *British shipping: Challenges and opportunities. Report of a joint working party chaired by the Secretary of State for Transport and President of the General Council of British Shipping.* London: HMSO, Department of Transport, 1990. 71 pp.

A reference study giving an agreed factual account of the position of the British shipping industry both domestically and industrially. Intensifies regulatory policy options, which could be undertaken by industry. Main obstacles to use of British flag and assesses benefits from encouraging its greater use.

2912 Great Britain. Committee of Inquiry into shipping. *Report (Chairman – the Rt. Hon. the Viscount Rochdale), Cmnd. 4337.* London: HMSO, 1970. 498 pp.

A report of the facets of the shipping industry and the industry's concern with international trade. The extent to which technological developments have transformed the nature of shipping operations and the demand for shipping services throughout the world. Seafaring careers. Personnel issues. Structure of command and of relationships aboard ship.

2913 Great Britain. Parliament. House of Commons. Transport Committee. *Decline in the U.K. registered merchant fleet.* London: HMSO, 1986/7.

Minutes of the evidence presented before the Transport Committee. A memorandum submitted by the National Union of Seamen, 'The crisis in the United Kingdom Merchant Shipping'. Examination of witnesses.

2914 Gubbins, E. J. *Implications in the decline of the British Merchant Navy.* Loughborough University of Technology, Department of Transport Technology, 1991. 39 pp.

2915 Ledger, G. and **Roe, M.** The decline of the UK merchant fleet: An assessment of government policies in recent years. *Maritime Policy and Management* 19(3), 1992, 239–51.

Outlines UK shipping policy since mid-1980s, as merchant marine declined. Analyses action of UK government to 1991 and assesses whether coherent policy was developed. Compares with EC rivals.

2916 National Union of Seamen. *British shipping: Heading for the rocks. A rescue plan from the National Union of Seamen.* London: National Union of Seamen, 1982. 32 pp.

Review of main factors which threaten shipping industry. Focus on need to extend public ownership. N.U.S. plan for recovery and survival.

2917 Newns, Ralph B. and **Parrington, Sheila.** *Double taxation relief for shipping.* London: Lloyd's of London Press, 1988. 354 pp.

Reference guide to judicious use of double taxation agreements in shipping sphere. Texts of Articles relevant to shipping in double taxation agreements between UK and other jurisdictions. Application. Useful addresses.

2918 NUMAST. *British shipping: Not a sunset industry.* London: National Union of Marine, Aviation and Shipping Transport Officers, 1991.

2919 *The ownership, management and use of shipping by nationalised industries.* London: HMSO, 1975.

2920 **Rich, C. A.** The role of the shipping organisation in the through transport of freight between the U.K. and Europe. Thesis submitted for MSc, University of Bath, 1976.

Western Europe and Scandinavia

2921 **Dunlop, Anthony.** *French shipping and ship finance, 1990.* Charter Newsletters, 1990. 63 pp.

2922 **Harlaftis, Gelina.** *Greek shipowners and the state, 1945–1975.* London: Athlone Press, 1993. 350 pp.

2923 **Norman, Victor D.** *Norwegian shipping in the national economy.* Bergen: Institute for Shipping Research, 1971. 67 pp.

2924 **Pantelidis, Evangelos T.** Greek shipping and the accession of Greece to the EEC. *Marine Policy* 3(1), January 1979, 3–20.

2925 **Smeets, G. C. A.** Special provisions for the taxation of Netherlands Antilles shipping and aviation companies. *Bulletin for International Fiscal Documentation* 26(8), August 1972, 311–16.

2926 **Smeets, G. C. A.** Special provisions for the taxation of Netherlands Antilles shipping and aviation companies: A clarification. *Bulletin for International Fiscal Documentation* 26(11), November 1972, 422.

2927 **Svendsen, Arnljot Stromme.** The maritime policies of the non-EEC Scandinavian countries and their impact on EEC shipping. In: George N. Yannopoulos (ed.) *Shipping policies for an open world economy.* London and New York: Routledge, 1989, pp. 176–87.

Analyses changing strategies of the non-EEC Scandinavian countries, Norway and Sweden. Non-EEC Scandinavian shipping industry restructuring by moving away from traditional shipping to maritime intelligence activities, offshore activities, aquaculture.

2928 **Tambakis, D.** An economic analysis of the factors affecting shipping in Greece. *Maritime Policy and Management* 11(1), 1984, 21–33.

Examines determining factors of each item making up shipping receipts in

Greek economy for period 1958–80, in order to show independent variables affecting each subcategory.

2929 **Theophanides, Stavros M.** The consumer firm: The popular shipping companies in Greece: A case-study of consumer participation in the creation of firms. *Economic Analysis and Workers' Management* 15(2), 1981, 231–50.

2930 **Verbeke, A.** and **Winkelmans, W.** The strategic search for sustainable country specific advantages: The case of the European shipping industry. *International Journal of Transport Economics* 17(1), February 1990, 57–75.

Central and Eastern Europe and Middle East

2931 **Bradshaw, Richard E.** The politics of Soviet maritime security: Soviet legal doctrine and the status of coastal waters. *Journal of Maritime Law and Commerce* 10(3), April 1979, 411.

2932 **Breitzmann, Karl-Heinz.** Transition from a planned to a market economy: Problems, tasks and first results in shipping of East Germany. In: K. M. Gwilliam (ed.) *Current issues in maritime economics.* Dordrecht: Kluwer Academic Publishers, 1993, pp. 22–35.

Examines the problems and structural changes that the shipping industry of Eastern part of Germany experienced up to the early 1990s. Analyses the functions and development of shipping in the former German Democratic Republic in order to understand such problems and changes.

2933 **D'Imperio, Robert** and **Niemiel, Joseph M.** Oceanographic efforts in Soviet maritime strategy. In: Curtis Cate (ed.) *The challenge of Soviet shipping.* London: Aims of Industry (in association with the National Strategy Information Center, New York), 1983, pp. 37–9.

Discusses importance of oceanographic research within Soviet merchant fleet. Soviet oceanographic organization. Soviet research development. Soviet research techniques/theory.

2934 **Dunlop, Anthony.** *The shipping and maritime industries of Poland: MRC overview.* MRC Marine, 1992. 85 pp.

2935 **El-Sayed, H.** *Maritime regulations in Saudi Arabia.* London: Graham and Trotman, 1987. 320 pp.

2936 **Glick, Leslie A.** *Trading with Saudi Arabia: A guide to the shipping, trade, investment and tax laws of Saudi Arabia.* Lanham (Maryland): Rowman & Littlefield, 1980. 620 pp.

2937 **Hill-Norton, Peter.** *World shipping at risk: The looming threat to the lifelines.* London: Institute for the Study of Conflict, 1979. 16 pp.

Considers the global impact of expansion of the Soviet maritime power.

2938 **Kwasniewski, K.** Shipping policy of the Comecon countries. *Marine Policy* 1, 1977, 110–17.

2939 **Levikov, G.** Co-operation among CMEA (Council for Mutual Economic Assistance) member countries in maritime transport. *Maritime Policy and Management* 17(1), 1990, 31–40.

Highlights activities relating to co-operation among CMEA members, e.g. joint planning and chartering.

2940 **Mott, William.** The Soviet maritime offensive. In: Curtis Cate (ed.) *The challenge of Soviet shipping.* London: Aims of Industry (in association with the National Strategy Information Center, New York), 1983, pp. 40–41.

Discusses author's belief of Soviet's aims to dominate Southern Africa. Relationship of Soviet destabilization efforts with Soviet maritime offensive in Southern Africa. Importance of area to Soviets.

2941 **Salatt, A. H.** Participation in shipping industry on the part of small countries: The case of Arabian Gulf States. Thesis submitted for MSc, UWIST, Cardiff, 1977.

North America

2942 **Agman, Robert S.** Competition, rationalization, and United States shipping policy. *Journal of Maritime Law and Commerce* 8(1), October 1976, 1–50.

Illustrates that the potential for abuse of monopolistic power can be restrained by non-governmental institutional arrangements. Suggests that these are at least as relatively satisfactory as the uncertain and possibly economically wasteful safeguards provided by US Shipping Act 1916.

2943 **Alexander, Lewis M.** and **Hawkins, Gordon R. S.** (eds) *Canadian–U.S. maritime problems.* Kingston: Law of the Sea Institute, University of Rhode Island, 1972. 88 pp.

2944 **Baker, R. Palmer (Jr.)** and **Fritzhand, Gary I.** United States federal income taxation of foreign-flag shipping earnings. *National Tax Journal* 26(4), December 1973, 537–51.

2945 **Blackwell, R. J.** Panel on U.S. shipping policy. In: Hans Ludwig Beth (ed.) *Liner shipping in the eighties: Report on the International Symposium, Bremen, October 24th–26th 1979.* Bremen: Institute of Shipping Economics, 1980, pp. 245–50.

Considers impact of Omnibus Maritime Bill. Review of historical developments leading to bill. Impact of bill upon functions and responsibilities of the Federal Maritime Commission.

2946 **Conley, M. W.** Canadian shipping policies and the United Nations Conference on Trade and Development: An analysis of UNCTAD V. *Maritime Policy and Management* 9(1), 1982, 35–43.

Looks at Canadian shipping policy *vis-à-vis* UNCTAD V. Stance was decided by fact of being major user of shipping without having a significant fleet of her own and her sympathy for Third World aspirations.

2947 **Daschbach, R. B.** Panel on U.S. shipping policy. In: Hans Ludwig Beth (ed.) *Liner shipping in the eighties: Report on the International Symposium, Bremen, October 24th–26th 1979.* Bremen: Institute of Shipping Economics, 1980, pp. 251–60.

US antitrust laws. Examination of shipping industry's criticism of regulatory policies – antitrust regulation.

2948 **Drewry Shipping Consultants.** *U.S. oil imports, policies and tanker shipping.* London: Drewry Shipping Consultants, 1978.

2949 **Fleming, Douglas K.** The Philadelphia story: 1975. *Maritime Studies and Management* 3(2), October 1975, 113–16.

Commentary on a national symposium on maritime transportation held in Philadelphia, April 1975. Symposium and report identified problems facing US Marine transportation system, progressing to new policy recommendations.

2950 Forsyth, C. Economic infrastructures and the development of state autonomy: A case study of U.S. maritime policy. *Sociological Spectrum* 8(4), 1988, 303–22.

2951 Frankel, E. G. Economic and commercial implications of the U.S. Shipping Act of 1984. *Logistics and Transportation Review* 22(2), June 1986, 99–114.

2952 Frankel, Ernst G. *Regulation and policies of American shipping.* Boston (Massachusetts): US Auburn House Pub. Co., 1982. 347 pp.

Analyses development and effectiveness of US shipping policy and regulation. Reviews structure of the industry and evaluates economic and regulatory aspects of US maritime policy. US international and inland shipping policy. Potential impacts of cargo sharing. Importance of government aid. Regulations. UNCTAD Code of Conduct for Liner Conferences.

2953 Friedman, Norman. *United States maritime strategy: Policy, programmes, tactics.* Coulsdon: Jane's Publishing Co., 1988. 288 pp.

2954 Gorter, Wytze. *United States shipping policy.* London: Greenwood Press, 1977. 230 pp.

2955 Harrington, J. W. A comparative view of US government subsidy to the ocean- and air-transport industries. *Maritime Policy and Management* 9(1), 1982, 23–8.

Compares nature and size of US government direct subsidy to the two industries, focusing on trends before the current 're-regulation' of commercial aviation.

2956 Hazard, John. A competitive U.S. maritime policy. *Transportation Journal* 22(2), Winter 1982, 32–62.

2957 Heaver, T. D. Canada and the evolving system and privatization in Canada: The forces for change. In: M. Zacher and C. Cutler (eds) *Canadian foreign policy and international economic regimes.* Vancouver: UBC Press, 1992, pp. 215–37.

Canadian policy should focus on efficiency in shipping services. The policies of the UK and European Union, the US, and developing countries in the UNCTAD Liner Code are reviewed.

2958 Heiman, Bruce J. and **Marshall, Rolf.** The Maritime Reform Act of 1992? *Journal of Maritime Law and Commerce* 23(4), October 1992, 507–46.

Review of interest of maritime sector. Outlines rationale for government support of a US flag fleet and programmes intended to do so. A 'new' maritime programme and development within Bush Administration. Provisions of the policy and implementing legislation.

2959 Heine, I. M. *The United States merchant marine: A national asset.* Washington (DC): National Maritime Council, 1976.

2960 Heine, Irwin M. *The U.S. maritime industry in the national interest.* Washington (DC): National Maritime Council, 1980. 397 pp.

A comprehensive historical and statistical reference, covering government policies, tables and charts, economic and defence aspects of the US maritime industry.

2961 Hilton, R., Pettavino, P. J. and **Ullman, H. K.** *U.S. Maritime industries: Down for the third time?* Washington (DC): Center for Strategic and International Studies, 1987.

2962 Jantscher, Gerald R. *Bread upon the water: Federal aids to the maritime industries.* Washington (DC): Brookings Institution studies in the regulation of economic activity, 1975.

2963 Kavass, Igor I. and **Blake, Michael J.** (eds) *United States legislation on foreign relations and international commerce: 1789 to date.* William S. Hein & Company, 1977.

2964 Lanthier, Allan R. Incentives for international shipping corporations: Canada. *Bulletin for International Fiscal Documentation* 45(6), June 1991, 269–75.

2965 Mangone, Gerard J. *Marine policy for America.* 2nd rev. edn. Bristol (Pennsylvania): Taylor and Francis, 1989. 400 pp.

2966 O'Brien, L. J. Panel on U.S. shipping policy. In: Hans Ludwig Beth (ed.) *Liner shipping in the eighties: Report on the International Symposium, Bremen, October 24th–26th 1979.* Bremen: Institute of Shipping Economics, 1980, pp. 261–8.

Problems with US maritime policy leading to drafting of the Omnibus Maritime Bill. Controversy surrounding the bill. Analysis of the bill.

2967 Rehm, John B. Proposed Trade Act of 1970: What direction U.S. foreign trade policy? *Journal of Maritime Law and Commerce* 2(2), January 1971, 189–321.

Analyses substance and direction of H.R. 1970 – then proposed Trade Act 1970. Initiatives in US foreign trade policy. Impact on international trade in the 1970s.

2968 Schmeltzer, E. Panel on U.S. shipping policy. In: Hans Ludwig Beth (ed.) *Liner shipping in the eighties: Report on the International Symposium, Bremen, October 24th–26th 1979.* Bremen: Institute of Shipping Economics, 1980, pp. 199–219.

General introduction to the panel statements on policy. Prospects for competitive structure of US liner trades. Open competition. Full rationalization. Regulation by government. Proposed US legislative changes.

2969 Singman, J. H. Panel on U.S. shipping policy. In: Hans Ludwig Beth (ed.) *Liner shipping in the eighties: Report on the International Symposium, Bremen, October 24th–26th 1979.* Bremen: Institute of Shipping Economics, 1980, pp. 220–28.

What US antitrust policy is in general; how it developed; what the policy means with respect to liner shipping in the 1980s.

2970 Transport Canada. *A shipping policy for Canada.* Ottawa: Ministry of Supply and Services, 1979.

2971 United States Department of Commerce. *U.S. ocean policy in the 1970's: Status and issues.* Washington (DC): US Government Printing Office, 1978.

2972 Whitehurst, Clinton H. *American domestic shipping in American ships: Jones Act costs, benefits and options.* Washington (DC): American Enterprise Institute, 1987.

2973 Wilson, Wesley W. and **Casavant, Kenneth L.** Some market power implications of the shipping act of 1984: A case study of the U.S. to Pacific Rim transportation markets. *Western Journal of Agricultural Economics* 16(2), December 1991, 427–34.

2974 Yang, K. C. The impact of the U.S. Shipping Act 1984 on the

Atlantic liner trades. Thesis submitted for M.Phil, University of Liverpool, 1988.

Considers the impact of the Act on the operation and competitive structure of the Atlantic liner trade. It attempts to distinguish the several different parties affected by the change in legislation and tries to separate short- and long-term gains and losses.

Japan

2975 Brown, Ronald G. and **Uesugi, Akinori.** Japanese regulation of ocean freight shipments. *Journal of Maritime Law and Commerce* 11(3), April 1980, 293–318.

Japanese shipping industry and its self-policing functions. Government efforts to restrain unfair business practices and monitor freight rate increases. Impact of anti-monopoly law and regulations. Problems in exerting extra-territorial control.

2976 Goto, Shigeya. *Japan's shipping policy.* Tokyo: Japan Maritime Research Institute, 1984. 28 pp.

Enactment of marine transportation law and subsequent developments. Role of Keikaku Zosen (government-sponsored shipbuilding programme). International co-operation in shipping. Future prospect.

2977 Sugiyama, Shinya. The impact of the opening of the ports on domestic Japanese industry: The case of silk and cotton. *Economic Studies Quarterly* 38(4), December 1987, 338–53.

Socialist Asia

2978 Heine, Irwin Millard. *China's rise to commercial maritime power.* London: Greenwood Press, 1989. 175 pp.

Examines development of China's maritime industry. The maritime policy. Use of maritime accords to advance China's maritime interests. Shipbuilding industry. Signatory countries to China's maritime accords. China's maritime relations with international organizations. Prospects.

2979 Taplin, John H. E. Economic reform and transport policy in China. *Journal of Transport Economics and Policy* XXVII(1), January 1993, 75–86.

First-hand information on how non-urban transport policy works at operator level. Assesses issues perceived by transport officials. Possible future directions. Covers *inter alia* shipping industry.

Africa

2980 Berridge, G. R. *Politics of the South Africa run: European shipping and Pretoria.* Oxford: Oxford University Press, 1989. 268 pp.

2981 Iheduru, Okechukwu C. A critical assessment of Nigeria's shipping policy and its implementation. *Journal of Maritime Law and Commerce* 23(4), October 1992, 547–84.

Summary of main provisions of 1987 legislation, 'The National Shipping Policy Decree 10 of 1987'. Evaluation of success of statutory objectives of National Maritime Authority policy. Consideration of defects of policy to be corrected to resuscitate Nigerian shipping industry.

2982 Iheduru, Okechukwu C. Merchant fleet development by legislation: Lessons from West and Central Africa. *Maritime Policy and Management* 19(4), October–December 1992, 297–317.

Examines some of cargo reservation policies and flag discrimination practices in West and Central Africa. Concludes these alone sufficient to build significant merchant fleet. Other problems to be solved for merchant fleet development.

2983 Mobereola, A. E. The establishment of a national shipping line – an asset or a liability? The case of Nigeria. Thesis submitted for PhD, UWIST, Cardiff, 1988.

2984 Solomon, V. E. The 'Open Market' in South African shipping: A forgotten controversy. *South African Journal of Economics* 47(3), September 1979, 266–76.

2985 United Nations Conference on Trade and Development Secretariat. *Maritime transport review for developing Africa 1986.* New York: United Nations, 1988. 41 pp.

Prepared by Shipping Division of UNCTAD. Material used is treated from the developing Africa maritime sector perspective. Uses of the information. Seaborne trade. Direction of trade. Cargo mix. Developing Africa merchant fleet. Freight costs. Landlocked and island countries. Port development. International Conventions.

Central and South America

2986 Economic Commission for Latin America and the Caribbean. *Basic concepts of maritime transport and its present status in Latin America and the Caribbean.* London: HMSO, 1988. 114 pp.

2987 Farrell, S. Brazilian merchant marine policy, 1958–80: A study in planning and development. Thesis submitted for PhD, University of Liverpool, 1984.

This describes the development of the Brazilian merchant marine. The conclusions are that shipping has made a positive but small contribution to economic growth. That the regulatory system was appropriate for the 'start-up' stage of development but should now be relaxed.

2988 Jhabvala, Farrokh (ed.) *Maritime issues in the Caribbean.* Gainsville (Florida): University Presses of Florida, 1989. 130 pp.

2989 Malins-Smith, R. L. An examination of the economic benefits of the establishment of a merchant marine by the countries of the Caribbean Free Trade Area. Thesis submitted for MSc, UWIST, Cardiff, 1974.

Asia, Australasia and Oceania

2990 Bolar, Malathi. *The shipping industry in India.* Geneva: International Institute for Labour Studies, 1984. 72 pp.

Reviews history of Indian shipping industry. Market conditions and business growth. The labour force. The role of Government and of legislation. International influences. Parties to the industrial relations system. Output of the system. Industrial relations system and action 1960–76.

2991 Buzan, Barry. Maritime policies in North-East Asia. *Marine Policy* 3(3), July 1979, 190–.

2992 Goss, R. Recent maritime work of the Australian Prices Surveillance Authority. (Review article.) *Maritime Policy and Management* 19(1), 1992, 1–17.

On four reports of PSA in 1990: 'Coastal shipping freight rates'; 'Congestion surcharge for cargo handling at Sydney'; 'Harbour towage charges'; 'Stevedoring and container depot charges'.

2993 Kim, G. Shipping investment and allocative efficiency in Korea. Thesis submitted for PhD, UWIST, Cardiff, 1991.

Here it is argued that the relative non-preference towards the shipping industry has resulted in an odd phenomenon – fleet growth in spite of the sustained non-profitability of shipping. This characteristic comes from the Government's preference towards externalities of the shipping industry in addition to its non-preference towards shipping itself.

2994 Kim, Gil-Soo. 'Government Preference' and its influence on the shipping industry in Korea. *Maritime Policy and Management* 19(4), October–December 1992, 265–77.

Suggests existence of 'government preference' has changed structure of Korean economy; how formation of export-dependent economy and shipping industry affected.

2995 Kokuryo, Hideo. *Korean shipping and shipping policy: Shipping industry rationalization plan.* Tokyo: Japan Maritime Research Institute, 1985. 40 pp.

A brief history of ROK shipping. Bright and dark sides of ROK shipping. Shipping industry rationalization plan and its future prospects.

2996 Lee, M. and **Pearson, R.** The expansion of the national fleet of the Republic of China on Taiwan 1960–1985: The policies adopted and conflicts created. *Maritime Policy and Management* 15(3), July—September 1988, 213–24.

Examines details of shipping policies implemented by government Republic of China *re* export expansion policy. Successful? Conflicts that implementation caused within shipping and shipbuilding sectors.

2997 Lee, T. W. Korean shipping policy: The role of governments. *Marine Policy* 14(5), 1990, 421–37.

2998 Lee, Tae-Woo. Some reflections on the causes of growth of Korean shipping. In: K. M. Gwilliam (ed.) *Current issues in maritime economics.* Dordrecht: Kluwer Academic Publishers, 1993, pp. 8–21.

Analyses causes of rapid growth of Korean shipping in the period 1962–81 in the context of the Gerschenkron model.

2999 *Ocean management – a regional perspective: Prospects for Commonwealth maritime cooperation in Asia and the Pacific.* London: Commonwealth Secretariat, 1984. 155 pp.

3000 *Problems of shipping and export strategy for 'eighties.* New Delhi: Indian Institute of Foreign Trade, 1980.

Protection

General and theory

3001 Ademuni-Odeke (Dr.) *Protectionism and the future of international shipping.* Dordrecht: Martinus Nijhoff, 1984. 446 pp.

The nature, development and role of flag discriminations and preferences, cargo reservations and cabotage restrictions, state intervention and maritime subsidies. Future prospects?

3002 Ademuni-Odeke. *Shipping in international trade relations.* Aldershot: Gower, 1988. 648 pp.

Interdisciplinary study of function of shipping in the economic, financial, legal, political and commercial relations between the trading nations of the world community. Examines rationales often given for state involvement in shipping industry of economics, employment, balance of payments, politics, geopolitics.

3003 Bohme, Hans. The economic consequences of restraints on transactions in the international market for shipping services. In: George N. Yannopoulos (ed.) *Shipping policies for an open world economy.* London and New York: Routledge, 1989, pp. 15–39.

Analyses in systematic way the economic consequences of three types of government-induced distortions: State subsidies, flag discrimination, price (freight rate) controls. Impact of restraints on competition in ocean shipping markets – effects on economic efficiency in shipping.

3004 Canadian Transport Commission. *Cargo reservation and liner conference shipping serving Canada.* Ottawa: Canadian Transport Commission, 1979.

3005 Claessens, E. M., Kamel, M. and **Shanker, A. K.** Traffic and trade projections on the eastbound Suez shipping route. *Maritime policy and management* 11(1), 1984, 45–55.

Applying traditional transport models, those mainly dealing with inland transportation, to maritime navigation to provide more empirical evidence as to maritime trade projections.

3006 Dickie, Paul. The Gulf crisis one year on: Problems for international traders. *Marine Policy* 16(2), March 1992, 107–10.

Describes the various UN resolutions issued during the 1990–91 Gulf crisis. Describes resolutions imposing trade and payment sanctions. Impact on trading.

3007 Drewry Shipping Consultants. *Protectionism and the bulk shipping industry.* London: Drewry Shipping Consultants, 1985. 62 pp.

Report – part of series covering world's bulk shipping. Developing countries and protectionism. Developing countries' dry bulk trade. Developing countries' shipping costs.

3008 Ellsworth, R. A., Schmitt, A. L. and **Zerby, J. A.** Regulating the rate practices of state controlled shipping companies: The development and implementation of the Ocean Shipping Act of 1978. *Journal of Maritime Law and Commerce* 12(4), July 1981, 467–83.

Examines development and ultimate implementation of the Act. Similarities between this 1978 Act and Anti-dumping Act 1921.

3009 Goss, R. O. Some economic aspects of flag discrimination. *Maritime Policy and Management* 3(3), July–September 1986, 245–50.

Paper considers decline in opposition to discriminatory measures from established maritime nations whose lines are members of shipping conferences. Why? Suggests (i) growth of ship management opportunities; (ii) modified attitudes towards increased freight rates.

3010 Langdon, Frank C. and **Ross, Douglas A.** (eds) *Superpower maritime strategy in the Pacific.* London and New York: Routledge, 1990. 256 pp.

3011 United Nations Conference on Trade and Development. *Protection of shipper interests: Guidelines in developing countries.* Geneva: UNCTAD, 1979.

3012 Yannopoulos, George N. World shipping between liberalism and protectionism: An introduction. In: George N. Yannopoulos (ed.) *Shipping policies for an open world economy.* London and New York: Routledge, 1989, pp. 1–14.

Introduction to general theme of entire book. Examines return to a market-based organization. A European framework for international shipping. Introduces papers by other contributors. Liberalization strategies.

Cargo preference; reservation

3013 Aspinwall, M. D. American cabotage law and outer continental shelf operations. *Maritime Policy and Management* 14(4), 1987, 313–20.

Cabotage (coastwise) maritime trade on USA coastline has long been reserved for US ships by system of cargo preference. Paper considers geographic scope of US coastwise trade law.

3014 Binkley, John and **Kesterman, Frank.** Economic analysis of the role of non-vessel operating common carriers in the U.S.–Puerto Rico trade. *Journal of Maritime Law and Commerce* 6(1), October 1974, 165–73.

A study providing an economic profile placing NVOCC services in perspective. Economic role of NVOCC in US–Puerto Rico trade. Benefits to public.

3015 Farrell, S. Protectionism in practice. *Maritime Policy and Management* 13(4), October–December 1986, 277–90.

Paper outlines experience in Brazil in respect of cargo reservation laws. Advantages/disadvantages. Favourable trade structure, pragmatic approach in Brazil.

3016 Frankel, E. G. Cargo reservation and restrictive shipping practices: Economic and operational impact. Thesis submitted for PhD, UWIST, Cardiff, 1985.

This thesis identifies the impact of measures limiting access to cargo and estimates the financial and economic effects on the use of resources, costs and freight rates, between developed and less-developed countries.

3017 Hearn, George H. Cargo preference and control. *Journal of Maritime Law and Commerce* 2(3), April 1971, 481–98.

Outlines cargo preference systems of US and some other countries and relates them to political and commercial conditions prevailing among maritime nations.

3018 Kilgour, John G. The Energy Transportation Act of 1974. *Journal of Maritime Law and Commerce* 7(4), July 1976, 557–72.

Examines campaign to extend cargo preference to oil and other commercial cargoes of which the proposed Energy Transportation Security Act 1974 is a part. (Following President Ford's veto of the proposed Act.)

3019 Lackman, Conway L. and **Heinzelmann, Raymond G.** Marine preference cargomarket: Opportunities for minority shippers. *Review of Black Political Economy* 11(3), Spring 1981, 365–74.

3020 United Nations Conference on Trade and Development. Cargo reservation and developing countries. *Journal of World Trade Law* 4(4), July/August 1970, 590–604.

Based on paper prepared by UNCTAD Secretariat for the fourth session of Committee on Shipping of Trade and Development Board. How to assure the developing countries a more equitable share in carriage of world seaborne trade.

Subsidy

3021 Kilgour, John G. The double-subsidy issue in shipping. *Journal of Maritime Law and Commerce* 6(3), April 1975, 395–408.

Examines opinion and order of Maritime Subsidy Board of the Maritime Administration in Docket No. S-244 Factfinding Hearing *Re*: Payment of Subsidy for Carriage of Preference Cargoes 1972. Evaluates double-subsidy issue.

3022 Moyer, R. C. Operating differential subsidies: Problems and alternatives for the United States. *Maritime Studies and Management* 3(3), January 1976, 153–62.

Paper suggests a proposal for a Competitive Rate of Return (CRR) based operating-subsidy system to be made as an alternative to the existing US foreign-cost parity operating differential subsidies system.

3023 Moyer, R. Charles. Maritime subsidies: Problems, alternatives and trade-offs. *Journal of Industrial Economics* 26(1), September 1977, 53–68.

This looks at three issues in the American maritime industry. Firstly, the rationale for current maritime support. Secondly, the weakness inherent in the foreign-cost parity approach to subsidies, and thirdly, the effect of trade-offs of various supporting programmes.

3024 Rouèche, Léonard. Notes on government subsidisation of ferry transport. *Journal of Transport Economics and Policy* XV(3), September 1981, 233–42.

A review of subsidies for ferries in Norway, Scotland, Canada and USA. Concentrates on two issues – the determination of appropriate subsidy level; the operation and management of ferry system.

3025 Tarpgaard, Peter T. U.S. shipping subsidies and ocean shipping of petroleum: Policy choices. *Contemporary Policy Issues* 4(4), October 1986, 79–92.

Trade embargo

3026 Boyd, Roy. Costs and distortions of U.S. regulations policy in intercoastal lumber shipping. *Journal of Policy Modeling* 5(2), June 1983, 293–307.

3027 Garson, John R. The American trade embargo of China: Recent developments. *Journal of Maritime Law and Commerce* 2(1), October 1970, 41–61.

Describes the particulars of various steps taken by American Administration in July 1969 to relax the trade embargo imposed against China. Mechanics of embargo; system in which changes to take place.

3028 Komarow, Gary. Foreign cargo vessels in the coastwise oil trade. *Journal of Maritime Law and Commerce* 10(2), January 1979, 271–87.

Focuses on 46 U.S. Coastwise Cargo Law S. 883 which provides for the forfeiture of any merchandise transported between points in US by vessels not documented under law of, and owned by citizens of, the US. Activities accompanying opening of Alaska pipeline. Impact of statute.

3029 Mukherjee, Neela. Multilateral negotiations and trade barriers in service trade: A case study of U.S. shipping services. *Journal of World Trade* 26(5), October 1992, 45–58.

Trade business in services. Accounting for trade barriers. Policy measures. Role of trade business in shipping services. Shipping services of the US. Trade business existing in the US shipping services. Producer subsidy equivalent – estimating in shipping services.

Flag preference

3030 Stemmer, Juan E. Oribe. Flag preference in Latin America. *Journal of Maritime Law and Commerce* 10(1), October 1978, 123–34.

General description of one of governmental policies used in allocation of cargoes, namely the flag preference policy, as practised by Latin American group of countries.

3031 Vanags, A. H. Flag discrimination: An economic analysis. In: R. O. Goss (ed.) *Advances in maritime economics.* Cambridge: Cambridge University Press, 1977, pp. 37–64.

The incentive to discriminate – the liner sector; the non-liner sector. Discrimination in a multilateral context. Balance of payment aspects of flag discrimination.

Deregulation

3032 Bancardi, C. G. Privatization: A recent example at the National Maritime Research Centre. *Maritime Policy and Management* 15(4), 1988, 275–8.

Analyses concept of privatization and its underlying philosophy exemplified by recent privatization of Computer Aided Operations Research Facility (CAORF) at National Maritime Research Centre, USA.

3033 Bennathan, Esra with **Escobar, Luis** and **Panakakos, George.** *Deregulation of shipping: What is to be learned from Chile.* World Bank Discussion Paper No. 67. Washington (DC): The World Bank, 1989. 109 pp.

Reviews and analyses regulations in force in Chile at end of 1970s. System resulting from deregulation, 1979. Further changes in the law, 1980s. Short-term consequences. Situation of shipping industry and users some nine years later. Analysis of determinants of freight charges in export trade with US, 1978 and 1986.

3034 Cathcart, Peter G. Transportation deregulation in the Canadian maritime industry. *Lloyd's Maritime and Commercial Law Quarterly,* February 1989, Part 1, 104–14.

Major changes in Canadian marine legislation in 1987 arising in part from philosophy in 'Freedom to Move' report from the Standing Committee on Transport. Examines: (i) Canada Shipping Act; (ii) Coasting Trade and Commercial Marine Activities Act; (iii) Shipping Conferences Exemption Act 1987.

3035 Eyre, J. Maritime privatization. *Maritime Policy and Management* 17(2), April–June 1990, 113–21.

Considers the public interest in attempts to privatize deep-sea shipping lines and ports. Comment on situation in year 2000. Review of present situation as reported in maritime press and observed in studies by authors.

3036 Gomez-Ibanez, José A. and **Morgan, Ivor P.** Deregulating international markets: The examples of aviation and ocean shipping. *Yale Journal on Regulation* 2(1), 1984, 107–44.

3037 Pollack, Gerald E. Promoting the U.S. flag merchant marine. *Business Economics* 26(2), April 1991, 45–50.

3038 Sirat, M. The federal ports of Malaysia and privatisation. Thesis submitted for M.Phil, UWIST, Cardiff, 1991.

This examines full privatization and three alternative regimes recommending regional control for privatized ports and submitting that ports are suitable subjects for a privatization geared to a landlord approach, based on the model of Rotterdam and Antwerp.

3039 *United Nations conference on conditions of registration of ships: A free market perspective.* Oslo: INTERTANKO, 1984. 12 pp.

Defence

General

3040 Corbett, Julian S. and **Grove, Eric** (eds) *Some principles of maritime strategy.* Oxford: Brassey's Defence Publishers, 1988. 351 pp.

3041 Gray, Colin S. *Maritime strategy, geopolitics, and the defense of the West.* New York: Ramapo Press, 1986. 96 pp.

3042 Grove, Eric. *Maritime strategy and European security.* Oxford: Brassey's Defence Publishers, 1990. 160 pp.

3043 Grove, Eric J. *The merchant fleet and deterrence: A study of the availability of shipping for NATO and national defence purposes.* London: British Maritime Charitable Foundation, 1988. 13 pp.

Considers the doubts about nuclear deterrence expressed by analysts and politicians from all sides of political spectrum. Deterrence into the twenty-first century. The North-Atlantic Link Ships. Issues for future study. The lessons of history.

3044 Hawkes, Kenneth G. *Maritime security.* Centreville (Maryland): Cornell Maritime Press, 1989. 353 pp.

3045 Hilditch, Peter J. Defence procurement and employment: The case of U.K. shipbuilding. *Cambridge Journal of Economics* 14(4), December 1990, 483–96.

Analyses the experience of the war-shipbuilding industry and the impact of competitive procurement policy. Also considers the conflicting roles that concerns over employment play in the implementation of procurement policies.

3046 Hill, J. R. *Maritime strategy for medium powers.* London: Croom Helm, 1986. 248 pp.

3047 Komer, Robert W. *Maritime strategy or coalition defence?* Bethesda (Maryland): University Publications of America, 1984. 234 pp.

3048 Luard, Evan. The law of the sea. *Royal United Services Institute for Defence Studies Journal* 119, June 1974, 15–21.

3049 Mattendorf, John and **Jordan, Robert S.** (eds) *Maritime strategy and the balance of power: Britain and America in the twentieth century.* New York: Macmillan Press, 1989. 400 pp.

3050 O'Connell, D. P. Transit rights and maritime strategy. *Royal United Services Institute for Defence Studies Journal* 123, June 1978, 11–18.

3051 Palmer, Michael A. *Origins of the maritime strategy, 1945–55.* Annapolis (Maryland): US Naval Institute Press, 1990. 192 pp.

3052 Smith, Robert W. An analysis of the strategic attributes of international straits: A geographical perspective. *Maritime Studies and Management* 2(2), October 1974, 88–101.

Paper examines question of if a special juridical regime be developed for certain straits, what criteria to be employed to determine which straits to be under jurisdiction.

3053 Till, Geoffrey. *Maritime strategy and the nuclear age.* 2nd rev. edn. New York: Macmillan Press, 1984. 304 pp.

3054 *To use the sea: Readings in sea power and maritime affairs.* 2nd rev. edn. Annapolis (Maryland): US Naval Institute Press, 1977. 390 pp.

3055 Tracy, Nicholas. *Attack on maritime trade.* Basingstoke, Macmillan Academic and Professional, 1991. 277 pp.

Reviews the motives for attack on maritime trade over five centuries and assesses the strategic utility of that form of naval action. Before First World War. The twentieth century.

3056 Tunander, Ola. *Cold water politics: Maritime strategy and geopolitics of the Northern Front.* London: Sage Publications, 1989. 208 pp.

3057 Young, Elizabeth. Policing offshore: Civil power or armed forces. *Royal United Services Institute for Defence Studies Journal* 122, June 1977, 18–22.

Military logistics

3058 Kemp, Paul. *Convoy protection: Defence of seaborne trade.* London: Arms & Armour Press, 1993. 128 pp.

3059 Olson, Richard E. and **Scrogin, Thomas W.** Containerisation and military logistics. *Journal of Maritime Law and Commerce* 6(1), October 1974, 119–46.

Examines impact that containerization has had as a technique in transport. Experience of US Department of Defense – technology; institutions; management; administration.

16 International Policy and Regulation

General

General

3060 Arroyo, Ignacio. *International maritime conventions.* Deventer, Netherlands: Kluwer Law and Taxation Publishers, 1991. 1568 pp.

An examination of international maritime conventions. Divided into five sections: International organizations; law of the sea; navigation and maritime traffic; seamen; maritime trade. Considers the 104 conventions approved at date of writing. (Excludes recommendations, resolutions, annexes and other complementary documents.)

3061 Bekiashev, Kamil A. and **Serebriakov, Vitali V.** *International marine organizations: Essays on structure and activities.* The Hague and London: Martinus Nijhoff, 1981. 578 pp.

Classifies international marine organizations; shows their structure and depicts their activities. Presents idea of volume and directions of international, economic, scientific, and technical co-operation in fields of exploration and use of world ocean.

3062 Butler, William E. (ed.) *Law of the sea and international shipping: Anglo-Soviet post UNCLOS perspectives.* New York and London: Oceana Publications, 1985. 432 pp.

Papers from the first Anglo-Soviet Symposium on the Law of the Sea and International Shipping, Moscow, 25–27 April 1983; the second Anglo-Soviet Symposium, London, 25–27 June 1984. Issues of concern to the international legal and shipping communities – implementation of 1982 UNCLOS; maritime jurisdiction; environmental protection and Law of the Sea.

3063 Comité Maritime International. *International conventions on maritime law.* Philadelphia: Coronet Books, 1987. 483 pp.

3064 **Farthing, Bruce.** *International shipping: An introduction to the policies, politics and institutions of the maritime world.* Colchester: Lloyd's of London Press, 1987. 204 pp.

A major study of the policies, politics and institutions of world-wide shipping community. Original minimal intervention in shipping by governments. European and Japanese maritime power. United Nations' efforts to redress balance of power in favour of Third World countries. Impact and pressures.

3065 **Frankel, Ernst G.** Shipping: Redefining functions. *Marine Policy* 8(2), April 1984, 165–70.

Reviews the international organizations which exist to affect, regulate or assist shipping. Considers the effect of increasing nationalism on shipping and associated policy development and questions whether it is necessary for existing international organizations to examine and redefine their functions.

3066 **Jackson, David C.** and others. *World shipping laws: International conventions.* New York: Oceana Publications, 1979.

3067 **Jagota, S. P.** Developments in the U.N. Conference on the law of the sea: A Third World review. *Third World Quarterly* 3, April 1981, 287–319.

This report traces the history and origins of the conference describing the position of various interest groups on the major issues involved and assesses the emergence of compromise solutions.

3068 **Mankabady, Samir.** *International Maritime Organization. Vol. 2. Accidents at sea and the law.* 2nd rev. edn. London and Sydney (*et al.*): Croom Helm, 1987. 480 pp.

Deals with consequences of accidents. Contains law and regulations on oil and chemical pollution; salvage; towage; general average; personal injury. Overall, comprehensive manual to the international law governing accidents at sea.

3069 **Reisman, W. Michael.** *Straight baselines in international maritime boundary delimitation.* New York: Macmillan Press, 1992. 210 pp.

3070 **Seifert, Gerald.** United States and European Community international maritime policies. In: George N. Yannopoulos (ed.) *Shipping policies for an open world economy.* London and New York: Routledge, 1989, pp. 139–44.

Compares developing common transport policy for the European Community with established set of standards in the US. Develops a basis for determining whether developing rules harmonious with regulatory scheme in US where conflicts may exist.

3071 **United Nations Conference on Trade and Development.** *An action programme.* New York: United Nations, 1975.

3072 **United Nations Conference on Trade and Development.** International legislation on shipping. *Journal of World Trade Law* 4(1), January/February 1970, 81–90.

Aims of the Working Group on International Legislation on Shipping. Review of economic and commercial aspects of international legislation and practices.

3073 **United Nations Conference on Trade and Development.** *Shipping investigation units.* Geneva: UNCTAD, 1975.

3074 **United Nations Conference on Trade and Development.** *What are shippers' interests?* Geneva: UNCTAD, 1975.

International Maritime Organisation

3075 **Henry, Cleopatre Elmira.** *The carriage of dangerous goods by sea: The role of the International Maritime Organisation in international legislation.* London: Frances Pinter, 1985. 181 pp.

Role of international organizations in enactment of international legislation; structure and mandate of I.M.O.; I.M.O. as an international legislator; regulation of carriage of dangerous goods by I.M.O.; International Maritime Dangerous Goods Code.

3076 **International Maritime Organisation.** *International conference on liability and compensation for damage in connection with carriage of certain substances by sea, 1984.* London: I.M.O., 1985.

3077 **International Maritime Organisation.** *International conference on limitation of liability for maritime claims, 1976.* London: I.M.O., 1977.

3078 **International Maritime Organisation.** *International conference on special trade passenger ships, 1971.* London: I.M.O., 1972.

3079 International Maritime Organisation. *Official records of the international conference on limitation of liability for maritime claims, 1976.* London: I.M.O., 1983.

3080 Lampe, Wilhelm H. The 'new' International Maritime Organisation and its place in development of international maritime law. *Journal of Maritime Law and Commerce* 14(3), July 1983, 305–29.

Historical development of I.M.O., from IMCO. Structural changes in 1982. Why I.M.O. now well suited to fulfil implementing role in field of international maritime law. Responsibilities. Options for the future.

3081 Mankabady, Samir. *International Maritime Organization. Vol. 1. International shipping rules.* 2nd rev. edn. London (*et al.*): Croom Helm, 1986. 450 pp.

Examines I.M.O. Conventions, Resolutions, Codes made up to end of 1985. Amendments to the Collision Regulations, Civil Liability Convention; the Fund Convention and Annex 11 of MARPOL 73/78 on chemicals, adopted December 1985.

3082 Smith, J. R. G. *IMO legislation to enter into force between 1989 and 1994.* London: Lloyd's Register of Shipping, 1989. 14 pp.

Discusses recent and impending legislation from the I.M.O. in respect of shipbuilding and ship safety.

3083 Srivastava, C. P. Safer ships and cleaner oceans: Thirty years' work of the International Maritime Organisation. *Transport Reviews* 9(1), January–March 1989, 45–58.

Objectives of I.M.O. to promote safety of shipping and to prevent pollution of seas from ships. International conventions, codes, recommendations – basis of shipping legislation. Improvements. Possible further improvements.

3084 Stemmer, J. E. O. A new economic order for postwar shipping and the liner conference system: The International Maritime Conference, 1948. *Maritime Policy and Management* 14(1), 1987, 5–26.

Examines negotiations leading to establishment of International Maritime Organisation from 1945 to 1948. Corrects some false impressions on functions of I.M.O. and UNCTAD.

Inter-Governmental Maritime Consultative Organization

3085 Inter-Governmental Maritime Consultative Organization. *The activities of the Inter-Governmental Maritime Consultative Organization in relation to shipping and related maritime matters.* London: IMCO, 1974. 47 pp.

A booklet which describes the work of IMCO, with reference to (i) promotion of maritime safety and efficiency of navigation; (ii) prevention and control of marine pollution; (iii) provision of technical assistance to developing countries in these fields.

3086 IMCO, Legal Committee. IMCO: The activities of the Legal Committee. *Journal of World Trade Law* 13(5), September/October 1979, 472–9.

Consideration by Committee of draft convention on liability and compensation in connection with carriage of noxious and hazardous substances by sea (other than oil). Additionally, draft conventions on air-cushion vehicles; civil jurisdiction on matters of collision at sea; drilling rigs; offshore mobile craft.

3087 Silverstein, Harvey B. *Superships and nation states: The transnational policies of the Inter-governmental Maritime Consultative Organization.* Boulder (Colorado): Westview Press, 1978. 251 pp.

General Agreement on Tariffs and Trade

3088 Von Raab, William. Pre-shipment inspections: Improved administration of an international trade regime. *Journal of World Trade Law* 25(5), October 1991, 87–98.

Examines the concept of and need for Pre-Shipment Inspection (PSI). Application to invoicing difficulties. Valuation and the GATT customs valuation code. PSI and transfer price determinations. Invoices are inadequate. Consistency of PSI with GATT. Need to modify code. PSI and automation. PSI in a non-tariff environment.

European Economic Community; European Community; European Union

3089 Bellis, J.-F., Vermulst, E. and **Musquar, P.** The new EEC regulation on unfair pricing practices in maritime transport: A forerunner of the extension of unfair trade concepts to services? *Journal of World Trade Law* 22(1), January/February 1988, 47–66.

Describes background of the 1986 EEC regulation No. 4057/86 which aims to protect community shipowners from unfair pricing practised by shipping lines of third countries. Analyses its procedural and substantive provisions. International ramifications. Tentative conclusions.

3090 **Bird, J.** Transport decision-makers speak: The Seaport Development in the European Communities Research Project – Part 1. *Maritime Policy and Management* 9(1), 1982, 1–22.

Questionnaire completed by shipowners and seaport managers showed contrasts between seaport planning approaches in UK, Ireland and Denmark *vis-à-vis* rest of EEC. Most important factor in ship route planning agreed to be regularity of cargo on offer.

3091 **Bird, J.** Transport decision-makers speak: The Seaport Development in the European Communities Research Project – Part 2. *Maritime Policy and Management* 9(2), 1982, 83–102.

Analyses responses in interviews of transport decision-makers to six sets of propositions. Concludes that transport decision-makers have a lot to offer in formulating future EEC seaport policy.

3092 **Bredima-Savopoulou, Anna** and **Tzoannos, John.** *The common shipping policy of the European Economic Community.* Amsterdam: North-Holland Pub. Co., 1990. 304 pp.

An interdisciplinary approach. Encompasses aspects of law, economics and policy-making. Analysis of development of the law and regulations governing common shipping policy of the EC and problems, prospects and implications of that policy.

3093 **Bredimas, A. E.** The common shipping policy of the EEC. *Common Market Law Review* 18, 1981, 9–32.

3094 **Bredimas, Anna E.** and **Tzoannos, John G.** In search of a common shipping policy for the E.C. *Journal of Common Market Studies* 20, December 1981, 95–114.

3095 **Brooks, M. R.** and **Button, K. J.** Shipping within the framework of a Single European Market. *Transport Reviews* 12(3), July–September 1992, 237–52.

Outlines main changes taking place in Europe which are directly related to maritime matters, and especially those stemming from actions associated with the move to the Single European Market.

3096 Cafruny, Alan W. Toward a maritime policy. In: Leon Hurwitz and Christian Lequesne (eds) *The state of the European Community: Policies, institutions and debates in the transition years.* Harlow, Essex: Longman, 1991, pp. 285–99.

A chapter which provides an account of the evolution of EC maritime policy since 1986, focusing particularly on developments during 1989–90. Brief historical review followed by survey of recent policies and proposals of the Council and Commission. Assesses their impact on the member states, the Community and the political economy of international shipping.

3097 Clifford Chance. *Unfair pricing practices in international cargo liner shipping: A guide to the EEC regulation.* London: Clifford Chance, 1989. 32 pp.

A summary of the European Community's regulation on unfair pricing practices in international cargo liner shipping. A general outline.

3098 Clough, Mark and **Randolph, Fergus.** *Shipping and European Community competition law.* London: Butterworth, 1991. 675 pp.

3099 Commission of the European Communities. *Progress towards a common transport policy: Maritime transport.* Luxembourg: Office for Official Publications of the European Communities, 1985. 68 pp.

Purpose is to develop coherent overall framework for Community shipping policy. Reviews Community actions to date of paper, and proposes measures to promote Community's trading and shipping interests.

3100 Cooke, Stephanie. Why U.K. shipping has a sinking feeling. *Management Today,* June 1993, 52–4.

This is the fourth of a series of articles on the uneven playing field. It concludes that the UK shipping industry has been blasted by its EEC competitors who receive state aid.

3101 Economic and Social Committee of the European Communities. *E.E.C. maritime transport policy.* Brussels: Economic and Social Committee of the European Communities, 1986. 89 pp.

3102 *E.E.C. customs union: What's the form? Joint conference proceedings. Confederation of British Industry – British Shippers Council.* London: CBI, 1977.

3103 *E.E.C. maritime transport policy.* Brussels: Economic and Social Committee of the European Communities, 1986. 89 pp.

3104 No entry.

3105 Erdmenger, Jürgen and **Stasinopoulos, Dinos.** The shipping policy of the European Community. *Journal of Transport Economics and Policy* XXII(3), September 1988, 355–60.

Summarizes scope and important provisions of the four regulations in the maritime package agreed on 26th December 1986 by European Community Ministers. Possible impact of community policy on community shipping fleet.

3106 *European Community transport legislation, 30 January 1992.* London: Department of Transport, 1992.

3107 Eussner, Ansgar. Industrial policy and southward enlargement of the European Community: The case of shipbuilding and repairs. *Journal of Common Market Studies* 22, December 1983, 147–72.

3108 Farrant, Stephen. Accession of Greece. EEC shipping: A new era looms. In: *EEC shipping: A guide to the sea transport industries of the European Economic Community.* Colchester: Seatrade Publications, pp. 19–21.

Impact of addition of Greek-registered tonnage to EEC fleets. Development of Greek-owned fleet.

3109 Gomer, Hilaire. Ireland: hopes pinned on oil and gas. In: Malcolm Wood (ed.) *EEC shipping 1978: A guide to the sea transport industries of the European Economic Community.* London: Seatrade Publications, 1978, pp. 199–208.

Growth potential for shipping industry. Impact of Irish offshore oil and gas exploration and its production. Shipbuilding industry. Fleet statistics. Ports details, services, etc.

3110 Gomer, Hilaire. Luxemburg: Free from maritime worries. In: Malcolm Wood (ed.) *EEC shipping 1978: A guide to the sea transport industries of the European Economic Community.* London: Seatrade Publications, 1978, pp. 243–4.

Registration of shipholding companies in Luxemburg. Impact of no corporation tax. Luxemburg as a finance centre.

3111 Goss, R. O. Sense and shipping policies. *Maritime Policy and Management* 14(2), 1987, 83–97.

Discusses ideas for shipping policy emanating from General Council of British Shipping, and EEC's document 'Towards a common transport policy: Maritime transport'.

3112 Hart, Paul. *Shipping policy in the European Community.* Aldershot and Avebury: Ashgate Publishing Co., 1993. 101 pp.

Examines the development of European Community maritime policy between 1988 and 1991. Contrasts and conflicts. State aid and the shipping industry. EUROS – the EC ship register.

3113 Hayman, Christopher. EEC shipping policy: The search for common ground. In: Malcolm Wood (ed.) *EEC shipping 1978: A guide to the sea transport industries of the European Economic Community.* London: Seatrade Publications, 1978, pp. 9–11.

Considers profile of shipping policy within Europe from the EEC perspective. Government interference in shipping.

3114 International Transport Workers' Federation. *Towards a common maritime transport policy: A report by the International Transport Workers' Federation.* London: International Transport Workers' Federation, 1985. 47 pp.

Suggests that decline of overall merchant fleet of Community has reached crisis point. Future of European shipping vital to both Community and Member States. Structure of merchant fleets. The European Commission's approach. Economic context. Problems facing shipping. Shipping sectors. Employment. Safety. Future developments.

3115 Kreis, Helmut W. R. EC competition law and maritime transport. *Antitrust Bulletin* 37(2), Summer 1992, 481–505.

3116 Lones, Trevor. Denmark: Last bastion of free enterprise in the EEC. In: Malcolm Wood (ed.) *EEC shipping 1978: A guide to the sea transport industries of the European Economic Community.* London: Seatrade Publications, 1978, pp. 67–97.

Impact of failure of cohesive EEC/European policies on shipping and shipbuilding. Relationship of shipowners and shipbuilders. Fleet statistics. Ports details, services, etc.

3117 Middleton, Ian. Italy: Shipping blazes expansion trail but builders face many problems. In: Malcolm Wood (ed.) *EEC shipping 1978: A guide to the sea transport industries of the European Economic Community.* London: Seatrade Publications, 1978, pp. 209–42.

Considers need for fundamental review of state industries. Ownership of Italian fleet – state ownership, private owners. Impact of Government's credit scheme. Increased activity in the ports. Fleet statistics. Ports details, services, etc.

3118 Middleton, Ian. West Germany: Economic strength is equally depressing for maritime industries. In: Malcolm Wood (ed.) *EEC shipping 1978: A guide to the sea transport industries of the European Community.* London: Seatrade Publications, 1978, pp. 141–98.

Problems of industry. Political developments. Impact on shipping and shipbuilding industries. Increased government involvement in shipping. Fleet statistics. Ports details, services, etc.

3119 The Netherlands: Support all round for owners and builders. In: Malcolm Wood (ed.) *EEC shipping 1978: A guide to the sea transport industries of the European Economic Community.* London: Seatrade Publications, 1978, pp. 245–92.

Government subsidies for shipbuilding industry and shipping companies. Business climate. Fleet statistics. Ports details, services etc.

3120 Power, Vincent. *European Community shipping law.* London: Lloyd's of London Press, 1991. 771 pp.

Describes and discusses EC law relating to shipping. Extracts from various documents, reports, communications, opinions and cases where the EC concerned with legal regulation of shipping industry.

3121 Renouf, Anthony. Finance and insurance: A question of harmony. In: *EEC shipping: A guide to the sea transport industries of the European Economic Community.* Colchester: Seatrade Publications, p. 45.

Reviews insurance and finance issues in respect of shipping industry and government shipping policies.

3122 Rich, Craig and **Thomas, Adrian.** Intra EEC trade: Dry cargo facing change and challenge. In: Malcolm Wood (ed.) *EEC shipping 1978: A guide to the sea transport industries of the European Economic Community.* London: Seatrade Publications, 1978, pp. 35–9.

Impact of increased competition faced by EEC trade from FOC vessels. Impact on intra EEC dry bulk trade.

3123 Seatrade Publications. *EEC shipping: A guide to the sea transport industries of the European Economic Community.* Colchester: Seatrade Publications, 1978, 164 pp.

A set of analytical articles by a group of specialist writers and academics on a broad cross-section of subjects – main liner trade routes, conferences and competition, ports and inland waterways, shipbuilding. Detailed factual and statistical sections.

3124 Seaward, Nicholas. Shipbuilding in the Community: The struggles of reducing capacity. In: Malcolm Wood (ed.) *EEC shipping 1978: A guide to the sea transport industries of the European Economic Community.* London: Seatrade Publications, 1978, pp. 13–30.

EEC Commission's efforts to impose general reduction in shipbuilding capacity on reluctant governments. Impact of reduction in shipbuilding. EEC and other statistics.

3125 Snoy, Yolande. Belgium: Struggling on all fronts. In: Malcolm Wood (ed.) *EEC shipping 1978: A guide to the sea transport industries of the European Economic Community.* London: Seatrade Publications, 1978, pp. 33–65.

Major problems facing shipowners. Impact of pressure from conference outsiders and competition from Comecon lines. Fleet statistics. Ports details, services, etc.

3126 *Towards a shipping policy for the E.E.C.: Seatrade Conference, Brussels.* London: Seatrade Publications, 1978.

3127 Tzoannos, John. The EEC common maritime transport policy and the liberalisation of world shipping markets. In: George N. Yannopoulos (ed.) *Shipping policies for an open world economy.* London and New York: Routledge, 1989, pp. 40–60.

Critical stand on the measures for co-ordinated action against other countries preventing EEC merchant fleets from having access to cargoes or permitting their shipping companies to indulge in predatory pricing.

3128 United Kingdom: Shipowners beginning to struggle. In: Malcolm Wood (ed.) *EEC shipping 1978: A guide to the sea transport industries of the European Economic Community.* London: Seatrade Publications, 1978, pp. 293–363.

Decline of UK flag fleet. British shipbuilding. Impact of joining EEC. Shipowning. Ports. Fleet statistics. Ports details, services, etc.

3129 Walker-Leigh, Vanya. France: Government only hope for future of owners and builders. In: Malcolm Wood (ed.) *EEC shipping 1978: A guide to the sea transport industries of the European Economic Community.* London: Seatrade Publications, 1978, pp. 99–140.

State of French shipping industry. Impact of competition from East European and FOC ships. French efforts at EEC level to harmonize social changes. Fleet statistics. Ports details, services, etc.

3130 Whitelegg, John. *Transport policy in the E.E.C.* London and New York: Routledge, 1988. 233 pp.

3131 Wood, Malcolm (ed.) *EEC shipping 1978: A guide to the sea transport industries of the European Economic Community.* London: Seatrade Publications, 1978. 368 pp.

Comprehensive guide to the sea-trading activities of the European Economic Community. First section – Community as a whole, discussing developments on EEC shipping – shipbuilding policies; trade and fleet figures; comparing EEC and other world sectors. Second section – comprises country by country breakdown of maritime industries.

Codes, Conventions and Reports

General

3132 *International conventions on maritime law: Texts.* Stockholm: Almqvist, 1987. 482 pp.

3133 Kushner, Lester. The 1966 International Load Line Convention: Compatibility of greater carrying capacity with safety of life and property. *Journal of Maritime Law and Commerce* 3(2), January 1972, 375–83.

Aims and achievements of 1966 Convention.

3134 Singh, Nagendra. *International conventions of merchant shipping. British Shipping Laws.* Vol. 8. 2nd edn. London: Stevens & Sons, 1973. 1697 pp.

A reference book on maritime conventions. Covers both private and public international law. Technical and operational conventions. Employment, welfare and status of seamen. The unification of private maritime law. Law of the sea, landlocked states, ports, canals and straits. International organizations.

3135 **United Nations Conference on Trade and Development.** *Economic co-operation in merchant shipping.* Geneva: UNCTAD, 1975, 1977.

3136 **University of Southampton, Institute of Maritime Law** (ed.) *The ratification of maritime conventions.* London: Lloyd's of London Press, 1990. 896 pp.

United Nations Conference on Trade and Development (not Liner Code)

3137 **Fabre, P. C.** Fleet restructuring in trades with developing countries. In: Hans Ludwig Beth (ed.) *Liner shipping in the eighties: Report on the International Symposium, Bremen, October 24th–26th 1979.* Bremen: Institute of Shipping Economics, 1980, pp. 134–44.

Assessment of development of LDCs' fleets in face of (i) the Code of Conduct and (ii) the bulk trade sharing scheme proposed by UNCTAD.

3138 **Haji, Iqbal.** UNCTAD and shipping. *Journal of World Trade Law* 6(1), January/February 1972, 58–118.

Aim of UNCTAD to devise policies in field of invisibles (e.g. shipping) to enable developing countries to increase earnings and decrease outflow of foreign exchange on this account. Freight rate; development and expansion of national merchant marines; subsidiary questions; UNCTAD vision of liner shipping.

3139 International legislation on shipping, 4th session. *Report: United Nations Working Group.* New York: United Nations, 1972.

3140 **Nicholson, A. A. N.** and **Bridger, M. W.** North/South co-operation in shipping: New approaches. *Transport Reviews* 8(2), April–June 1988, 145–60.

Examines North/South confrontation following Second World War. Impact of inter-governmental organizations – UNCTAD, and associated attempts to regulate commercial practices. International Chamber of Commerce and Action Programme 1981. Centre for Maritime Co-operation, 1985.

3141 **Seaward, Nick.** Liner trades: Opportunity for a common policy on all aspects of shipping. In: *EEC shipping: A guide to the sea transport industries of the European Economic Community.* Colchester: Seatrade Publications, 1979, pp. 23–5.

Impact of accession to the UNCTAD Code of Conduct for Liner Conferences by EEC. Opportunities for EEC Commission.

3142 Strange, Susan and **Holland, Richard.** International shipping and the developing countries. *World Development* 4(3), March 1976, 241–51.

This describes the structure of international shipping, the way in which liner conferences operate, the efforts of UNCTAD to get a better deal for the Third World, and the options open to developing countries to improve their relative economic position in this industry.

3143 Sweeney, Joseph C. The UNCITRAL Draft Convention on Carriage of Goods by Sea (Part I). *Journal of Maritime Law and Commerce* 7(1), October 1975, 69–125.

Analysis of Draft Convention for replacement of Hague Rules. Debates surrounding proposals. First of five parts. Examines: Harmonization of international maritime law; UNCTAD, UNCITRAL; substantive issues – period of carrier responsibility, deck cargo and animals, jurisdictional provisions, liability of carrier, arbitration.

3144 Sweeney, Joseph C. The UNCITRAL Draft Convention on Carriage of Goods by Sea (Part II). *Journal of Maritime Law and Commerce* 7(2), January 1976, 327–50.

Analysis of Draft Convention for replacement of Hague Rules. Debates surrounding proposals. Second of five parts. Continues examination of substantive issues – unit limitation of liability and subsidiary doctrines.

3145 Sweeney, Joseph C. The UNCITRAL Draft Convention on Carriage of Goods by Sea (Part III). *Journal of Maritime Law and Commerce* 7(3), April 1976, 487–503.

Analysis of Draft Convention for replacement of Hague Rules. Debates surrounding proposals. Third of five reports. Continues examination of substantive issues – delay, documentary and geographic scope of Convention.

3146 Sweeney, Joseph C. The UNCITRAL Draft Convention on Carriage of Goods by Sea (Part IV). *Journal of Maritime Law and Commerce* 7(4), July 1976, 615–70.

Analysis of Draft Convention for replacement of Hague Rules. Debates surrounding proposals. Fourth of five parts. Continues examination of substantive issues – elimination of invalid clauses in bills of lading, reconsideration of deck cargo, animals, definitions, legal effects of documents, letters of guarantee/indemnity.

3147 Sweeney, Joseph C. The UNCITRAL Draft Convention on Carriage of Goods by Sea (Part V). *Journal of Maritime Law and Commerce* 8(2), January 1977, 167–94.

Analysis of Draft Convention for replacement of Hague Rules. Debates surrounding proposals. Last of five parts. Concludes examination of substantive issues – liability of shipper, dangerous goods; notice of loss provisions, derogations from the Convention.

3148 Tetley, William. Canadian comment on the proposed UNCITRAL Rules: An analysis of the proposed UNCITRAL text. *Journal of Maritime Law and Commerce* 9(2), January 1978, 251–68.

Comparison of UNCITRAL Draft Convention 1976 to replace the Hague Rules and the Hague Rules as amended by Visby. Outlines advantages and disadvantages of the UNCITRAL text over Hague Rules and Visby Rules. Canada's participation.

3149 United Nations Conference on Trade and Development Secretariat. *Multinational shipping enterprises: Report.* New York: United Nations, 1972. 24 pp.

Embodies the results of study which UNCTAD Secretariat carried out in response to resolution of Committee on Shipping: Resolution 12(11) on development of merchant marines of developing countries, and establishment of multinational shipping enterprises.

3150 United Nations Conference on Trade and Development. *Shipping: Participation of developing countries in world shipping and the development of their merchant marines.* Geneva: UNCTAD, 1978.

3151 United Nations Conference on Trade and Development. Success in shipping. *Journal of World Trade Law* 4(4), July/August 1970, 586–90.

Policy crisis among traditional maritime nations. Resolutions adopted relating to freight rates and conference practices; finance for purchase of ships by developing countries; assistance to shipping.

3152 United Nations Conference on Trade and Development. The UNCTAD Board and shipping. *Journal of World Trade Law* 5(2), March/April 1971, 225–9.

A report on two developments in UNCTAD: One concerning legal status of the Committee resolutions; the other concerning a draft text on shipping and parts approved by the UNCTAD Board at its 10th session.

3153 United Nations Conference on Trade and Development. *World developments in shipping, ports and multimodal transport.* Geneva: UNCTAD, 1977. 134 pp.

A review which examines statistical evidence and other information with regard to the development of international maritime transport. Particular attention is given to factors and developments affecting the trade and shipping of developing countries.

3154 United Nations Economic and Social Commission for Asia and the Pacific. *Impact of containerisation on shipping in selected developing countries of the ESCAP region.* Bangkok: United Nations ESCAP, 1982.

International Labour Organization

3155 Goldberg, Joseph P. ILO tightens standards for maritime safety. *Monthly Labor Review* 100(7), July 1977, 25–30.

3156 *International Labour Conference. 55th (Maritime) Session, Geneva, 1970. Report of the Director General.* Geneva: I. L. O., 1970. 75 pp.

3157 International Labour Organization. *Maritime labour conventions and recommendations.* 2nd rev. edn. Geneva: International Labour Office, 1988. 238 pp.

3158 International Labour Organization. *Joint Maritime Commission, 23rd Session, 1980. Geneva. The minimum basic wage for able seamen.* Geneva: International Labour Office, 1980. 26 pp.

New Organizations

3159 Couper, A. D. New maritime organisations. (Editorial.) *Maritime Studies and Management* 3(1), July 1975, 3–4.

Looks at question of why oil-producing countries of the Arab world most vigorous advocates of new marine transport investments.

Name Index

Aandahl, C. L. 658
Abbott, Charles 1874
Abdalla, A. A. F. 563
Abecassis, David W. 831, 1875
Abol al Shamat, M. F. 2128
Abou-el-Hassan, A. A. 1620
Abrahamsson, Bernhard J. 166, 576, 586, 1418, 2853
Abu-El-Hassan, Aly 1621
Abu Hakmeh, Juma 9
Adamczak, Wojciech [Weseiech] 2366, 2407
Adams, George 1211
Adams, J. N. 457
Adams, John 1997
Ademuni-Odeke 956, 957, 1936, 3001, 3002
Agapos, A. M. 381
Agarwal, Vinod B. 768, 1210, 1486
Agman, Robert S. 2942
Ahern, John R. 2645
Ahmed, A. E.-S. 1513
Ahnish, Faraj Abdullah 1876
Aila, M. T. M. 1045
Ainah, E. O. 1275
Akgul, B. U. 2730
Akhtar, Saleem 2276
Akiyama, Tomomasa 1046
Al-Hassan, O. S. A. 1697
Al-Kazily, J. 1047
Alabere, A. D. 1276
Alba-Barriga, T. E. 482
Albrecht, Hans-Christian 2200, 2277, 2464
Albury, P. F. 1741
Aldcroft, Derek H. 1, 2905
Alderton, Patrick M. 70, 231, 324, 1309
Alexander, Bruce, E. 2332, 2379
Alexander, L. 1796
Alexander, Lewis M. 2943
Alexandrakis, A. J. 659
Alfred, I. 867
Allbritton, Jack L. 2201, 2421
Allen, Betty B. 2178
Allen, J. C. 2620, 2621
Allen, Joseph M. 2447
Altahuri, A. M. 2588
Alty, S. 1174
American Bar Association, Tort and Insurance Practice Section 2589
American Bureau of Shipping 2754
American Society of Naval Engineers 1838
Aminoff, Jan 2278
Amorouayeche, F. 2854
Anand, J. C. 868
Anders, Przemyslaw 1811
Anderson, Andrew W. 1877
Andreassen, J. A. B. 331
Antoun, Y. Y. 2722
Apostolides, Michael 798
Apostolidis, A. V. 104
Apostolopoulos, Y. N. 382
Appleby, George 1998
Arab–British Chamber of Commerce 614
Argiroffo, Enrico 1653, 2834
Ariel, A. 332, 660, 661
Arlt, W. H. 1372
Armstrong, Terence 2723
Armstrong, Timothy J. 1048
Arnold, G. 1419
Arnold, R. A. H. 2333
Arnott, D. 1212
Arnould, Sir Joseph 2590
Arria, Luis Coja 1988
Arroyo, Ignacio 2465, 3060
Arthur D. Little Limited 656
Artz, John C. 731
Ashar, A. 638
Asokan, Govindarajalu 2656
Aspinwall, M. [D.] 2835, 3013
Assaad, A. F. 1250
Asser, T. M. C. 2129, 2367
Associated British Ports 1213
Association of Ship Brokers and Agents 866
Association of Singapore Marine Industries 2

Asteris, Michael 1214
Astle, W. E. 1878, 1999, 2000, 2001, 2086, 2087, 2130, 2179, 2368, 2490
Auchter, Gerard 2369
Audo, S. 1310
August, Raymond S. 2202
Australian National Maritime Association 2778
Australian Shippers Council 869
Avi-Itzhak, B. 247
Azzee, Baha Bin Hussein 167, 2724, 2725

Babbage, Ross 2779
Badger, Dennis 2591
Baer, Herbert R. 2523
Bagwell, David Ashley 1565
Baker & McKenzie 2279
Baker, C. C. R. 1357
Baker, R. Palmer Jr. 2944
Bakr, M. 3
Ballantyne, W. M. 2280
Bancardi, C. G. 3032
Banda, C. A. 1879
Bang, H.-S. 1296
Bank, R. K. 870
Bankston, W. 1772
Bankston, W. B. 1797
Bannister, J. E. 2684
Baptist, C. N. T. 2334
Bar-Lev, Joshua 958
Baran, Metin 2281
Barker, Simon 186
Barnes, B. I. 1692
Barnes, L. 1654
Barry, David 169
Barry, Michael 1673
Bartlett, Paul 732
Barton, J. R. 390
Barty-King, Hugh 699
Bashford, A. S. 2592
Bashir, Ahmed 1952
Basilico, Rocco 391
Bassis, Michael S. 1698, 1713, 1714, 1840
Bassoff, Joel M. 2370
Bateman, Sam 2779
Bauchet, P. 349, 1311
Baumel, C. Phillip 1643
Bauer, R. Glenn 1957, 2002, 2180
Baxter, Richard 1633
Bayer, Nathan J. 2181
Bayley, A. Gordon 799
Baylis, Arthur E. 1880
Be Vier, Michael J. 985
Beatson, J. 2003
Beaulieu, Jeffrey 1643
Beddard, Ralph 2371
Beenstock, Michael 223, 587, 662, 733, 2653
Behnam, A. 901, 2855
Behrendt-Poulsen, E. 2203
Bekiashev, Kamil A. 3061
Belcher, Gary L. 1634
Bell, Andrew P. 2131
Bellis, J.-F. 3089
Benabi, B. 441
Bendall, Helen B. 1401–1403, 1805
Benford, H. 350–352, 1825
Bennathan, Esra 871, 1312, 1404, 3033
Bensch, Erhard 26
Bentil, J. Kodwo 2583
Bentley, Anthony M. 2204
Beresford, A. K. C. 1197
Berg, Jack 2088
Bergstrand, Simon J. 657, 2726, 2727, 2797
Berlingieri, Francesco 2089, 2205, 2282, 2491
Bernard, H. Russell 1790, 1791
Bernstein, Ronald 2492
Berridge, G. R. 2980
Bertschuger, P. P. 2448
Bes, J. 663
Bescoby, J. H. 413
Besman, John 2090
Bess, H. David 105
Bessler, David A. 155
Beth, Hans Ludwig 106, 564, 776, 872, 873, 1590, 1591
Betz, John 2004
Biagini, E. 1237
Bierman, Don Edward 1635
Billy, George J. 4
BIMCO 701, 2091
Binkley, James K. 155, 664
Binkley, John 3014
Bird, James H. 1198, 1313, 1314, 1436, 1847, 1848, 3090, 3091
Birnie, P. W. 2545
Bjornland, D. 1142
Black, Jan Knippers 1605
Black, John N. 1420
Blackwell, R[obert] J. 2755, 2945
Blake, Gerald H. 2728
Blake, Michael J. 2963
Bland, Gail 1847
Blaney, Harry C. 2546
Blauwens, G. 1437
Bloom, Murray A. 383
Blumenhagen, D. 1049
Bobrovitch, Dov 1438
Bockrath, Joseph 2335
Boczek, Boleslaw Adam 2798
Boe-Oystein 340
Boffey, T. B. 1050
Bohme, Hans 534, 588, 2729, 2856, 3003

Bolar, Malathi 2990
Bolton, Rita V. 5
Bomchil, M. 2283
Bond Smith, J. 2182
Bonnick, Stanley 2024, 2092, 2093, 2127
Bonwick, George J. 2697
Booker, Mark D. 1051
Bookman, Charles A. 2336
Booz-Allen and Hamilton Inc. 615
Borrie, G. 1175
Boschken, Herman L. 1297
Bosselaar, B. C. 1533
Bouayad, A. 2857
Bouzoubaa, A. F. 2858
Bowman, Robert J. 1134
Bowman, Roy G. 1958
Boyd, Roy 3026
Boyd, Stewart C. 2149
Bradley, M. F. 1176
Bradshaw, M. R. 1514
Bradshaw, Richard E. 2931
Branch, Alan E. 6–9, 71, 276, 1515, 1812, 1849–1851
Brann, Scott R. 1962
Breakfield, James W. 768
Bredima-Savopoulou, Anna E. (Bredimas, Anna E.) 3092, 3093, 3094
Breitzmann, Karl-Heinz 2932
Bremer, W. M. 734, 759
Brice, Geoffrey 2337, 2338, 2372
Bridger, M. W. 3140
Briggs, Adrian 1881
British Maritime Charitable Foundation 2698–2700
British Maritime League 2906
British Shippers' Council 442
British Shipping Federation 1655, 1742
British Transport Docks Board 1215
Brittin, Burdick H. 2547
Brock, Denis 2593
Brockel, Harry C. 1429
Brodefors, Rigmor 126
Brodie, Peter R. 10–17, 464
Bromich, M. 1315
Bromley, Raymond Curtis 986
Brookfield, H. C. 1052
Brooks, Mary R. 874, 875, 959, 1053, 1054, 1967, 2780, 3095
Bross, Steward R. 72
Brown, E. D. 2548, 2859
Brown, Robert H[enry] 18, 1841, 2594–2597, 2606, 2671, 2678–2681
Brown, Ronald G. 2975
Brown, W. R. 422
Browne, Michael 657
Brownsword, Roger 1997
Brunck, Arthur E. 2609
Bryan, F. J. 5
Bryan, Ingrid A. 465, 987, 988
Bryant, D. 1765
Bryon, R. V. 19
Bryon, T. N. 19
Buchan 296
Buchwald, Todd 1959
Buck, E. F. G. 1216
Buckley, D. J. 1421
Bugg, Ralph 1852
Buglass, Leslie J. 2449, 2450
Bunel, Jean Claude 903
Bureau, Lisa C. 1964
Burger, W. 187, 1699
Burke, Christopher 1963
Burrell, D. C. E. 423
Burrick, Robert S. 2253
Burroughs, R. H. 1143
Burton, V. C. 1766
Bush, Marian 162
Bushell, George E. 209
Butcher, John 188
Butler, Eamonn 1316, 1439
Butler, William E. 2493, 2494, 3062
Butterfield, C. S. J. 80
Button, K. J. 817, 2827, 3095
Buxton, I. L. 248, 320, 437, 2730
Buzan, Barry 2991
Buzek, F. J. 2432
Byington, Russell 876

Cadwallader, F. J. J. 639, 1719, 1792, 1937–1940, 2094, 2206, 2373, 2374, 2584
Cafruny, Alan W. 2860, 2861, 3096
Cahill, Richard A. 2422, 2423
Caillol, R. 199
Calder, Douglas 2284
Calder, Korella J. 1989
Calvert, J. 20
Cameron, R. M. 168
Campbell, Donald 2533
Campbell, M. R. 384
Canadian Transport Commission 877, 989, 3004
Canfield, J. L. 1251
Canna, Elizabeth 1144, 2687
Canner, W. H. P. 589, 1196
Cargo Systems International 1055–1057
Carlisle, Rodney 2326, 2799
Carmichael, Arthur J. 1440
Carnilla, Phyllis D. 2213
Casas, J. 325
Casavant, Kenneth L. 1317, 2973
Case, Leland S. 1636
Cashman, J. P. 353
Cashmore, Christopher 2005, 2006, 2017
Cassegrain, Jean 2095
Casson, Mark 169, 590, 878
Cate, Curtis 2731
Cathcart, Peter G. 3034

Centre d'Études de Transport pour la Méditerranée Occidentale (CETMO) 1238
Cervone, John B. 1960
Chaballe, L. Y. 60
Chadwin, Mark L. 1058
Chakradaran, C. 1952
Chalmers, Sir Mackenzie D. 2598
Chambers, J. D. C. 1391
Chameides, Steven B. 2424
Champness, Michael 735
Chandler, E. J. 1767
Chandler, George F. 2007, 2132, 2495
Chang, S[emoon] 1318, 1631
Chang, W. H. 1422
Chaplin, Peggy 2174
Chapman, J. B. 1294
Chapman, Paul K. 1656
Chapman, Stephen E. 250
Chappell, D. 935
Charemza, W. 216
Charity, David E. 1882
Charles, Alison 66
Charlier, J[azques] J. 1177, 1193, 1239
Chartered Insurance Institute 2599–2603
Cheka, C. N. 1199
Chen, C. V. 1990
Chen, H. 592
Cheng, Phillip C. 281
Chesterman, Philip J. 2285
Chida, Tomohei 392
Chinese Institute of Navigation 2762
Chiu, T. N. 1274, 1298
Chizea, B. I. 151
Choi, Chong Ju 1951
Choo, Chung Kek 1441
Chorley, Robert Samuel Theodore 2207
Chowdhury, Khabirul Haque 1178
Chrispeels, Erik 2023, 2133
Christopher, M. 1824
Chrzanowski, Ignacy H. 82–84, 251, 252
Chu, David K. Y. 1274, 1298, 1299, 1516
Chudleigh, Peter 990
Churchill, Robin 2242
Cimokowski, Edwin W. 2134
Claessens, E. [M.] 180, 1319, 1826, 3005
Clapp, I. C. 1374
Clark, M. 1442
Clark, Peter D. 2585
Clark, R. B. 820
Clarke, A. P. 2524
Clarke, Malcolm [A.] 354, 2008, 2451, 2604
Clarkson Research Studies 483, 736, 832, 833
Cleary, Camilla 571
Clegg, Paul 108
Clement, David A. 1646
Clifford Chance 3097
Clingan, Thomas A. 2549
Clough, Mark 3098
CMI *see* Comité Maritime International
Coffee Trade Federation 2009
Coghlan, F. P. 1534
Coghlin, Terence 2126, 2339
Cohen, D. 253
Cohen, Michael Marks 2097, 2164, 2496
Cole, Stamford Darley 1883
Colinard, R. J. 2191
Colinvaux, Raoul 1945, 2010, 2208
Collins, J. W. F. 1637, 1638
Collins, Lawrence 2497
Collins, Michael G. 2340
Collis-Squires, N. 1535
Collison, F. M. 879, 880
Colvin, Michael 2907
Comité Maritime International (CMI) 1884, 2091, 3063
Commission of the European Communities 355, 3099
Committee on effect of smaller crews (for National Research Council) 1720
Compton, M. 1487
Conley, M. W. 2946
Conlon, R. M. 65, 636, 881, 953–955, 1044
Connell, Raymond 2498
Conner, M. Timothy 2375
Connolly, D. J. 1536
Connolly, John 1583
Container Operations and Shipping Conference 991
Conway, Barbara 2532, 2550
Cook, T. C. C. 1219, 1517
Cooke, Julian 2098, 2463
Cooke, Stephanie 3100
Cooper, J. J. 2003
Cooper, M. J. M. 1059
Coopers and Lybrand Associates 1375
Corah, R. 2457
Corbet, A. G. 1885
Corbett, Julian S. 3040
Corkhill, Michael 170, 171, 834–836
Corlett, Ewan 109
Correa Chairez, R. 1504
Costabel, Attilio [M.] 2286, 2466
Cotton, A. R. 992
Cottrell, Alvin J. 2885
Coulson, E. C. 737
Coulthard, Peter 2341
Couper, Alastair D. 22, 73, 110, 1300, 1320, 1423, 1505, 1842, 3159
Couper, Frank E. 2425
Couraux, M. B. 1842
Cousen, B. 1376
Coveney, A. 1721
Cowap, G. R. 1827
Cox, J. C. 558, 1484

Coyne, J. 443
Craig, R. S. 111
Cranfield Institute of Technology 1217, 1218
Crawshaw, B. W. 1220
Creigh, S. W. 1537
Crew, J. G. 1256
Cross, Edward M. 610
Crowley, J. A. 1153
Croxson, R. C. 702
Cullinane, K. P. B. 254, 703–705
Cullison, A. C. 1750, 1751
Cunningham, John C. 2056
Cunningham, John R. 2209, 2287, 2467, 2499
Cuny, P. J. 23
Curtis, Jeffery 2732
Curtis, Simon 356
Curwen, Nicholas 2099
Curzon, L. B. 1886
Cuzan, A. G. 1200
Czapski, Waldemar 1146

D'Arcy, Leo 2011
D'Este, G. M. 1180, 1181
D'Imperio, Robert 2933
Dabrowski, K. 616
Dale, M. 2733
Dalenburg, Douglas R. 1266, 1354
Daley, James M. 1266, 1354
Dally, H. K. 1359, 1377, 1405, 1424
Daltung, S. 960
Damigos, E. 665
Danish, A. M. A. F. R. 1277
Danter, John 1465
Darling, Gerald (Q.C.) 2342, 2343, 2500
Darvas, Robert 1920
Daschbach, R. B. 2947
Datz, I. M. 255
Davies, Donald 2100, 2165
Davies, J[ohn E.] 882–894, 993–995
Davies, M. B. 1941
Davies, Marian 2728
Davies, Martin 1991, 1992, 2101, 2135, 2210, 2376
Davies, Michael 2702
Davies, Peter N. 134, 135, 392
Davies, Warwick 1316
Davis, D. 1538
Davis, H. C. 1321
Davis, J. G. 112
Davis, P. V. 200
Davison, Richard 1887
Dawes, B. 1378
De Boo, Arie 188
De Borger, B. 666
De Gaetano, George A. 2288
De Jong, J. 1639
De Kerchove, René 24
De Koning, J[ohn] P. M. 2211, 2289, 2468
De La Pedraja Toman, René 113
De Lombaerde, P. 1322
De Monie, G. 1592
De Neufville, R. 1406
De O. Goncalves, Franklin 617
De Orchis, M. E. 1060
De Sieno, Timothy B. 2212
De Weille, Jan 1379
Deakin, B. M. 996, 997
Deakin, F. R. 1240
Dean, C. J. 2719
Debattista, Charles 2136, 2137, 2221
Debievre, A. 1765
Dedial, Jurg 2734
Deere, Derek H. 601
Degenhardt, Henry W. 25
DeGurse, John L. 2012
Denholm, I. 2908
Denning, Stephen M. 2138
Denniston, Thomas 1853
Denny, Maurice E. V. 2452, 2686
Desai, H. B. 2781
Devanney, J. W. 999
Di Pilato, Vincent 1460
Diamond, Anthony 2102
Dias, Mateus Andrade 2290
Dicer, G. N. 1330
Dick, Howard W. 1061, 1147–1149, 2782, 2783
Dickey, Anthony 2210
Dickie, Paul 3006
Dickinson, B. 1278
Diehl, Niel N. 1632
Dillon, Stephen D. 1961
Din, M. A. E. 2735
Dinwoodie, John 1443
Dipl-ling, V. 26
Diplock, Kenneth 640, 2183
Dobler, Jean-Pierre 357
Dobson, H. W. 1197
Dockray, Martin 2013
Doganis, R. S. 2800
Doganis, Rigas 657, 2727
Dolman, Antony J. 1200a
Domke, Martin 2501
Donaldson, Sir John 2502
Donaldson, J. F. 2453
Donn, Clifford B. 1743–1746, 2756, 2846, 2847
Dooley, G. 1695
Douglas, B. W. 943
Douglas, P. S. 236
Douglas, R. P. A. 2327
Dover, Victor 2606
Dowd, T. J. 1380, 1407
Down, Paul M. 2607
Downard, John M. 277, 326
Doxley, J. M. 1696

Doyer, S. 895
Drewry Shipping Consultants 85, 149, 156, 172, 217, 224–228, 237, 279, 280, 297–299, 327, 334, 358–365, 385, 414–418, 435, 436, 438, 444–446, 471–481, 484–512, 523–531, 535–544, 565, 602–604, 618, 667–681, 707–709, 716, 720–727, 738–742, 742a-749, 777–787, 800–802, 807–812a, 821, 822, 837–854, 864, 865, 896, 1062–1064, 1114–1116, 1120, 1135, 1136, 1170, 1323–1327, 1622, 2688, 2763, 2771, 2784, 2801, 2802, 2862, 2863, 2948, 3007
Driehuis, W. 897
Driesen, George B. 1747
Drury, Charles 300
Drzal, Michael P. 2213
Du Cann, Edward 2909
Dubner, B. H. 2551
Duff, Peter 114
Dunbar-Nobes, Anne C. 1444, 2803
Dunlop, Anthony 2921, 2934
Duran, Oyden Ortega 1606
Dutton, C. M. J. 552
Dyer-Smith, M. 1765
Dykes, Andrew 2377

Eadie, J. D. 2175
ECMT *see* European Conference of Ministers of Transport
Economic and Social Commission for Western Asia 2785
Economic and Social Committee of the European Communities 2804, 3101
Economic Commission for Latin America and the Caribbean 2986
Economist Intelligence Unit 2805
Egiyan, G. S. 2806, 2830
Ekpenyong, S. 1279
El Noshokaty, S. M. 1381
El-Sayed, H. 2935
Elawad, B. A. M. 1280
Elborne, Robert 2344
Ellen, Eric [F.], 2533, 2552, 2566, 2567
Ellman, Mark 2525
Ellsworth, Robert A. 898, 1000, 3008
Elton, C. 1607
Elvey, Lisa A. 770
Embiricos, Philip 823
Emerson, E. C. 2777
Emery, Chris 28
Engerrand, Kenneth G. 1962
Erdmenger, Jürgen 3105
Erez, Miriam 1700
Eriksen, I. Erik 717
Erlund, Jan 2469
Ernst Robert Lindley & Sons, B. V. 2454
Escobar, Luis 3033
Essien, E. E. 2772
Ettinger, Jan Van 1200a
European Communities 1640
European Conference of Ministers of Transport (ECMT) 1151, 1641, 1642
Economic Research Centre 1150
European Shippers Councils, Council of European and Japanese National Shipowners Associations 256
Eussner, Ansgar 3107
Evangelista, Rafael E. 2291
Evans, J. J. 249, 335, 386, 899–901, 995
Evans, John 86
Evans, Malcolm D. 1888
Everett, J. 750
Ewart, W. D. 321
Ewart, W. O. 710
Ewing, Keith 2848
Eyre, J. [L.] 641, 642, 1065, 2831, 3035
Ezzat, H. 1623

Fabre, P. C. 3137
Fadem, J. A. 1539
Faghfoori, M. 2162
Fairplay Publications 157, 301, 2470
Fajemirokun, Bola 2214
Falowo, I. O. 1281
Falvey, Patrick J. 2189
Fariday, Edward K. 855
Farrant, Stephen 3108
Farrell, S[heila] 1495, 2807, 2987, 3015
Farris, Martin T. 105
Farrugia, R. 2215, 2471
Farthing, Bruce 3064
Farthing, R. B. C. 961
Fasbender, Karl 1001
Fastenau, R. 1506
Faust, Peter 619, 689
Fawcett, F. C. 1002
Fearnresearch 447, 591, 620, 682
Federation of National Associations of Ship Brokers and Agents (FONASBA) 2091, 2103
Feltham, J. D. 2014
Feng Li Qi 1983
Fernandes, Rui M. 2608
Fernandez, M. 1963
Ferraz, Joao Carlos 393, 394
Fettner, Lee 1808
Ffooks, Roger C. 856, 857
Filani, M. O. 1282
Finc, Mark J. 1964
Fink, Ze'ev S. 2216
Finnie, Alasdair 2015
Fisher, J. P. 1540
Flatman, A. J. 1152
Fleming, D[ouglas] K. 201, 244, 643, 1003, 1257, 1328, 1329, 1768, 2949
Fletcher, B. 1748

Flitner, Arthur L. 2609
Florida Bar 2217
Flyntz, Frank J. 2426
Fogarty, Aengus 1889
Foggin, J. H. 1330
FONASBA *see* Federation of National Associations of Ship Brokers and Agents
Force, Robert 2218
Ford, C. R. 1331
Ford, Michael 2610
Forsyth, C. J. 824, 1658, 1749–1751, 1769–1773, 1797, 1798, 2950
Forward, C. N. 1258
Fossey, John 1089
Foster, Leslie A. 63
Fox, Nancy Ruth 902, 1004
Franck, Bernard 903
Franco, G. Robert 1659, 1810
Frankel, Ernst G. 116, 117, 181, 218, 448, 553, 904, 1332, 1425, 1445, 1507, 1813, 1828, 1829, 2864, 2865, 2951, 2952, 3016, 3065
Franklin, P. 1814
Frazer, R. J. 1201
Fricke, Peter H. 1650, 1660, 1722–1724, 1793
Friedell, Steven F. 2219
Frieden, Luc 2292
Friedman, Norman 2953
Fritzhand, Gary I. 2944
Fulford, K. G. 422
Fuller, Stephen 521, 1368, 1608

Gajawira, B. P. 2866
Galbraith's Shipping Advisory Services 2703
Gallagher, F. D. 906
Gallagher, M. 1153
Gallagher, Thomas J. 301a
Gallimore, William 1608
Ganado, Max 2016, 2220
Gandhi, D. S. 118, 1890
Gandy, Dick 2139
Gardiner, Paul 2714
Gardiner, Robert 110
Gardner, Bernard M. 189, 302, 336, 337, 577, 686, 907, 908, 929, 1066, 2867, 2868
Garey, D. 1569
Garnett, H. C. 1074, 1508
Garratt, M[ike G.] 1154, 1171, 1182, 2704
Garrett, K. S. 136
Garrod, P. V. 1155
Garrod, Peter 605
Garson, John R. 3027
Gaskell, N[icholas.] J. J. 2221, 2378
Gates, Edward T. 190
Gautrat, Patrick 2869
GCBS *see* General Council of British Shipping
Geen, G. K. 2327
Gemmel, Andrew W. 1649
General Council of British Shipping 1408, 1661, 1662, 1830, 2091, 2705, 2708
Gentle, N. F. 328
Georgandopoulos, E. A. 729, 962
George, James A. 1566, 1567, 2526
Georghadjis, A. 2222
Ghang, J. H. 683
Gibbs, Stephen R. 1609, 1615
Gibney, R. F. 1067
Gifford, J. Morris 1202
Gilman, Jonathan C. B. 2636, 2637
Gilman, S. 202, 238, 567, 606, 1068–1070, 1117, 1333, 1382
Gino, Paul 2786
Gitsu, M. A. 2293
Giziakis, K. 191
Glaskowsky, Nicholas A. 2328
Glass, David A. 2017, 2104
Gleerup, Alf 1759
Glen, D. [R.] 751, 788, 789
Glennie, Angus J. 2488
Glick, Leslie A. 2936
Gloersen, Erik 513
Goddard, K. S. 2427
Goghlin, T. G. 2658
Gold, E[dgar] 25, 192, 1701, 1891, 1892, 2345, 2870
Goldberg, Joseph P. 1774, 1775, 3155
Goldie, C. W. H. 2018, 2659
Goldman, Steven E. 1965, 2186
Goldrein, Iain S. 2472
Goldstein, A. M. 2690
Gomer, Hilaire 3109, 3110
Gomez-Ibanez, José A. 3036
Goodacre, J. Kenneth 2455, 2611–2614, 2672
Gooding, K. 1856
Goodman, Allen C. 1446
Goodwin, E. M. 32
Gooneratne, S. G. 1421
Gordon, Paul 1776
Gorman, Francis J. 1568, 2019, 2379
Gorter, Wytze 2954
Gorton, Lars 1831, 1870, 2020–2022, 2105, 2223
Goschka, Dennis A. 2428
Goss, R. O. 74, 87, 182, 189, 193, 194, 219, 232, 239–241, 282, 302, 338, 554, 684, 909, 1005, 1259, 1301, 1334–1338, 1383, 1492, 1496, 1518, 1519, 1799, 1806, 1807, 2706, 2871–2874, 2910, 2992, 3009, 3111
Goto, Shigeya 2976

Gough, Neville 1893, 2615
Gould, H. R. 803
Goulielmos, A. 220
Grafton, R. G. T. 158
Graham, M. G. 119
Graham, Thomas 2023, 2133
Graillot, A. 1497
Gram, Per 2024
Gramling, R. 1769
Grammenos, C. Th. 303
Gray, James W. 304, 305, 711
Gray, R[ichard] 1861, 2538
Great Britain, Central Office of Information 75, 1221
Great Britain, Committee of Inquiry into Shipping 2912
Great Britain, Department of Employment 1541
Great Britain, Department of the Environment 1222
Great Britain, Department of Trade – Shipping Policy Division 1071
Great Britain, Department of Trade and Industry 1593, 2707
Great Britain, Department of Transport 578, 910, 1156, 1169, 1183, 1223, 1663, 2654, 2708, 2911
Great Britain, Parliament. House of Commons. Transport Committee 2913
Green, Meana Fernando 2294
Greenway, Ambrose 2732, 2736
Greenwich Forum Conference on Britain and the sea 2709
Grey, Michael 857a, 1118, 1651, 2617, 2875
Grey, R. J. M. 861
Griffiths, J. D. 1384, 1385
Griggs, J. G. R. 2224, 2473
Griggs, Patrick 2380
Grime, Robert P. 2225, 2381, 2382, 2625
Grimsley, R. H. 1857
Grobey, Harold A. 1447
Gronfors, Kurt 2025, 2184
Gronicki, M. 216
Grosdidier de Matons, Jean C. 1493, 1520
Grove, Eric 3040, 3042, 3043
Grover, Steven F. 2066
Grundy, Ken 2808
Grzelakowski, Andrzej S. 1386
Gubbins, Edmund J. 76, 2914
Gunaratne, M. H. 1006
Gunatillake, D. S. 1302
Gunton, G. Paul 33, 34, 1702
Guzhenko, T. 2737
Gwilliam, K. M. 88, 89

H. Clarkson and Company 752
Ha, T. M. 1664
Ha, Yeong Seok 183
Habeeb, William Mark 1610
Haberbeck and Galloway 1953
Hader, Arnulf 106
Haellquist, Karl Reinhold 2738
Hagberg, Lennart 2226, 2295, 2296, 2474
Haines, Stephen W. 2568
Haji, Iqbal 3138
Hakiki, M. A. 2527
Hale, C. 229, 712
Hall, Geoffrey Michael 2616
Hallet, G. 1448
Hallock, D. S. 963
Hamilton, D. O. 2297
Hampton, Michael J. 159
Hanappe, Paul 1339
Hansen, Bent 1624
Hansen, H. 2876
Hansen, Victor V. 2298
Hanson, Charles 1542
Hanson, Philip 2739
Haralambides, Hercules E. 233, 306, 579, 580, 1777
Harding, A. S. 242
Harlaftis, Gelina 2922
Harley, S. T. 2408
Harlow, Bruce 2553, 2569
Harper, J. C. 1942
Harrer, Bruce 664
Harrington, J. W. 2955
Harris, Nigel 1894
Harrison, R[ichard] T. 366, 395
Hart, Paul 3112
Harte, M. 1594
Hartwell, N. M. 1543
Harvey, D. A. 339
Hassan, E. M. 1625
Hattori, Katsuhiko 164
Hauser, Robert J. 1643
Havens, George R. 2570
Hawdon, David 790–792, 812b
Hawkes, Kenneth Gale 1125, 3044
Hawkins, J. E. 581, 1509
Hawkins, Gordon R. S. 2943
Hayman, Christopher 2740, 3113
Hayter, D. M. 120
Hayuth, Y[ehuda] 644, 645, 1252, 1361, 1387, 1388, 1428
Hazard, John 2956
Hazelwood, Steven J. 1895, 2227, 2228, 2383, 2540–2543, 2554, 2660, 2661
Head, G. I. 1157
Heal, David W. 35
Healy, Nicholas J. 2026, 2219, 2429–2431, 2503
Hearn, George H. 1752, 3017
Heaton, P. 1753
Heaver, Trevor D. 90, 329, 607, 621, 658,

911, 912, 924, 1007–1009, 1340, 2757, 2877, 2957
Hecker, G. 1411
Heere, Wybo P. 1934
Heggie, Ian 1341
Heiman, Bruce J. 2958
Heine, Irwin Millard 2959, 2960, 2978
Heinemann, Guy L. 2849
Heinzelmann, Raymond G. 3019
Heller, Paul P. 2384
Helmick, Jon S. 2328
Hemson, D. 1585
Henderson, Lewis E. 2185
Henrion, F. M. 514
Henry, Cleopatre Elmira 3075
Herber, Rolf 2385
Herman, Amos 1010
Herman, Lewis 2186
Hershey, R. 1693
Hershman, Marc J. 1595
Hewer, Christopher 2571, 2617
Hicks, R. M. 2673
Higgs, Andrew 1948, 2032
Higham, J. R. 203
Hilditch, Peter J. 396, 3045
Hill, C. J. 2027
Hill, Christopher 2229–2232, 2661
Hill, D. J. 1858, 2409
Hill, J. M. M. 1800
Hill, J. R. 3046
Hill, S. 1544
Hill, Stephen 1522
Hill-Norton, Peter 2937
Hilling, David 1203, 1224, 1283–1286, 1344, 1449, 2773
Hills, W. 753
Hilton, Chris 2528
Hilton, R. 2961
Hjort, Erling C. 2299
Hlaing, Maung Tin 1703
Hodges, S. L. H. 1665
Hodgin, W. R. 1666
Hodgson, Godfrey 137
Hoffman, Stanley 1896
Hogwood, Brian W. 387, 397
Holder, L. A. 1704, 1725
Holdert, H. M. C. 2432
Holland, Richard 3142
Holloway, Ronald 1342
Holman, Michael 2106
Holt, J. A. 1426
Honan, William J. 2028
Hong Kong Register of Shipping 2787
Hong Kong Shipowners' Association 2788
Honko, Jaakko 2716
Honnold, John O. 2029
Honour, J. P. 2030
Hope, Einar 91, 92, 340
Hope, Ronald 121, 121a, 221, 1667, 1705
Hopkins, F. J. 1897
Hopwood, Warren H. 1706
Horn, K. K. 1256
Horner, Dennis 169
Hornick, Robert N. 1993
Horton Rogers, W. V. 2534
Houdlett, Ronald R. 2662
Howard, Alex T. 1966
Howard, C. A. 545
Howard, Tim 2031
Howarth, Stephen 138
Howell, Paul 1644
Hoyle, B. S. 1203, 1241, 1260, 1286, 1303, 1343, 1344, 1449–1452
Hoyle, Mark D. W. 1898
Huang, C. T. 2618
Huang, Mao-Ching 2410
Hubbard, Vincent K. 2809
Hudson, A. H. 2083
Hudson, G. F. 1832
Hudson, Geoffrey N. 2346
Hudson, N. G. 2456, 2619–21
Hudson, Robert 1545
Hughes, C. 1117
Hughes, C. N. 93, 1726, 1815
Hughes, D. E. 2878
Hughes, D. O. 119
Hughes, G. S. 2457
Humphreys, Gordon 1948, 2032
Hunter, C. M. 1287
Hunter, Ormunde 2329
Hwang, H.-D. 398
Hyslop, I. R. 2555

I.C.C. Business Ratios 257
ICHCA *see* International Cargo Handling Co-ordination
I.M.O. *see* International Maritime Organisation
Ignarski, Jonathan 2556, 2572
Iheduru, Okechukwu C. 2981, 2982
Ihre, Rolf 1870, 2022
Imakita, Junichi 1346
IMCO *see* Inter-Governmental Maritime Consultative Organization
Immediata, M. J. 1758
Imoyin-Omene, O. W. 2774
Ingpen, B. D. 122
Institute for Shipping Research, Bergen 913, 914, 2720
Institute of Chartered Shipbrokers 1871
Institute of London Underwriters 2622
Institute of Shipping Economics and Logistics, Bremen 913, 914, 2720
Insurance Institute of London 2685
Inter-Governmental Maritime Consultative Organization (IMCO) 204, 3085

Legal Committee 3086
International Association of Independent Tanker Owners 2107–2109
International Cargo Handling Co-ordination (ICHCA) 1072, 1345, 2140, 2187
International Chamber of Commerce 2110, 2535, 2536
International Chamber of Shipping 1409
International Labour Office 1668–1670
International Labour Organization 3157, 3158
International Maritime Organisation (I.M.O.) 3076–3079
Library Information Services 36
International Maritime Seminar, Seoul 2879
International Shipping Federation 1671
International Tanker Nominal Freight Scale Association 866
International Tanker Owners' Pollution Federation 825
International tanker time charterparty (INTERTANKO) 793, 794, 813, 814, 2111, 2188
International Transport Workers' Federation 2850, 3114
INTERTANKO *see* International tanker time charterparty
Ircha, M. C. 1389, 1569, 1843
Ivamy, E. R. Hardy 37, 2033, 2034, 2049, 2084, 2234, 2598, 2623, 2624
Ivković, Dorde 2300

Jackson, David C. 1899, 1900, 2035, 2235, 2387, 2411, 3066
Jackson, John 1727–1729
Jackson, Michael P. 1546
Jacobs, Jon 1859
Jagota, S. P. 3067
Jankowski, W. B. 915, 1011
Jansen, Willem 2301
Jansson, Jan Owen 165, 608, 916–921, 1121, 1122, 1347
Jantscher, Gerald R. 2962
Japan Container Association 1073
Jaramillo, D. I. 922, 934
Jarvis, Robert M. 2236, 2586
Jayawickrama, S. S. 466
Jeans, C. 1707
Jeffrey, Angus Cuthbert Trevelyan 804
Jenkins, Brian M. 2573, 2574
Jenkins, Gilbert 735, 754
Jensen, V. H. 1547
Jhabvala, Farrokh 2988
Jixiong, Piao 411
John, M. 1860
Johnsen, Erik 513
Johnson, A. 2039
Johnson, Edward L. 2302, 2303
Johnson, K. M. 1074
Johnson, Patrick 2458
Johnson, R. A. 1832
Jon, J. S. 2790
Jones, Sir A. 134
Jones, C[olin] D. 219, 684, 1126
Jones, James R. 1634
Jones, Philip N. 546, 1390
Jordan, Robert S. 3049
Joshi, P. B. 1809
Juda, Lawrence 964, 965

Kačić, Hrvoje 2388
Kahn, L. M. 1570
Kaijo Hoan Kokusai Kenkyukai 1488
Kalindaga, Y. C. 755
Kamel, M. 3005
Kan, Vun-Ping 2237
Kanjilal, S. M. 234
Kapoon, Peter 39, 2537, 2538
Kappel, R[obert] 106, 230, 555
Karatzas, Theodoros 1949
Katz, Harold 1808
Kavass, Igor I. 2963
Kaye-Kucharzewski, S. L. 636
Keane, C. P. 2504
Kearsley, Harold 123
Kegels, Tony 2238, 2304, 2475
Kelly, T. 1617
Kemp, J. F. 32
Kemp, Paul 3058
Kendall, Lane C. 77
Kendall, P. M. H. 609
Kenyon, James B. 1348
Keriakosk, Mary Jane 2239
Kerr, Donald A. 2347, 2348
Kesterman, Frank 1075, 1304, 3014
Khajuria, Sajay 2653
Khan, E. 1571
Khanti-Akom, Sophon 756
Kieding, Nicolay 2663
Kilgour, John G. 2758, 2810, 3018, 3021
Killworth, Peter D. 1791
Kim, Gil-Soo 2993, 2994
Kimball, John D. 2036, 2126, 2505
Kindleberger, Charles P. 258
Kindred, Hugh M. 1967, 2016, 2037
Kingsley, Jeremy 2664
Kissling, Christopher C. 2791
Klaassen, Leo N. 1498
Klausner, R. 342
Klein, A. 1596
Kleyn, L. 1485
Klinghoffer, Arthur Jay 2539
Knight, H. Gary 1204
Knott, John 2529
Knudsen, O. 2880
Koh, Peter S. K. 2112

Koide, Shuzo 2741
Koishi, Y. 1730
Kokolo, T. 2775
Kokuryo, Hideo 2995
Komarow, Gary 3028
Komer, Robert W. 3047
Kondratowicz, L. J. 646
Koo, Won W. 467
Korsakov, E. 1586
Koster, John C. 2429
Kotowitz, Y. 988
Kotsovilis, V. 2682
Kouladis, Nicholas 2625
Kovats, L. J. 1194, 2113
Kozolchyk, Boris 2141
Kreis, Helmut W. R. 3115
Krommenacker, Raymond J. 2881
Kruse, H. J. 259
Kryvoruka, Kenneth J. 1012
Krzyzanowski, M[aciej T.] 84, 923
Kubota, Takeo 2657
Kuo, Tu-Cheng 1262
Kushner, Lester 3133
Kwak, K. S. 1083
Kwan Natividad, B. 2291
Kwang, Peter K. 2038
Kwasniewski, K. 2938

La Feber, Walter 1611
La Saponara, Francesco 67, 1157a, 1510
Labys, Walter C. 522
Lackman, Conway L. 3019
Lael, R. L. 1612
Laffer, K. 2851
Laing, E. T. 283, 468, 622, 1013, 1014, 1076, 1410, 1411
Lambert Brothers Shipping 515, 517, 1077, 2742
Lambert, Leroy 2254
Lambeth, R. J. 2626
Lampe, Wilhelm H. 3080
Lamson, Cynthia 2900
Landau, E. 2305
Lane, D. E. 924
Lane, Tony 1801
Langdon, Frank C. 3010
Langenskiold, Henrik 2240, 2476
Lanier, R. A. 2349
Lanthier, Allan R. 2964
Lapper, R. 1184
Larkin, J. E. 1158
Larner, R. 623
Larsen, [Paul] B. 966, 2189, 2389
Larson, Donald W. 467
Laskey, Steve R. 547, 728, 1137
Laube, Jan A. 209
Laughlin, James P. 2241
Laurence, C. A. 1833
Lauriat, George 124
Lave, Lester B. 1636
Lavery, H. I. 1834
Lavon, B. 321a
Law, Stuart C. 367
Lawrence, D. P. 1802, 1832
Lawrence, S. A. 94
Lawson, Guy 2376
Lax, Michael D. 2166
Lay, S. Houston 2242
Lazaratos, G. 2811
Leader, P. G. F. 2665
Ledger, G. 2915
Ledwith, Frank 2666, 2667
Lee, Chung H. 2792
Lee, D. C. 1901
Lee, John S. 2306
Lee, M. 125, 2996
Lee, Tae-Woo 2997, 2998
Leeper, John 1453
LeFevre, C. N. 1172
Leloir, Nicole 2628
Lemon, Robert T. 2668
Lennie, J. M. 2142
Leon, P. P. 1295
Leontief, Wassily 1349
Leschine, T. M. 1380
Levikov, G. 967, 2743, 2939
Lewis, Caroline 1904
Lewis, Charles 1943
Lewis, N. 1944
Liang, G. Y. 2764
Liao, S.-P. 2143
Liberian Shipowners' Council 2836
Liebman, P. 1168
Lin, B. 592
Lindeborg, Karl H. 1634
Linden, Rigmor 127
Line, G. 2717
Liu, S. 1078
Livanos, V. M. 999
Livesey, Robert C. 1225
Llorca, Manuel R. 1015
Lloyd's List International 139
Lloyd's of London Press 926, 2114, 2145, 2350
Lloyd's of London 42, 2629, 2812
Lloyd's Register of Shipping 44, 2307
Lloyd's Shipping Economist 195, 568
Lloyd, Anthony 2144
Lobosky, Reginald H. 2308
Logan, J. A. 685
Lones, Trevor 389, 2813, 3116
Long, D. M. 2744, 2745
Loose, V. W. 1159
Lopez, Norman J. 48
Lorange, Peter 283a, 284, 343, 1836
Lorenz, Edward H. 399
Los, Matheos D. 2718
Low, K. M. S. 260

Lowe, A. V. 1903, 1968
Lowe, R. 1226
Lowenfield, Andreas F. 985
Lowry, Nigel 261
Lu, W. G. 569
Luard, Evan 3048
Luddeke, C. F. 2039
Luke, D. F. 1288
Luks, Krzysztof 84
Lukvic, Branimir 2391
Lundy, C. C. 1227
Lux, Jonathan 2630
Lydolph, Paul E. 2746
Lyons, R. A. 2459
Lysenko, Vladil 1778

MacAlister, Terry 196
McArdle, A. D. 2244
McCalla, R. J. 1265
McCaul, James R. 1808, 2814
McClain, Ellen Y. 2181
McConnell, John W. (Jr.) 1969
McConville, James 20, 1455, 1755, 1756
McCormick, B. J. 1548
McCowen, Peter 1673
McCune, D. Thomas 2115
Macdiarmid, K. J. 1391
McDonnell, John J. 1017
McDonnell, Paul 1305
McDorman, T. L. 49, 1970
McDowell, Carl E. 2675
McGoldrick, James 400
McGuffie, Kenneth C. 2245, 2433
McHose, John C. 2246, 2478
McIntosh, David A. 2507
Mackay, J. J. D. 927
McKee, David L. 1129
Mackenzie, C. H. C. 49
McKenzie, J. S. 205
McLaughlin, Susan 2058
McMahon, John P. 2116, 2509
McTamaney, Robert A. 1572
Madge, Tim 198
Madigan, Richard E. 173
Maguire, F. J. 1392
Mahassni, Hassan 2309
Major, F. 1376
Makovskaia, A. A. 1954
Makus, Larry 1608
Malek, Ali 1904
Malins-Smith, R. L. 2989
Mallon, Paul 2631
Maltenfort, George G. 1079
Manca, Plinio 1905, 1906
Mancroft, Lord 928
Mangone, Gerard J. 1907, 2965
Mankabady, Samir 1350, 2040, 2041, 2146, 2176, 2190, 2434, 3068, 3081
Mannheim, Bilha F. 1803, 1804
Mansell, Chris 1185
Marcus, Henry S. 553, 1263, 1412, 1816
Marine Technology Society 1672
Maritime Institute 2720
Maritime Law Association of South Africa 2310, 2477
Maritime Transportation Research Board 647
Market Development Dept. B & W Diesel A/S 152
Marks, Jonathan 2907
Marlow, Peter B. 86, 302, 308, 337, 686, 929, 2867, 2873, 2882–2884
Marsh, Allan G. 1817
Marsh, Arthur Ivor 1754
Marsh, James Barney 2793
Marshall, J. K. 1393
Marshall, R[olf] 624, 2958
Marston, Geoffrey 2311
Marti, Bruce E. 1080, 1081, 1127, 1128, 1264, 1295, 1352, 1489, 2759
Martin, Michael V. 1646
Martinez, Robert D. 1908
Maruno, Y. 309
Marx, Daniel (Jr.) 1016
Maslov, G. A. 2747
Mason, Herbert B. 50
Massenburg, J. W. 2674
Massey, Eugene A. 1909
Matear, S. 1189, 1861
Mather, Graham 1454
Mathews, John 1910
Mattendorf, John 3049
Matthew, Paul 2243, 2506
Mattis, Brian 2412–2414
Matz, James R. 561
Mavrakis, J. A. 2632
Maybourn, Ralph 827
Mayer, H. 174
Mayo, A. H. B. 1082
Mazzone, F. 2575
Mbanefo, L. N. 1984, 1985
Mead, Ray 2633
Meanley, John A. 51
Meek, M. 422
Meeson, Nigel 1911, 2247
Meisel, Frank 1998
Melis, D. 1818
Mellish, M. 1549
Mendelsohn, Allan I. 2042, 2147, 2392
Mendizabal, Marcial J. 2393
Menefee, Samuel P. 2557, 2558, 2576
Meng, Tan L. 2043
Menon, P. K. 1844
Mentzer, Marc S. 1130
Meredith, B. 1456
Merren, H. Orren 2085
Mesrati, S. al-A. 2634

Metaxas, B. N. 68, 95, 687, 688, 2800, 2815, 2816
Metwally, H. I. 1779
Meyrick, S. 906, 1180
M'Gonigle, R. Michael 826
Middleton, Bruce W. 2794
Middleton, Ian 3117, 3118
Mikelis, N. E. 2719
Milklius, Walter 605
Miller, Anthony R. 2351
Miller, E. Willard 52
Miller, Michael D. 2676
Miller, Ruby M. 52
Miller, William H. 1131
Mills, C. P. 2044
Mills, G. 1511
Mills, Herb 1573
Milner, Simon 1550
Mitchell, C. Bradford 1394
Mitchell, William Harry 140
Mitchellhill, Alan 2148
Mitra, B. C. 2045
Miyajima, M. 1083
Miyashita, Kunio 344, 625, 713
Mobereola, A. E. 1084, 2776, 2983
Mocatta, Alan Abraham 2149
Mohammed, F. J. 2635
Molina, E. L. 1845
Moloney, G[erald] J[ohn] 2248, 2312, 2479
Möncke, Helmuth 689
Monopolies and Merger Commission 1186
Montgomery, T. M. 1647
Moodie, M. 2885
Moon, J. R. 1205
Moore, C. G. 160
Moore, John C. 2046
Moore, Michael 1708
Moreby, D. H. 245, 1674, 1675, 1709, 1710, 1780, 2710, 2886, 2887
Morgan, Ivor P. 3036
Morgan, M. 1187
Moriarty, James 1228
Morris, Albert 2117, 2435
Morris, Charles 1613
Morris, J. H. C. 2047
Morris, Michael A. 2888
Morris, R. 1587, 1757
Morrison, W. S. G. 1711
Mort, Derrick 1
Morton, Donald S. 1523
Moses, Eli 968
Moskowitz, Daniel B. 1018
Moss, Donald C. 2849
Mossack, Jürgen R. D. 2313
Mott, William 1574, 2940
Motte, R. 206
Moukhtar, G. A. 1712
Mountjoy, Alan B. 1626
Moustakas, P. 419
Moyer, R. Charles 3022, 3023
Moyse, Andrew 171
Mukherjee, Neela 3029
Mukherjee, Proshanto K. 2436
Muldowney, T. 1676
Muller, E. J. 595, 930, 1353
Mulley, Corrine 1191
Murphy, Paul R. 1266, 1354
Murphy, Thomas Francis 2437
Musquar, P. 3089
Mustill, Sir Michael John 2149, 2510, 2636, 2647
Myburgh, Paul 2249

Nadel, Ernest 571, 2894
Naess, Erling D. 141, 2837
Nagatsuka, Seiji 161, 368, 369, 401–403, 439, 449, 450, 516, 714, 757, 2748
Nagi, A. A. 420
Nagorski, Bohdan 1457
Nakagawa, Keiichiro 142, 148
Nakagawa, Yoshiaki 1819
Naschitz, P. G. 2150, 2250, 2415
National Economic Development Office (NEDO) 1160, 1229
National Ports Council 1395, 1458, 1459
National Research Council 1720
National Union of Marine, Aviation and Shipping Transport Officers (NUMAST) 1680, 2918
National Union of Seamen 2838, 2916
Nautical Institute 197, 269, 1677, 1820
Naya, Seiji 2792
NEDO *see* National Economic Development Office
Neff, Stephen C. 969
Nersesian, Roy L. 78
Netherlands Maritime Institute 2720
Nettle, Stanley 1355
New, Christopher 2817
Newborne, M. J. 1862
Newman, D. L. 58
Newman, N. 1230
Newns, Ralph B. 2917
Ngantcha, Francis 1912
Nicholson, A. A. N. 3140
Nicholson, Sir John 582
Nicholson, T. A. J. 1821
Nicolich, Lillian 270
Nicoll, C. C. 2048
Niemiel, Joseph M. 2933
Nilsson, Bertil 2689
Ninomiya, Seijiro 2314
Noakes, Robert 974
Nolah, D. C. 1002
Nolan, Bryan 1731, 1732
Nonneman, W. 666, 1356

Nordquist, Myron 2242
Nordvik, Helge N. 1781
Norman, Victor D. 283a, 284, 343, 690, 1836, 2923
Norris, Martin J. 1678, 1679
North, John 546, 1390
Northrup, H. R. 1758, 2852
Norton, Roger D. 207
Norwegian Trade Council 2721
NUMAST *see* National Union of Marine, Aviation and Shipping Transport Officers
N.Y.K. Research Chamber 1085

O'Brien, K. H. 1460
O'Brien, L. J. 2966
O'Connell, D. P. 1913, 1914, 3050
O'Kane, Gerry 1863
O'Keefe, Patrick 2191
O'May, Donald 2638
O'Shea, James A. 2639
O'Sullivan, Patrick 2640
Oceans Institute of Canada 2889
Oda, Hisashi 648, 1086
Oditah, Fidelis 2416
OECD *see* Organization for Economic Co-operation and Development
Offerhaus, J. P. S. 1934
Office of the Defence Attaché, Permanent Mission of Nigeria to the United Nations 2559
Ogawa, Yoichi 2480
Ogundana, [T. T. Babafemi] 1289, 1290, 1291
Ohashi, Mitsuo 2511
Ohsuga, Hiroshi 931
Oil Companies International Marine Forum 1206
Oiyu, Wei 2765
Olev, Kulno 53
Olin, Gary 876
Olson, Richard E. 3059
Oram, R. B. 1357
Organization for Economic Co-operation and Development (OECD) 96, 2818
Otimong, S. 222
Oum, Tae Hoon 153
Owen, David R. 2438
Owen, M. 789

P.A. Management Consultants Limited 421
Pace, M. H. 932
Pace, Malcolm N. 626
Packard, William V. 79, 278, 451, 596, 730, 2118, 2167, 2481
Packer, Bernard 2819
Padelford, Norman J. 1614, 1615
Padfield, Peter 143
Paine, Frank 310
Pak, M. S. 556
Palmer, Michael A. 3051
Palmer, N. E. 1524, 1525
Panagopoulos, P. C. 2251, 2315, 2482
Panakakos, George 3033
Pantelidis, Evangelos T. 2924
Panunzio, Vito 1242
Papakidis, N. 597, 598
Pardini, Juan 2252
Parks, Alex L. 2394, 2641, 2642
Parmer, Jean M. 54
Parmer, Jerome F. 54
Parrington, Sheila 2917
Parrish, Robert B. 2192
Parritt, Brian A. H. 2560
Partridge, Christopher 860
Patrick, Bruce R. 2316
Paulsen, Gordon W. 2253, 2254
Payne, William 2049
Peacock, J. Hoke 2050
Pearson, Peter 790
Pearson, Roy 933, 1019, 1087–1089, 2996
Peavy, Robert 1093
Peck, Francis 404
Pedrick, John L. 758
Peebles, D. R. 453
Peeters, C. 1168
Palaez, Alfred S. 2352
Pemberton, Barry 1161
Pender, David R. 1371
Pengilley, Warren 1020
Pentti, Frank 2561
Perakis, A. N. 330, 597, 598, 734, 759, 922, 934
Perkins, R. J. 328
Perry, Nick 1733, 1794
Peters, H. J. 454
Peterson, M. J. 2562
Pettavino, Paul J. 572, 2961
Peugniez, Dominique 2749
Peyrefitte, Léopold 2051
Phelan, G. 1745, 2847
Phelan, Patricia J. 2488
Philip, Allan 2255
Philip, Craig E. 1632
Phillips, Gordon 1551
Phillips, S. W. 2814
Pike, J[on] 1152, 1162
Pilcher, William W. 1575
Pinckernelle, Erdewin 2643
Pinder, D. A. 1241, 1243, 1451
Pineus, Kaj 345, 1915, 2439, 2460–2462
Pirie, Madsen 1439, 1461
Pitfield, David E. 817, 2827
Plant, G. 208
Pocock, Pamela 1526
Pollack, Gerald A. 270

Pollack, Gerald E. 3037
Pollock, Eric E. 1313, 1462, 1463
Pomrehn, H. P. 160
Pond, P. J. 1357a
Poole, M. 1795
Pope, James A. 610, 611, 1058, 1103
Porter, Robert W. 1397
Posgate, Ian 137
Posner, J. 1188
Potter, Douglas 2587
Potts, A. 413
Potts, T. H. A. 2256, 2317, 2483
Power, Michael E. 49
Power, Vincent 3120
Powles, David G. 1916–1919, 2052, 2168, 2169, 2540–2544
Powrie, P. J. 1552
Poynder, Anthony 946
Practicing Law Institute 2257
Prescott, John Robert Victor 2890
Price, Richard 1955
Pritchard, S. Z. 829
Pritchett, Russell W. 2119, 2258, 2531
Proctor, N. C. 2655
Psillaki, Efti 2318
Pullat 691
Pullen, R. D. 1821
Punekar, S. D. 1782
Purvis, Rodney N. 1920

Quinn, Alonzo De F. 1358
Quon, Tony K. S. 209
Rajwar, L. M. S. 2891
Ramberg, Jan 2053
Ramsden, W. A. 1986
Ramwell, Dave 198
Randall, James E. 1267
Randall, P. 1448
Randolph, Fergus 3098
Ranken, M. B. F. 2711, 2892
Ratcliffe, Mike 760
Ray, Anandarup 1379
Rayner, Joe 1553
Ready, N[igel] 1949, 2820
Rechell, R. E. 2814
Redgwell, C. J. 2353
Reed, Peter B. 2681
Rees, D. J. 144
Regul, Rudolf 1244, 1245
Rehm, John B. 2967
Reid, A. B. 1359
Reilly, Danny 1776
Rein, Alex 2259, 2484
Reisman, W. Michael 3069
Rembe, N. S. 1987
Renouf, A[nthony] 761, 3121
Reynardson, W. Birch 2677
Rich, Craig A. 1163, 1189, 2920, 3122
Richard, M. 1745
Richards, Edward 1550
Richardson, John 2054, 2120
Richardson, P. 336
Ricks, Vincent E. 2208
Riddle, Ian 370
Rifas, B. E. 2894
Riley, N. T. 1306
Rimmer, P. J. 1272
Rinman, Thorsten 126, 127
Ritchie, J. D. 1527
Roach, D. K. 1603
Roberts, W. E. 1090
Robertson, Bill 2661
Robertson, David [W.] 1576, 1783, 1971
Robins, P. 1021
Robinson, M. D. 1784
Robinson, Pamela 1499
Robinson, Ross 612, 1091, 1307
Robinson, Susan K. 2260
Robol, Richard T. 2261
Rodda, William H. 2644
Rodger, Alan 2151
Roe, M. 2915
Rogers, Andrew 1921
Rogers, David N. 1922
Rogers, Nigel C. 2645
Rogers, Phillip 518
Roggema, J. 1681, 1736, 1785
Rohart, Jean-Serge 2319
Rohreke, H. G. 2439
Rokison, Kenneth 2121
Rol, Y. 1428
Romlov, Robert 2262, 2320, 2485
Ronen, D[avid] 210, 1822
Ronzitt, Natalino 2577
Rose, F[rancis D.] 1022, 2163, 2193, 2330, 2357, 2512
Rosengren, W[illiam R.] 1698, 1713, 1714, 1840
Rosenstein, E[liezer] 1803, 1804, 2690
Ross, Douglas A. 3010
Ross, P. 1577
Rossi, James A. 285
Rossmere, Andrew E. 2683
Roueche, L[eonard R.] 1159, 1190, 3024
Rowan, Richard L. 1758, 2852
Rowe, R. P. 1273
Rowley, Charles 1191, 1192
Rowlinson, Mervyn P. 692, 2839
R. S. Platou Economic Research AS 627
Rubenowitz, Sigvardi 1759
Rubin, R. 1616
Rudolph, James L. 2354
Rutter, J. I. 1092
Ryan, P. 1578
Ryan, Victoria 1754
Ryden, Inger 2893
Ryder, S. C. 242, 935

Saaban, Hairani 311
Sagger, R. K. 1123
Salatt, A. H. 2941
Salmans, S. 1786
Sampson, D. H. 2263
Sampson, Gary P. 469
Sandevarn, Arne 1870
Sapsford, David 1561
Sasaki, H. 1760
Sassoon, David M. 2055, 2056, 2085, 2194
Savy, Michael 1339
Sawiczewska, Z. 2750
Sawyer, L. A. 140
Saxena, K. B. C. 1809
Sayers, C. R. 936
Schaefer, Heinz 371
Scheaker, Eric 1429
Schechter, Joel 211
Scheller, Douglas M. 2379
Schiels, Kenneth T. 2177
Schiller, Thomas S. 2578, 2579
Schirach-Szmigiel, Christopher Von 948, 2893
Schlichting, Mathias P. 2264
Schmeltzer, Edward 937, 1093, 1094, 2968
Schmidt, Folke 2821
Schmitt, A. L. 3008
Schmitthoff, Clive M. 457, 2057
Schoenbaum, Thomas J. 2265
Schofield, John 2171
Schönknecht, Rolf 128
Schrank, Robert 1734
Schrier, Elliot 571, 2894
Schulze, G. 1360
Schwampe, D. 2122
Scobie, W. E. 1529
Scott, Robert 861
Scowcroft, Jerome C. 2395
Scrogin, Thomas W. 3059
Seatrade 322, 323, 2751, 3123
Seatrade Conference Proceedings 286–294
Seaward, Nicholas 3124, 3141
Sedano, Alejandro G. 2396
Sedgwick, Stanley 145
Seifert, Gerald 3070
Selvadurai, P. 2317
Selvig, Erling 2195, 2397, 2398, 2646
Serafetinidis, M. 2691
Serebriakov, Vitali V. 3061
Seward, R. C. 2647
Seward, T. 997
Shachar, Yoram 1095, 2058
Shah, M. J. 970–972, 2059, 2355
Shanker, A. K. 3005
Sharp, C. H. 97
Sharpe and Spicer 2266
Sharpe, P. 1796
Sharpe, Thomas 1787
Shashikumar, N. 1023, 1024, 1972, 1973
Shaw, B. P. 973
Shaw, Gordon C. 1649
Shaw, Richard 2356, 2399
Shell Briefing Service 766
Shepherd, Diana C. 59
Sheppard, Edward J. 1094
Sher, Stanley O. 1579
Sherar, Mariam G. 1684
Shim, Duke 1430
Shimojo, Tetsuji 98, 175, 583, 613, 628–630, 815
Shipping Economist Conference 767, 767a, 938
Shneerson, Dan 165, 253, 321a, 584, 608, 917–921, 939, 940, 1025, 1122, 1347, 1431, 1512, 1685
Shore, P. 2712
Shorter, James Russell 1580
Shujian, Liu 2513
Shum, Clement 2514
Siemss, John 55
Silard, Stephen A. 2060
Silverstein, Harvey B. 3087
Simmonds, K. R. 1923, 2061
Simmons, Kevin P. 2417
Simon, Jeffrey D. 2580
Simon, Seymour 1096–1098, 2062, 2418
Sinan, I. M. 2840
Sinclair, Clive 1464
Singels, A. 941
Singh, Avtar 2063
Singh, Nagendra 1716, 1924, 1925, 1945, 2440, 3134
Singman, J. H. 2969
Sirat, M. 3038
Sivewright, Sandy 146
Sjostrom, William 130, 942, 1026–1028
Skinner, George 974, 1029
Slabotzky, Albert 2515
Slack, Brian 1864
Sletmo, G. K. 624, 1030, 1031, 2692
Sloggett, J. E. 313
Smedresman, Peter S. 2516
Smeets, G. C. A. 2321, 2925, 2926
Smillie, Ian 1465
Smith, Brian D. 2895
Smith, Christopher 2343
Smith, J. R. G. 3082
Smith, Leigh 796
Smith, M. F. 1994
Smith, M. H. 1681, 1736
Smith, Robert W. 3052
Snelson, Anthony 1887
Snoy, Yolande 3125
Social Surveys (Gallup Poll) Limited 1686

Society of Maritime Arbitrators (New York) 2517
Society of Marine Port Engineers of New York 1838
Society of Naval Architects and Marine Engineers 1837, 1838
Soffer, A. 1253
Söhmen, Helmut 263, 1687
Sokoloff, Valentin A. 2766
Solomon, V. E. 2984
Soman, S. 2901
Somner, Graeme H. 147
Song, Yann-Huei Billy 2767
Soondrum, S. 519
Soper, R. T. 264
Southwood, Peter 1554
Sparks, A. 532
Sparshott, K. J. 346
Spathi, Sophia K. 408
Speidel, Volker 371
Spencer, C. F. 176
Spicer, W. Wylie 2267, 2489
Sprake, Raymond F. 145
Springett, Peter 2710
Spruyt, John 1839
Srivastava, C. P. 3083
St. Seidenfus, Hellmuth 1246
Staniland, Hilton 2268
Stanley, W. R. 1466
Stanton, L. F. H. 1926
Stapleton, J. 1865
Starr, J. T. 1268
Stasinopoulos, Dinos 3105
Stav, Chaim 2581
Steeg, Robert M. 1581
Steel, David [W.] 1947, 2208, 2357
Stemmer, Juan E. Oribe 3030, 3084
Stent, A. F. 1401–1403, 1805
Stephenson Harwood 314
Stern, E. 1361
Stevens, Edward F. 56, 80
Stevens, Paul 790
Stewart, R. J. 999
Stokes, James C. 2269
Stokes, Peter 300, 315
Stonham, P. E. 1032, 1432
Stopford, Martin 81, 316
Stopford, R. M. 372, 390, 459
Stott, K. L. 943
Strandenes, Sir I. Pettersen 805
Strange, John 518
Strange, Susan 3142
Strasburger, John H. 2270
Strath, Bo 373, 374
Strevens, Peter 1688
Stuart, Keith 1555
Stubbs, P. 2795
Stuchtey, Rolf W. 1164, 1598–1600
Studd, Brian 518
Studer, Keith R. 607
Sturley, Michael F. 1927, 2064–2066, 2648
Sturmey, S. G. 99, 131, 177, 944, 976–978, 2713, 2822
Sturt, Richard H. B. 2441
Suez Canal Authority 1627, 1628
Sugiyama, Shinya 2977
Sullivan, Eric 57
Summ, G. H. 1617
Summerskill, Michael Brynmor 2172
Susilowidjojo, Henry 2842
Suthiwartnarueput, K. 470
Suykens, Fernand 1414, 1500, 1601
Svendsen, Prof. A. S. 91, 92, 182, 268, 271, 340, 690, 717, 805, 945, 1781, 2927
Svensen, T. E. 248
Swatton, R. J. 1397, 2221
Swayne, [Sir] R[onald] 272, 2752, 2896
Swedish International Development Authority 1604
Sweeney, Joseph C. 1974, 2067, 2189, 2430, 2431, 3143–3147
Swinbank, C. 1099
Sychrava, Lev 162
Symeonides, N. 317
Tabak, Herman D. 1100
Tabbush, S. J. 2271
Tabernacle, J. B. 1477
Tae-Woo, L. 2897
Takel, R. E. 1233
Talley, Wayne K. 611, 768, 1058, 1101–1103, 1398, 1478
Tambakis, D. 2928
Tampoe, Moira 2768
Tan, Jin Oon Francis 1104
Tangsubkul, Phiphat 1995
Taplin, John H. E. 658, 2979
Tarpgaard, Peter T. 3025
Tatham, Nigel, 946
Taylor, A. J. 178, 631–633
Taylor, Leslie Granville, 1362, 1399
Taylor, R. 375
Taylor, Zbigniew, 1254
Teewia, John, 2303
Terazaki, Naomichi H. 2563
Terminal Operators Ltd 816
Terry, J. P. D. 2649
Tetley, William 1928, 1975–1980, 2068–2072, 2152, 2400, 2419, 2420, 2442, 3148
Tettenborn, A. M. 1929, 2197
Textor, James M. 2198, 2518
Theophanides, Stavros M. 2929
Theunis, Jan 2272
Thomajan, Robert 2519
Thomas, A[drian J.] 693, 3122

Thomas, B. J. 1415, 1556, 1602, 1603, 1843
Thomas, D. Rhidian 1946, 2358, 2520, 2521
Thomas, Michael 1947
Thomas, William B. 2443
Thommen, Jistice T. 2153
Thompson, Dennis 2823
Thompson, Peter 2322
Thompson, Stanley R. 467
Thorburn, Thomas 100
Thuong, Le T. 634, 649, 770, 1033, 1582, 2841
Tiberg, Hugo 2173
Till, Geoffrey 3053
Timmermans, W. A. 2073
Tinsley, David 1165, 1173
Titz, M. A. 1490
Tiyaphorn, Anurat 2323
Tkporukpo, C. O. 1282
Toda, Mitsuhiro 2401
Todd, D[aniel] 376, 377, 405
Todd, Paul 2074, 2154–2156
Toh, Rex S. 2842
Tolofari, S. R. 817, 2824–2827
Tomassoni, Mark 1453
Tombari, Henry A. 1105
Tomlinson, D. S. 246
Tongzon, J. 1588
Totten, George O. 406
Touche Ross and Company 1207
Tourk, Khairy 1624
Tourkolias, A. 265
Townsend, Alan 404
Tracy, Nicholas 3055
Trade Union Research Unit 2843
Trakman, Leon E. 1930, 1931
Transport Canada 2970
Trappe, J. 2123
Trappe, Johannes 1932
Treitel, G. H. 2157
Tresselt, D. 1788
Trew, Antony 771
Trotz, Norbet 2402
Tsunokawa, K. 1406
Tsurumaki, Yasuo 407
Tubielewicz, Andrzej 1501
Tuin, J. D. Van Der 58
Tukur, B. M. 1106
Tunander, Ola 3056
Tuppen, J. N. 1479
Turnbull, Peter J. 1502, 1557–1561
Turner, Harold A. 2650
Tye, William B. 947
Tyrer, H. 135
Tzamouzakis, Frangiscos G. 408
Tzoannos, John 3092, 3094, 3127

Ueland, Svein 2669
Uemura, Masahiro 1166
Uesugi, Akinori 2975
Uhm, Ihr H. 1649
Ullman, Gerald H. 650, 1866, 1867
Ullman, H. K. 572, 2961
UNCITRAL *see* United Nations Commission on International Trade Law
UNCTAD *see* United Nations Conference on Trade and Development
Underwood, David 1761
Underwood, Eugene 2444
Ungar, A. 1480
United Nations Commission on International Trade Law (UNCITRAL) 2075, 2158, 2159
United Nations Conference on Trade and Development (UNCTAD) 101, 132, 133, 179, 243, 273, 460, 520, 548–550, 573, 635, 651–653, 979–982, 1034, 1107, 1208, 1234, 1363, 1365, 1365a, 1400, 1433, 1434, 1491, 1494, 1530, 1604, 1629, 1873, 2160, 2651, 2695, 2828, 2832, 2844, 2898, 2985, 3011, 3020, 3071–3074, 3135, 3149–3153
United Nations Economic and Social Commission for Asia and the Pacific 1366, 3154
United Nations. Economic Commission for Asia and the Far East 1846
United States Department of Commerce 2971
United States Department of Justice 1981, 2899
United States Maritime Administration 1269
University of Southampton, Institute of Maritime Law 2386, 3136
University of Wales, Institute of Science and Technology, Department of Maritime Studies 1035
Urhoma, I. K. 1762
Uyeno, D. 924

Vafai, G. H. 1956
Valencia, Mark J. 59
Validakis, J. E. 1124
Vallega, A. 461
Vambery, Robert G. 409, 562
Van Alstyne, R. W. 1618
Van Beuningen, John 2359
Van De Voorde, E. 1437
Van der Feltz, F. 2273, 2486
Van Der Meer, R. 789
Van Notten, Michael 1481
Van Poelgeest, F. M. 830
Van Rijn, Hendrik 637
Van Voorden, W. 1689

Vanags, A. H. 184, 185, 194, 229, 712, 3031
Vandenberghe, J. P. 60
VanderZwaag, David L. 2900
Vanplantinga, John R. 1531
Vardi, Michael 1418
Veldhuizen, Paul J. 718
Veldman, Simme J. H. 212, 1108
Vella, Edwin A. 1950
Ventris, F. M. 2124
Verbeke, Alain 1247, 1248, 1322, 1596, 2930
Vergottis, Andreas 587, 599, 662, 733
Verhoeff, J. M. 1367
Verhoeven, W. 2324
Vermulst, E. 3089
Verzijl, J. H. W. 1934
Vetterick, V. 966
Vidali, M. W. 266
Vielvoye, Roger 772
Vigarie, A. 1503
Villar, Roger 2564
Villareal, Dewey R. 1982, 2199, 2445
Villeneau, Jacques 2274, 2487
Vincenzini, Enrico 2360, 2361
Viscencio-Brambila, Hector 521, 1368
Visscher, Sue L. 634
Vitek, J. 2845
Vlachos, Gerasimos M. 2403
Vleugels, R. L. M. 1249
Volk, B. 347, 564, 574
Von Raab, William 3088
Vongkiatkachorn, Suparb 2323

Waaler, Claus 797
Wada, Keiji 1738
Waddams, S. M. 2362
Wagner, Wolfgang 1001
Wainwright, David 1763
Wakeford, G. W. 1789
Walden, L. 1532
Waldron, A. J. 949, 2077
Walkate, Jaap A. 2565
Walker, A. G. 1868
Walker, A. R. 1255
Walker, Fred M. 378
Walker, Jim 1235
Walker-Leigh, Vanya 3129
Wallace, I. 1109
Walsh, Robert G. 348
Walters, A. A. 102, 871, 950, 1209, 1404
Walters, A. S. 1369
Walters, Brian 1482
Walters, Milklius 605
Walz, C. J. 1823
Wan, H[ua] 1739, 2769, 2796
Wanhill, S. R. C. 1036, 1132, 1292, 1483
Ward, Les 2404
Warf, B. 558, 600, 1270, 1484, 1485
Waring, Anne 2125
Warner, David W. 267
Warsash Nautical Bookshop, 61
Wassermann, Ursula 1630
Waters, C. D. J. 2901
Waters, D. M. 1718
Waters, R. C. 1370
Waters, W. G. 62
Watkins, G. G. 1897
Watson, Andrew 2770
Watson, Mary-Anne L. 1694
Weale, John 695
Webster, G. H. 68
Webster, P. J. K. 1236
Weeks, Eldon E. 1317
Weeks, John 1619
Wei, J. A. 645
Weil, Prosper 2405
Weiner, George J. 937
Wellard, D. J. 1359
Wellman, David 1573
Wenger, Larry B. 64
Wentworth, F. 1824
Wepster, A. 274
Wergeland, Tor 268
West, A. C. 1562
Westinform 103, 551, 696, 1037, 1690
Weston, S. 1502, 1559, 1560
White, Brian 806
White, Lawrence J. 559
White, M. W. 1996
Whitehead, Geoffrey 1852, 1869, 2591
Whitehurst, Clinton H. 410, 1935, 2760, 2972
Whitelegg, John 3130
Whiteside, Noel 1551
Whittaker, J[ulie R.] 1110, 1191
Whyte, I. M. 773
Wiese, B. 1293
Wijkman, P. M. 983
Wijnolst, Ir. N. 1168
Wiktor, Christian L. 63, 64
Wild, Peter 1141
Wildeboer, Ina H. 2363
Wilder, Ronald P. 1371
Wilford, Michael 2126
Wilkie, Roy 1727–1729, 1733
Wilkinson, Paul 2582
Williams, B. K. 2078
Williams, David McI. 2275
Williams, Ernst W. 1031
Williams, G. F. 1069, 1333
Williams, Harvey 2127
Williams, J. T. 1435
Williams, Richard 2380
Williamson, G. A. 1117
Willingale, M. C. 213–215
Willingham, R. 1195
Wilmot, Nicholas 2406

Wilson, D. F. 1563
Wilson, J. F. 2079–2081, 2463
Wilson, Peter 1564
Wilson, V. E. 1039
Wilson, Wesley W. 2973
Wing, John F. 575
Wingfield, M. 1154, 2704
Winkelmans, W. 1247, 2930
Winston, C[lifford] 150, 154
Winter, W. D. 2652
Winters, Tobey L. 1271
Wishart, J. 1312
Wistner, E. J. 318
Wiwatpattarakul, P. 2161
Wohlmuth, Karl 560
Wojewokda, Czeslaw 69, 2753
Wood, Malcolm 3131
Wood, N. J. 585
Wood, Thomas W. 1133
Wooder, James B. 2364
Woodhead, Gregory 379
Woods, James R. 1111
Wooler, R. G. 862
Wormmeester, Ir. G. J. 1533
Worth, R. W. 2325
Wright, A. J. 1416
Wright, D. 1860
Wright, G. 163, 698
Wright, Joseph P. 1652
Wurfbain, J. W. 2365
Wydra, H. N. 1388

Yamada, Hideo 655, 1041, 1042, 1112
Yamajo, T. 1417
Yamamoto, Hiromasa 1589, 1691, 1740, 1764
Yang, Chin Wei 522
Yang, K. C. 2974
Yannopoulos, George N. 2833, 2902, 3012
Yeats, Alexander J. 462, 469, 2903
Yeshou, A. B. 2904
Yi, Seong Deog 1951
Yinghui, Min 411
Yochum, G. R. 1210, 1486
Yoshida, Shigeru 380, 412, 463, 533, 863, 2761
Young, Elizabeth 3057
Young, James 2522
Ysern, M. Gorina 319
Yui, Tsunehiko 142, 148

Zachariades, A. 248
Zachcial, M. 1164
Zacher, Mark W. 826
Zammit, Joseph P. 2331
Zamora, Stephen 984
Zannetos, Zenon S. 774, 775
Zapf, Robert J. 2446
Zaphirious, G. A. 2082
Zartman, I. William 1610
Zerby, J. A. 65, 636, 881, 952–955, 1043, 1044, 1308, 3008
Zimbalist, Andrew 1619
Zurek, Janusz 275

Subject Index

accidents 190, 191, 1663, 1668, 1678, 1765, 1784
'Achille Lauro' hijacking 2557, 2580
admiralty law 2062, 2200–2275, 2417
Africa 134, 1275–1293, 1984–1987, 2771–2776, 2980–2985
agencies 2174–2177
American shipping *see* Canada; United States (US)
Antwerp 1596, 1601
Arabian Gulf states 1304, 1697, 1712, 2280, 2941
  see also flag and fleet development, Middle East
arbitration, law of 2490–2522
Arbitration Act (1979) 2502, 2512, 2516
Arctic 2900
Argentina 2283, 2393
arrest of ships 2226–2228, 2230–2232, 2237–2238, 2240, 2243, 2244, 2246, 2248, 2250–2252, 2255–2256, 2259–2260, 2262–2264, 2271–2274
  uniformity lack 2220, 2222
ASEAN (Association of South-East Asian Nations) 260, 906, 1995
Asia 149, 1091, 1653, 2738
  socialist *see* China
  South-East 59, 1703, 2546
  see also flag and fleet development; state and policy
associations, professional *see* trade unions
Australia 1161, 1174, 1799, 1805, 2211, 2289, 2468
  exorbitant jurisdiction 1881
  liner trades 869, 884, 906, 936, 1032
  maritime law 1989, 1991, 1992, 1996
  Prices Surveillance Authority (PSA) 2992
  Trade Practices Act (1974) 1020
  unions 2847, 2851
  see also flag and fleet development; ports

balance of payments 576–585
barge carriers 601, 689, 1631, 1632
barratry 1895
Bathside Bay (Harwich) 1225
Belgium 1239, 1247, 1248, 1437, 1491, 2238, 2304, 2475, 3125
berths, deep-sea multi-user 1419, 1424
BIFFEX (Baltic International Freight Futures Exchange) 700, 703–705, 711
BOXTIME charterparty 2104, 2120
Brazil 393, 394, 2987
Bruges Week (1970) Conference 1244, 1245
Brussels Package 956–984
bulk trades 155, 172, 344, 447, 537, 602, 604, 1169–1196
  carriers 226, 708, 724, 831–863, 1326, 1327
    ferries and ro-ro 1174–1192
    hovercraft 1193–1195
    Panamax 721, 726, 742a
    rescue systems 1196
    see also ULCCs; VLCCs
  dry 482–522, 525–527, 659–730, 722, 2862
    demand 716–718
    freight rates/markets 699–715
    general and theory 659–698
    regulation 729, 730
    supply 719–728
  fleets 217, 219, 235, 721, 722, 2801
  'mini' and 'neo' 523–533
  Norwegian 340
  ports 1323
  and protectionism 3007
  wet 471–481
    demand 798–806
    freight rates/markets 776–797
    general and theory 731–775
    pollution 820–830
    product carriers 831–863
    regulations 864–866
    supply 807–819
  see also cargo; costs
bulkers *see* bulk trades, carriers
bunkers 320–323, 2028
business of shipping 77, 78, 80

cabotage law 3013
Canada 153, 988, 2287, 2467, 2943, 3034
  admiralty law 2209, 2267
  flag and fleet development 2757
  hazardous goods 186
  insurance law 2607, 2608
  laws and policy 1963, 1970, 1975–1980
  longshoring industry 1571
  Merchant Marine 1781
  ports 987, 1259, 1260, 1265
    container 1109, 1389
  security 296
  state and policy 2070, 2943, 2946, 2957
  training 1701
  and UNCTAD Liner Code of Conduct (1974) 959, 974
  see also North America
canals and waterways 1605–1649
car transport 523, 529, 534, 545, 546, 547
cargo 443, 451, 452, 828, 867, 2878

claims 1928
delays 1967, 2016
dry 227, 327, 587, 1722, 1724, 3122
insurance 2591, 2678–2686
liner shipping 1120–1124, 3097
preference/reservation 3004, 3013–3020
sharing 1080
shipping costs 283
see also bulk trades
cargo-handling 11, 1308, 1372–1403, 1410, 1505, 1536
Caribbean 444, 804, 2777, 2986, 2988, 2989
carriage of goods 1997–2199
Carriage of Goods by Sea Acts (1971 & 1992) 2031, 2032, 2052, 2057, 2065
cartels 986, 1004, 1016, 1018
casualties/injuries 1663, 1678, 1765, 1784
cement trade 530, 531, 533
Central America 1295, 2777, 2986, 2988, 2989
chartering 158, 702
tanker 731, 738, 781–783, 793, 794
charterparties 2086–2127, 2149
chemicals 171, 476, 544, 832, 835, 837–839a, 841, 857a, 859
China 124, 569, 1739, 1983, 2410, 3027
arbitration 2513, 2514
flag and fleet development 2762–2770, 2796
port development 1274
shipbuilding 401, 412
state and policy 2978, 2979, 2996
CIS (Commonwealth of Independent States) 2748
Cityport 1451, 1452
CMEA (Council for Mutual Economic Assistance) 2939
CMI (Comité Maritime International) (1981) 2344, 2346, 2363
coal 552, 1388, 1453, 2761
transport 482, 483, 485, 495, 501, 518, 519, 522
coking 488, 490, 496, 497
steam 491, 494, 506, 509, 510
coastal/short-sea trades 1142–1168, 1169–1173
collective bargaining *see* trade unions
colleges, nautical 1708, 1710
collisions, law on 2421–2446
combined carriers *see* bulk trades, wet
Comecon 2736, 2938
communication, at sea 1780, 1790–1795
companies 42, 137–147
conferences 49, 130, 1200, 2389, 3067
congestion 184–185, 1159, 1418, 1421, 1422, 1433
containers/containerization 21, 38, 119, 641, 991, 1395, 1397
bills of lading 2142
China 569
developing countries 1047, 3154
insurance and liability 2675, 2684, 2685
law 1096–1098, 2418
and military logistics 3059
Nigeria 1282
optimum size 1101, 1108
ports/terminals 1307, 1374–1378, 1380, 1389, 1392, 1393, 1477, 1864
US 1261, 1262, 1407
US-Canada flows 2759
contracts 1653–1691, 1997–2082, 2515
conventions 3060, 3063, 3066, 3132, 3133, 3134, 3136
costs 177, 235–243, 882, 2824, 2827
bulker 235, 283, 484, 725
bunker 320–323
human resources 1653–1691, 1807
operating 324–330
ports 1205, 1207, 1209, 1315, 1359, 1365a, 1369
tanker 810, 812b
transport 464–470, 484
see also pricing
credit, letters of 1920
crew 118, 1657, 1720, 1807
criminal activity 2532–2587
law on 2583–2587
crude oil *see* oil
cruise trade, liner 1125–1133
Cyprus 2318

damages 2100, 2201, 2589, 3076
dangerous goods 186, 192, 3075
death, wrongful 1962, 1965
decasualization 1546, 1547
defence 3040–3059
delays 1967, 2016, 2089
demand 149–165, 2983
bulk trades 716–718, 798–806
demolition 437
demurrage 2166, 2171, 2173
Denmark 2203, 2298, 2469, 3116
deregulation 3032–3039
developing countries 556, 2635, 2876, 2891, 3142, 3150
and cargo reservation 3020
co-operation among 2857
containerization 1047, 3154
flag and fleet development 117, 462, 2774, 2904, 3137
freight rates 468, 1036
liner trades 868
management and training 1842, 1843
ports 1292, 1312, 1345, 1404, 1457, 1592
shipbroking 1873
tariffs 561
trade barriers to 2903
transport costs and policy 462
see also individual names; less developed countries (LDCs); Third World
development, shipping 104–133, 128
Devlin modernisation programme 1540, 1549, 1561
dismissal/redundancy (employment) 1662, 1691, 1830
disputes/conflicts 1696, 1741, 1749–1751, 1792, 1920
dockers (dockworkers) 1522, 1554, 1563, 1564, 1585
dry bulk shipping *see* bulk trades
dry cargo *see* cargo
drydocking 415

Eastern bloc *see* Europe, Central and Eastern
EC *see* European Community (EC)
economic theory 164–165, 552–575, 1309–1371
economics 65, 68, 74, 81, 82–103, 84
education and training 1552, 1578, 1670, 1697–1718, 1725, 1891
management 1840–1846
EEC *see* European Economic Community (EEC)

Egypt 1252, 1779
Energy Transportation Act (1974) 3018
England 1199, 2297
enterprise 269–275, 637
entrepreneurs 134–147, 271
environment 195, 196, 449, 1423, 1490, 2875
Esso, British 140
Europe 453, 878, 1170, 1185, 1477, 1583, 2690
  coastal/short-sea trades 1150, 1157a, 1164, 1168
  North-West 187
  shipbuilding/repair 349, 350, 371, 373, 374
  Western 350, 371, 373, 1948–1951, 2921–2930
  see also flag and fleet development; ports; state and policy
European Community (EC) 40, 1951, 2522, 2719, 3070, 3107
  law 3098, 3106, 3115, 3120
  ports 1249, 1313
  shipping policy 3096, 3105
European Economic Community (EEC) 3089, 3102, 3123
  and flags of convenience 2804, 2843
  ports 1413, 1584
  shipbuilding 382, 389, 3124
  shipping policy 3092–3094, 3113, 3126, 3127, 3130, 3131
export 1847–1868

Federal Arbitration Act (1925) 2498, 2503
ferries 1153, 1171, 3024
  bulk trades 1174–1192
  cross-Channel 1179, 1183, 1186, 1188
fertilizers 487, 489, 504
finance 236, 281–295, 296–319, 1492–1494, 3121
Finland 2240, 2278, 2476, 2716
fire 1957, 2370
flag and fleet development
  general 2687–2796
    Africa 2771–2776
    America
      Central and South 2777
      North 2754–2760
    Asia 2779, 2780, 2785, 2793, 2972
      socialist 2762–2770
    Australasia 2778–2796
    Europe
      Central and Eastern 2723, 2726, 2727, 2729–2734, 2736–2739, 2741–2753
      Western 2714, 2715, 2718–2720, 2811
    Japan 2761
    Middle East 2722, 2724, 2725, 2728, 2735
    Oceania 2778–2796
    policy and theory 2687–2695
    Scandinavia 2716, 2717, 2721
    United Kingdom 2696–2713
  open registry 2797–2852
  Soviet 2723, 2726, 2727, 2729, 2731–2734, 2737, 2739, 2741–2749, 2751–2752
  see also developing countries
flagging out 2830–2833
flags
  discrimination 3009
  preference 3030, 3031
flags of convenience 2326, 2798–2829, 2834, 2838–2842, 2845
fleets 151, 330, 597, 598, 740, 1037
  bulk trades 217, 219, 235, 721, 722, 2801
  development *see* flag and fleet development
forecasting, international trade 563–575
forest products 526, 527, 528, 539, 667
France 2274, 2319, 2369, 2487, 2921, 3129
  ports 1503, 1520
fraud 2532–2544, 2610, 2656, 2657
freeports 1316, 1439, 1454, 1460, 1461, 1464, 1465, 1481
freight 98, 120, 153, 154
  costs 464–470, 484
  fixtures 2162, 2163
  forwarders 1847–1868
  inland waterways 1637, 1638
  insurance 2686
  rates 468, 1036, 2824
    liner 872, 897, 899–901, 920, 921, 932, 1019
      stability 993
    and markets 614–636, 699–715, 776–797

General Agreement on Tariffs and Trade (GATT) 2881, 3088
general average 2447–2463
Germany 2200, 2264, 2464, 2643
  Democratic Republic 2402, 2932
  Federal Republic 2277, 2385, 3118
Ghana 1283, 1284
government, and shipbuilding 381–390
grain 492, 493, 507, 516, 521, 664, 714, 1368, 2515
Great Britain 75, 108, 114, 121, 131, 1169, 1516
  shipbuilding 375, 387, 396, 399, 404, 405
  see also England; Scotland; Wales
Greece 104, 265, 579, 1818, 2251, 2315, 3108
  flag and fleet development 2714, 2718, 2811
  law 1949
  and Limitation Convention (1976) 2403
  ship sale and purchase 2482
  shipbuilding/repairing 382, 408, 419
  state and policy 2691, 2922, 2924, 2928
Greenwich Forum Conferences (1980 & 1983) 2709, 2711, 2892
Gulf Arab region *see* Arabian Gulf states
Gulf War (1991) 2578, 3006

Hague Rules 1999, 2008, 2012, 2029, 2065, 2116, 2129, 2648
  revision 2023, 2036, 2059, 2075, 2082
  Visby amendments 2000–2002, 2042, 2054, 2055, 2068, 2646
Hamburg Rules 2000, 2018, 2039, 2041, 2046, 2060, 2067, 2069, 2070, 2077
  and conflicts of law 2035
  Convention leading to 2061
  deviation under 2056
  effect on insurance 2078, 2588, 2646, 2648, 2659
  and New Zealand 2048
  non-contractual claims under 2025
  v. Hague-Visby regime 2002, 2004, 2029, 2040, 2056
harbours, law of 2327
Harter Act (1893) 1974, 2007, 2130
Hawaii 1155
hazardous goods *see* dangerous goods
health and safety 1520, 1528, 1532
history 111, 126, 127, 142, 148
Hong Kong 1298, 1299, 2322, 2787, 2788
hovercraft 1194, 1195
hull insurance 2645, 2671–2677

human resources 1650–1810

India 1654, 1692, 1782, 2781, 2990, 3000
Indian Ocean 1994
Indonesia 1148, 1149, 1993, 2782, 2783, 2878
industrial relations *see* trade unions
information 4, 5, 19, 36, 274
information technology 1416, 1860
injuries *see* casualties/injuries
inland waterways 1633–1648
insurance 2122, 2355, 2450, 2588–2686, 2670, 3121
  premiums 2653–2655
Inter-Governmental Maritime Consultative Organization (IMCO) 3085–3087
intermodalism/multimodalism 638–655
International Labour Organization (ILO) 3155–3158
International Load Line Convention (1966) 3133
International Maritime Organization (IMO) 3068, 3075–3084
international seaborne trade 440–585
  bulk *see* bulk trades
International Transport Workers' Federation (ITWF) 2850, 2852, 3114
INTERTANKO (International Tanker Time Charterparty) 793, 794
investment 224, 225, 331–348, 808, 1404, 2883, 2884
  ports 1504–1512
  in shipping 577, 580, 584
Iran 1956
Ireland 2248, 2312, 2479, 3109
iron ore 488, 490, 496–498, 512, 514, 520, 1385
iron and steel 35
island economies 1305, 1444
Israel 1685, 1700, 1803, 2150, 2250, 2305, 2415, 2690
Italy 67, 391, 1237, 1242, 2282, 2286, 2466, 3117

Japan 463, 809, 1738, 1740, 1760, 1819, 2314, 2401
  flag and fleet development 2761
  maritime safety 1488
  ports 1272, 1273, 2977
  ship sale and purchase 2480
  shipbuilding 368, 369, 376, 380, 391, 392, 406, 407
  state and policy 2975–2977
jetfoils 1193
job satisfaction 1796–1804
joint ventures 2223
Jonglei Canal 1644

Kobe, port of 1589
Korea 183, 398, 411, 683, 1296, 2790, 2897
  state and policy 2993–2995, 2997, 2998
Kuching 1430
Kuwait 1835

lading, bills of 1095, 2059, 2064, 2128–2161
Lagos 1275
Lanasse Rule 2668
Latin America 1295, 2651, 2807, 2986, 3030
law 37, 63, 80, 3120
  Africa 1984–1987, 2263, 2264
  America
    Central and South 1988
    North 1957–1982
  arbitration 2490–2522
  Asia 1983, 1990, 1995
  Australasia 1989, 1991–1993, 1996
  carriage of goods 1997–2199
  coastal/short-sea trades 1146
  container 1096–1098, 2418
  criminal activity 2532–2587
  deep draft harbour facilities 1204
  Europe
    Central and Eastern 1952–1956
    Western, and Scandinavia 1948–1951
  general 1874–1996
  IMO 3082
  insurance 2590, 2607, 2608, 2636–2638, 2642
  international 3072, 3132
  maritime law 2200–2463
  Middle East 1952, 1953, 1955, 1956
  Oceania 1994
  ports, harbours and docks 1199
  seamen's rights 1665, 1677, 1679, 1783
  seaworthiness 2523–2531
  ship sale and purchase 2464–2487, 2488, 2489
  United Kingdom 1936–1947, 2224
  see also European Community (EC)
law merchant 1930, 1931
law of the sea 118, 1890, 1910, 1912–1914, 1923, 1987, 2872, 3048, 3062
lay-up 231–234
laytime 2164–2172
leasing 236, 319
less developed countries (LDCs) 1441, 1456, 3137
liabilities 2178–2199, 2495, 2592, 2648, 3076
  limitation 2366–2406, 3077, 3079
Liberia 1466, 2303, 2799, 2812
liens, maritime 1946, 2407–2420
life salvage 2335
lifeboats/liferafts 189
Limitation Convention (1976) 2369, 2371, 2381, 2387, 2388, 2397, 2399, 2401
limitation of liabilities *see* liabilities
liner trades 49, 264, 1114–1119, 1134–1141, 3004, 3141
  conference/non-conference 985–1044, 1103
  containers 1045–1113
  conventional cargo liners 1120–1124, 3097
  cruise trade 1125–1133
  general and theory 867–955
  liner code 956–984
  services 1582
  Soviet 2747
  terminals 242
Liverpool 135, 1212
Lloyd's 45, 47, 137, 2307, 2628, 2629
  Form of Salvage Agreement 2342, 2351
  Open Form (LOF)
    1980 2339, 2341–2344, 2346, 2351
    1990 2343, 2356
LNG (liquified natural gas) shipping 474, 535, 541, 806, 1409, 1447
longshoremen 1523, 1566–1568, 1572, 1574, 1576, 1580, 1581
loss, ship 152, 423, 436, 438, 439, 2631
loyalty contracts, shipper 992, 1019
LPG (liquified propane gas) shipping 544, 831, 834–835, 839, 840, 843a, 849, 853–857, 862–863
Luxemburg 2292, 3110

Malaysia 3038
Malta 1950, 2215, 2288, 2471
management 34, 59, 69, 144, 250–280, 1590–1604, 1809, 1811–1873
Manchester Ship Canal 1647
manpower 1513–1604, 1655, 1661
  see also human resources
Mareva injunctions 1882, 1898, 1904, 1917–1919, 1921, 1922
Marine Insurance Act (1906) 2598, 2623, 2624, 2625, 2628
market structures 586–1196
  bulk trades *see* bulk trades
  coastal/short-sea trades 1142–1168
  general 601–658
  liner trades *see* liner trades
Mauritius 1292
Mediterranean 28, 42, 199, 461, 1238, 1876
Melbourne Dockers War (1971–3) 1587
merchant fleet 131, 267
  British *see* United Kingdom
Merchant Marine Act (1970) 1659, 1666, 1810, 2755
Merchant Navy 121
merchant seamen *see* seamen
Merchant Shipping Acts, (1864, 1970 and 1988) 1936, 1942, 1944, 1947, 2368
Middle East *see* flag and fleet development; law; ports; state and policy
military logistics 3058, 3059
minerals and ores (incl. bauxite and alumina) 499, 500, 538, 551
Morflot 1787, 2729
mortgages 296, 301, 1911, 2415
multimodalism *see* intermodalism/ multimodalism

National Dock Labour Scheme 1502, 1535, 1538, 1541, 1542, 1553, 1554, 1559, 1560
National Union of Seamen 1754
nationalism, economic 562
Netherlands 1240, 1243, 1689, 2273, 2324, 2486, 3119
Netherlands Antilles 2301, 2321, 2925
New York Produce Exchange (NYPE) forms 2026, 2112, 2119
New Zealand 1741, 2325
Nigeria 867, 1106, 1282, 2214, 2559, 2776
  maritime law 1984, 1985
  ports 1275, 1276, 1278, 1279, 1281, 1290, 1512
  state and policy 2981, 2983
North America *see* flag and fleet development; law; ports; state and policy
North/South perspectives 2888, 3140
Northern Ireland 395, 1227
Norway 340, 2259, 2299, 2484, 2717, 2721, 2923
  sailors 1781

obsolescence 435, 436
ocean shipping 72, 116
Oceania 2784, 2786, 2791, 2792, 2794, 2999
officers 1660, 1692–1694
oil 825, 1324, 2267, 2633, 2948
  pollution 1875, 2377
  tankers 348, 739, 751, 754, 756, 760, 766, 774, 775
  trade 444–446, 800–804, 816, 842, 851, 852
    bills of lading 2161
    Caribbean 444, 804
    transport 471–473, 475, 477–480
on-board systems 1719–1740
open registry 2797–2852, 2834–2845
optimum ships 601–613, 935
organization 144, 1785
organizations 3159
  international 29, 3061, 3075–3084, 3159

P & I (Protection and Indemnity) 2030, 2339, 2658–2669, 2677
P & O (Peninsular and Oriental Line) 143
Pacific Forum Line (PFL) 2794
Pacific Ocean 1007, 1078, 1444, 2999
  microstates 2791, 2792, 2803
  ports 1297–1300, 1300, 1302, 1305, 1306
  South 2786, 2794
  Western 1994
  see also Oceania
Pacific Rim 450, 1297, 2784
package limitation 2071, 2395
Pakistan 2276
Panama 2252, 2313
Panama Canal 1605–1619
parcel tankers 845, 857a
passenger ships 1935, 3078
Pennsylvania Rule (1873) 2442, 2446
Peru 1294
petroleum 1270, 2518
Philippines 2291
phosphates 505, 513
pilotage 1219, 1517, 2327–2331
piracy 2545–2565, 2566
Poincare gold franc 2384, 2392
Poland 275, 1254, 1501, 1811, 2366, 2407, 2750, 2934
policy 59, 63, 69, 84, 105, 881
  and regulation
    international 3060–3159
      codes, conventions and reports 3132–3159
      general 3060–3131
  see also state and policy
pollution 820–830, 1490, 1875, 2377, 2859
ports 30, 43, 58, 222, 455, 456, 1864
  Africa 1275–1293, 1512
  America
    Central and South 1294, 1295
    North 1256–1271, 1389, 1407, 1565–1589, 2759
  Asia 1274, 1296
  Australasia 1301, 1303, 1307, 1308
  Canada 987, 1109, 1259, 1260, 1265, 1389
  economic theory and price 1309–1491
  Egypt 1252
  employers 1529
  Europe
    Central and Eastern 1251, 1254
    Western, and Scandinavia 1237–1249
  finance 1492–1494
  general 1197–1210
  investments 1504–1512
  Japan 1272, 2977
  manpower and management 1513–1604
  Mediterranean 199
  Middle East 1250, 1252, 1253, 1255
  Oceania 1297–1300, 1302, 1305, 1306
  purpose and congestion 1418–1435
  and regional development 1436–1486
  routeing 213, 214, 215
  safety 1350, 1487–1491
  subsidies 1495–1503

United Kingdom 1211–1236, 1534–1564
United States 1348, 1368, 1371, 1478, 1485, 1489, 1520
see also Cityport; costs; freeports
Portugal 2290
pricing
conferences 1013, 1014, 1025
EC 1951
liner shipping 999
ports 1205, 1209, 1404
secondhand 223–230
unfair practices 3089, 3097
see also costs
privitization 3032, 3035, 3038
product carriers *see* bulk trades; tankers
productivity 180–183, 1401–1417, 1805–1809
professional associations *see* trade unions
protection, from state and policy 3001–3039
protectionism 3001, 3007, 3012, 3015

radar 1908, 2426, 2437
radioactive materials 207, 1489
ratings 1695, 1696
reefers 1134–1141
registers 33, 34
registration 2276–2325, 2693, 2694, 2809, 2820
registry, open 2797–2852
regulation 821, 994, 2853
bulk trades 729, 730, 864–866
see also policy and regulation
reinsurance 2597, 2616, 2649, 2654, 2681
rescue at sea 1196, 2204
research 66–69, 244–246
risk 305, 306, 440, 1884, 2602
investment 331, 332, 342, 343
war 2675–2677
ro-ro (roll on/roll off) 1021, 1114–1119, 1174–1192
Rochdale Inquiry (1970) 2871, 2905, 2912
routeing 199–215, 1223

safety 186–198, 1328, 1488, 1680, 1682, 1683, 3083
and crew size 1657, 1720
criminal penalties 2587
shipbuilding 422
tanker 824
terminal 1487
sale
of goods 2083–2085, 2137
and purchase, ship 2464–2489, 2544
salvage 2332–2365, 2500
see also Lloyd's, Open Form (LOF)
Salvage Convention (1989) 2338, 2343, 2347, 2356, 2364
Saudi Arabia 1953, 2309, 2724, 2725, 2935, 2936
Scandinavia 283a, 284, 343, 374, 1948–1951
see also flag and fleet development; state and policy
Scotland 1235, 2151, 2316
scrapping 152, 436, 438, 439
scuttling 1895, 2536, 2584
seafaring life, deprivation 1732
seamen (seafarers)
American merchant 1658, 1684, 1695, 1770, 1771, 1773
home lives 1766
Soviet 1778
stereotype 1769, 1772
see also human resources
Seaport Development 3090, 3091
seaworthiness, law on 2523–2531
security 296, 2560
service 2, 44, 51, 1024, 1582
Shell 138, 2109
shipbrokers 1870–1873, 2117
shipbuilding 161, 216, 334, 393, 394, 422
EC/EEC 382, 389, 3107, 3124
national 391–412
and repair 349–380, 381–390, 408, 413–421
United Kingdom 397, 421, 3045
United Kingdom 375, 387, 396, 399, 404, 405, 2702, 3045
see also Japan
shipowners 118, 134–147, 2094, 2373, 2374
liability 2376, 2386
and management 250–280
UK 1945, 3128
shipping pools 730
shiprepairing *see* shipbuilding
shop stewards 1534
short-sea trades 1142–1168, 1169–1173
Singapore 2, 311, 2043, 2256, 2317, 2483
socialist countries 2753
see also China
South Africa 122, 2310, 2477, 2777, 2980, 2984
law 1986, 2263, 2264
South America 1294, 1988, 2986, 2987
Southampton 1226, 1232, 1766
Soviet shipping 637, 1787, 2293, 2723
see also flag and fleet development; state and policy
Spain 2294, 2365, 2396
speed 247–248, 321a
Sri Lanka 963, 1006, 1302
state and policy
Africa 2980–2985
America
Central and South 2986–2989
North 2070, 2942–2974
Asia 2978–2979, 2990, 2991, 2993–3000, 2996
Australasia 2992
defence
general 3040–3057
military logistics 3058, 3059
Europe
Central and Eastern 2931–2934, 2937–2940
Western 2921–2924, 2928–2930
general 2853–2904
Japan 2975–2977
Middle East 2935, 2936, 2941
Oceania 2925, 2926, 2999
protection 3001–3039
Scandinavia 2923, 2927, 2929
Soviet 2931, 2933, 2937, 2940
United Kingdom 2905–2920
United States 2942, 2944, 2945, 2947–2956, 2958–2963, 2965–2969, 2971–2974
statistics 1, 4, 20, 32, 35, 67, 85, 102
steel 511, 532, 540, 1172
stevedores 1524, 1525, 1531, 1556, 1580
strikes, seamen's (1960) 1696, 1792
subsidies
differential 1810, 3022
ports 1495–1498
state 2955, 3021–3025
Sudan 563

Suez Canal 130, 1620–1630, 3005
sugar 543, 1303
supertankers 770, 771, 807
supply 100, 882, 2893
  bulk trades 719–728, 807–819
  economic theory 216–249
  finance, cost and investment 281–348
  general 166–215
  loss and scrapping/demolition 423–439
  shipbuilding/repairing 349–422
  shipowning/management 250–280
Supreme Court Act (1981) 2235
Sweden 1759, 2262, 2320, 2485

Taiwan 125, 1422, 1990, 2306, 2996
tankers 279–280, 735–738, 741, 742, 746–802, 807–819, 2519
  chartering 731, 738, 781–783, 793, 794
  chemical 171
  fleet 138
  freight 156, 159
  markets 587, 733
  owners 279
  product 836, 844, 846–848, 860, 861
  regulations 864–866
  safety 824
  scheduling optimization system 734, 759
  scrapping age 439
  secondhand 226
  sub-standard 830
  trade, forecasting 564, 565, 568
  voyage charterparties 2107, 2108, 2109, 2111, 2124, 2127
  see also oil; supertankers
tariffs 561
taxation 173, 301a, 2917, 2944
Tayside 1220
technology 218, 272, 1711, 1785
terminals 1401–1417, 1959
terrorism 2566–2582
text books 70–81
Thailand 470, 2323
Third World 642, 1325, 2863
  see also developing countries; less developed countries (LDCs)
Tokyo 2511
tonnage 164, 170, 2858
tortfeasors 2421, 2444
towage 1887
trade embargo 3026–3029
trade unions 1546, 1550, 1577, 1741–1764, 2846–2852
training *see* education and training
tramp shipping 687, 688, 695, 697, 698, 713, 2177
transhipment 656–658
Transport & General Workers' Union (T& GWU) 1540, 1546
Turkey 46, 2281
turnaround time 607, 612

ULCCs (ultra-large crude carriers) 798, 812a, 812
United Arab Emirates (UAE) 1952
United Kingdom (UK) 120, 453, 1066, 1167, 1391, 2383
  balance of payments 578, 582, 585
  merchant fleet decline 2699–2701, 2712, 2908, 2909, 2913–2916, 3100
  ro-ro vessels 1192
  ship sale and purchase 2473
  shipowners 1945, 3128
  see also England; flag and fleet development; Great Britain; law; Northern Ireland; ports; Scotland; shipbuilding; state and policy; Wales
United Nations Commission on International Trade Law (UNCITRAL), Draft Convention on Carriage of Goods by Sea 3143–3148
United Nations Conference on Trade and Development (UNCTAD) 3137–3154
  liner Code of Conduct (1974) 956–984
United Nations (UN), Convention on Registration of Ships (1986) 2311
United States (US) 105, 559, 1081, 1125, 2943
  Canadian free-trade agreement (FTA) 558
  admiralty law 2201, 2219, 2246
  anti-trust laws 1002, 1012, 1015, 1033
  Construction-Differential Subsidy (CD) programme 383
  container terminals 1261, 1262, 1407
  Effective United States Control (EUSC) programme 2810
  freight forwarders 1866, 1867
  Government bill of lading 2134
  inland waterways 1636
  international policy 3070
  international trade forecasting 572
  and Limitation Convention (1976) 2394
  liner trades 870, 876, 895, 908
    conferences 1002, 1012, 1017, 1030, 1031, 1033
    and containerization 1105
  lumber shipping regulations 3026
  management 1829
  Merchant Marine Act (1936) 367, 383
  merchant shipping 113
  Mortgage Act (1989) 2275
  oil and tanker supply 787
  Pacific Basin 1264
  ports
    political economy 1200
    West Coast container 643
  and post-war piracy 2558
  registration 2302
  ship sale and purchase 2478
  shipbuilding 402, 409, 410
  Shipping Act (1984) 1969, 1972, 1973, 1981, 2951, 2973, 2974
  subsidies 3025
  tankers 850
  tort liability 2375
  Trade Act (1970) 2967
  waterfronts 1595
  see also longshoremen; North America; ports; state and policy
USSR (Union of Soviet Socialist Republics) 1954, 2493, 2494
  see also Soviet shipping

value, ship 345
Vancouver 1258, 2499
Vanuatu 2809
Venezuela 1988, 2280
venture capital 2696
vessel traffic services 208, 209, 1885
vessels 79, 103, 176, 2488, 2489
violence at sea *see* terrorism
Visby amendments *see* Hague Rules
Vladivostok 1201
VLCCs (very large crude carriers) 780, 798, 811, 812a, 813, 818, 1324

Volkswagenwerk decision (1968) 1579

wages 1548, 1659, 1664
waiting-to-service-time (W/S) ratios 1411
Wales 1233, 1448, 2297
war risk 2675–2677
Washington tanker law (1975) 1961
waterfront development 1452, 1595
waterways, sea and inland 1633–1648
waybills 1948, 2072, 2145
wet bulk shipping *see* bulk trades
work 1551, 1730
  conditions 1669, 1670
  stoppages (dock strikes) 1537, 1545, 1558
  see also dismissal
World Bank 1440, 1493
World Maritime University (WMU) 1844
world shipping 112, 117, 132
wrecks 820, 2185

York-Antwerp Rules 2449, 2453, 2455, 2456, 2463
Yugoslavia 2300, 2388

Zaire 2775